Ecology of the Brain

Ecology of the Brain
The Phenomenology and Biology of the Embodied Mind

Thomas Fuchs
Karl Jaspers Professor of Philosophical Foundations
of Psychiatry, Psychiatric Clinic, University of Heidelberg,
Germany

OXFORD
UNIVERSITY PRESS

Great Clarendon Street, Oxford, OX2 6DP,
United Kingdom

Oxford University Press is a department of the University of Oxford.
It furthers the University's objective of excellence in research, scholarship,
and education by publishing worldwide. Oxford is a registered trade mark of
Oxford University Press in the UK and in certain other countries

Originally published as *Das Gehirn – ein Beziehungsorgan*
by W. Kohlhammer GmbH, 2007

© W. Kohlhammer GmbH 2018

The moral rights of the author has been asserted

First Edition published in 2018

Impression: 1

All rights reserved. No part of this publication may be reproduced, stored in
a retrieval system, or transmitted, in any form or by any means, without the
prior permission in writing of Oxford University Press, or as expressly permitted
by law, by licence or under terms agreed with the appropriate reprographics
rights organization. Enquiries concerning reproduction outside the scope of the
above should be sent to the Rights Department, Oxford University Press, at the
address above

You must not circulate this work in any other form
and you must impose this same condition on any acquirer

Published in the United States of America by Oxford University Press
198 Madison Avenue, New York, NY 10016, United States of America

British Library Cataloguing in Publication Data

Data available

Library of Congress Control Number: 2017955391

ISBN 978-0-19-964688-3

Printed and bound by
CPI Group (UK) Ltd, Croydon, CR0 4YY

Oxford University Press makes no representation, express or implied, that the
drug dosages in this book are correct. Readers must therefore always check
the product information and clinical procedures with the most up-to-date
published product information and data sheets provided by the manufacturers
and the most recent codes of conduct and safety regulations. The authors and
the publishers do not accept responsibility or legal liability for any errors in the
text or for the misuse or misapplication of material in this work. Except where
otherwise stated, drug dosages and recommendations are for the non-pregnant
adult who is not breast-feeding

Links to third party websites are provided by Oxford in good faith and
for information only. Oxford disclaims any responsibility for the materials
contained in any third party website referenced in this work.

Preface

This book is based on a German precursor entitled *Das Gehirn—ein Beziehungsorgan* [*The Brain—A Relational Organ*], which recently appeared in its fifth edition (Fuchs 2016). Its continued success has encouraged me to make it available to the international public, albeit in a completely revised and extended form. The new title, *Ecology of the Brain*, indicates the crucial role of the *Umwelt* for understanding the human brain, namely as an organ of relation, interaction, and resonance: with the body itself, with the immediate environment of the organism, and with the social and cultural environment of the lifeworld.[1] Of course, this essential relatedness applies to the embodied mind as well—an idea that is illustrated by the front cover of this book: being immersed in an enveloping milieu, which serves as a carrier for the diver and at the same time offers flexible resistance to his motion.

However, an ecological approach to the brain requires, first of all, an autonomous concept of *life*. Up to now, the neurosciences have largely neglected that the brain is primarily an organ of the living being, not of the mind. The life sciences, too, are far away from grasping life as a phenomenon of its own, displaying a self-organization and inwardness that is irreducible to mere physical processes. In contrast, embodied and enactive approaches have considerably advanced our understanding of the living organism. Only on this condition does it become possible to overcome the direct "short circuit" of brain and mind that is still pervasive in the neurosciences.

Certainly, in a time of a growing number of publications on embodied and enactive cognition, a book like this should also indicate what its particular focus is. I only wish to highlight some aspects here:

- This book reformulates the traditional mind–brain problem, presenting it as a *dual aspect of the living being*: both as a lived or subjective body and as a living or objective body. The resulting "body–body problem" has already

---

[1] The reference to Gregory Bateson's *Steps to an Ecology of Mind* (1972) is of course intended. I only give one significant quotation here: "we may say that 'mind' is immanent in those circuits of the brain which are complete within the brain. Or that mind is immanent in circuits that are complete within the system, brain *plus* body. Or, finally, that mind is immanent in the larger system—man *plus* environment" (Bateson 1972, 317; emphasis in original).

been pointed out in Evan Thompson's seminal book *Mind in Life* (2007), but I put more emphasis on the mediation of both aspects, in particular on the transformative and integrative functions of the brain. The investigation of these aspects requires a dual approach, combining the phenomenology of subjectivity with ecological biology and the philosophy of life.

- Embodied and enactive approaches are usually based on the sensorimotor interaction of body and environment, or in the phenomenological tradition, on our bodily "being-towards-the-world" (Merleau-Ponty). While this is doubtlessly a crucial dimension of embodiment, it should not be forgotten that *the primary locus of self-awareness is the body itself*. According to the concepts presented here, basal consciousness or the "feeling of being alive" emerges deep inside the organism, only to then direct itself, on higher levels of integration, towards the environment.

- There has been up to now no detailed account of how the brain functions as a mediating organ, both in our sensorimotor interaction with objects and our social interaction with others. This book develops such an account, based on thorough analyses of the brain as an *organ of mediation, transformation, and resonance*. Its central concept of consciousness as an *extended integral* of the ongoing relations between the brain, the body, and the environment may not be completely novel, but it is still a far cry from mainstream neurophilosophy, thus warranting a comprehensive explanation. For this, a thorough *critique of representationalist concepts* in current neuroscience and neurophilosophy is of particular importance.

- A further focus of this work lies on the peculiar causality of the living organism, which is conceived as *circular causality* both in the vertical (upward/downward causation within the organism) and in the horizontal dimension (loops of interaction reaching into the environment). This leads to the concept of an integral causality of the living being—as well as serving as a basis of a libertarian account of free will ("embodied freedom").

- The ecological approach will also be applied to the *socially and culturally scaffolded development of the human brain*, especially in early childhood. Intercorporeality and interaffectivity are shown to provide the basis for the incorporation of habits and skills as well as for higher forms of intersubjectivity, including the acquisition of language through shared embodied practices.

- Finally, I investigate the consequences of an ecological conception of the brain for my own primary discipline, namely *psychiatry and psychological medicine*, where reductionist paradigms of mental disorders are currently prevailing. In contrast, the paradigm of the brain as a relational organ may serve as the basis of a holistic approach that regards subjectivity and intersubjectivity as crucial dimensions for diagnosis, etiology, and treatment, without the risk of falling back into a dualistic framework.

All this results in understanding the brain as an organ which does not produce the mind like a gland produces its secretions,[2] but which mediates our bodily, emotional, and mental relations to the world. The processes of life and experiencing life are inextricably linked: it is not the brain but the living human being who feels, thinks, and acts.

---

[2] Such views are by no means overcome today—to give just one example: "The product of the interaction of all these billions of neurons is called 'mind'. Just as kidneys produce urine, the brain produces mind …" (Swaab 2014, 5).

Acknowledgments

An earlier version of my German book has been translated by Nuala Hughes, Rudolf Müllan, and Susanne Kirkbright. Although I have completely revised the book and rewritten major parts since then, I still owe them a debt of thanks for providing me with the basis of this final revision in English. Many of the thoughts and concepts revealed in this book originated from two European Research Training Networks that I coordinated, namely "Disorders and Coherence of the Embodied Self" (DISCOS, 2007–2011) and "Towards an Embodied Science of Intersubjectivity" (TESIS, 2011–2015). Many senior researchers and fellows from both projects, among them Dan Zahavi, Josef Parnas, Shaun Gallagher, Dan Hutto, Vittorio Gallese, Andreas Roepstorff, Peter Henningsen, Hanne De Jaegher, Sanneke de Haan, and even Thomas Metzinger have helped me to sharpen my concepts through the continuous debate between similar or quite contrary viewpoints. It has always been fun to cooperate with them, and I already miss the time we devoted to these projects.

Finally, a book like this is not possible without an "ecology" of its own, especially when the author is a philosopher, psychiatrist, and clinician. The Clinic of General Psychiatry headed by Sabine Herpertz, and my Research Section of Phenomenological Psychopathology were both wonderful environments in which to work and develop my theories. Special thanks must go to Christian Tewes and Rixta Fambach who provided the necessary support in the Section. The same of course applies to my wife Gabriele and my family who are the most agreeable environment I could imagine.

During the production of the volume I was happy to collaborate with Martin Baum, Charlotte Holloway, and Nic Williams from Oxford University Press who provided all necessary support. Charlotte in particular was most helpful at times when I needed encouragement during the extended revision of my book—my special thanks go to her.

Thomas Fuchs
Heidelberg, August 2017

Contents

Introduction *xiii*

Part 1 **Criticism of neurobiological reductionism**

1 Cosmos in the head? *3*
2 The brain as the subject's heir? *29*

Part 2 **Body, person, and the brain**

3 Foundations: Subjectivity and life *69*
4 The brain as an organ of the living being *107*
5 The brain as an organ of the person *173*
6 The concept of dual aspectivity *209*
7 Implications for psychiatry and psychological medicine *251*
8 Conclusion *279*

References *293*
Index *325*

Introduction

Neurobiology and the overthrow of the lifeworld

Since the brain and its activity during mental processes may be observed in increasingly finer detail, neuroscience is prepared to "naturalize" human consciousness and subjectivity, that is, to explain it in neurobiological terms. Mental activity, so it seems, can be localized inside the head, indeed even be visualized in detail with imaging techniques. Perceiving, feeling, thinking, and planning seem to take place at specific locations, being observable in vivo through the color-coded "lighting up" of brain structures. A stream of books with titles such as "The brain show," "We are our brains," "The brain and its self," or "How our brains become who we are"[3] paint a picture of the brain as an information-processing machine that in its convolutions and networks constructs a monadic inner world and a subject caught up in deceptions. At the same time, a deluge of articles in popular science journals teach us the real, namely neuronal and hormonal, causes of our feelings, perceptions, thoughts, and actions.

Undeniably, neurobiology has revealed abundant insights into the biological foundations of mind, experience, and behavior but also of mental illness, from which fruitful applications may be derived. On the other hand, this approach privileges a brain-centered view of human beings that is becoming more and more influential in medicine, psychology, and social sciences. In psychiatry, for example, the trend for the neurobiological paradigm suggests that mental illnesses are ultimately material processes within the brain, thus isolating them from a person's reciprocal relationships with their environment. Similarly, in educational science, learning and attention difficulties are increasingly traced back to organic causes inside the brain.

The neurobiologically informed concept of human beings affects the lifeworld and changes our self-understanding in everyday life. As a result of a gradual process of self-reification we start to see ourselves less as human beings taking decisions based on reason or motives, but rather as agents of our genes, hormones, and neurons. Equally, neuroscience calls into question our experience as authors of our own actions, thus casting doubt over whether we are in control of our lives at all. A person's will, it seems, emerges too late in the

---

[3] See Nitsan 2017, Swaab 2014, Knoll 2005, LeDoux 2003, or Churchland 2013.

day when neuronal processes underlying decisions have already run their full course. In this scenario, freedom is merely experienced as a result of the brain's self-deception, leaving us with a sense of mastery and self-control, while in fact the neurons have long made the decisions on our behalf.

Many neuroscientists or neurophilosophers arrive at a similar conclusion about the functioning of consciousness: it only mirrors the mechanisms of neuronal information processing of which we are unaware in principle. The brain machinery running in the background produces only the illusion of a permanent self. Long ago, the search for an "ego-center" or an "entry portal" to the mind within the brain was called off, which Descartes still supposed he had discovered in the pineal gland. The brain seems well capable of performing its computational tasks without any involvement of the human subject. In the words of neurophilosopher Thomas Metzinger: "We are mental self-models of information-processing bio-systems [...]. If we are not computed, we do not exist" (Metzinger 1999, 284).

Clearly, the validity claims of neuroscience are anything but modest. Its dominance is confirmed, too, by the popularity of its opening prefix in connection with other disciplines: "neuro-philosophy," "neuro-ethics," "neuro-pedagogy," "neuro-psychotherapy," "neuro-theology," "neuro-economy," and "neuro-law" claim dominance over other branches of science. Neurobiological terminology is infiltrating our self-descriptions and taking precedence over subjective and intersubjective experiences. The language of our lifeworld, which is still characterized by self-attributions and psychological terms, is now being remodeled, step by step, into an objectivizing, scientific idiom.

A citadel under siege

This overthrow of the lifeworld is the inherent logic built into the scientific program begun in early modern times. The program's underlying principle is reductionism. Its objective is to achieve a conception of nature, which eliminates all qualitative, holistic, that is, not discretely countable properties as merely subjective or anthropomorphic ingredients. It is achieved by breaking down what were originally experiences in the lifeworld into a physical, quantitative and a subjective, qualitative component: the former is accessible to experimental and explanatory research; the latter is displaced to a subjective inner world. For instance, the phenomenon of "heat" is divided, on the one hand, into physical particle movements and, on the other hand, into subjective sensation. A scientist therefore redefines heat by carving off its phenomenal aspect and shifting it as a "sensation of heat" into the subject. The same applies to color, sound, smell, and taste: as of now, they are merely subjective additions to actual reality. Conversely, the constructs originally developed for purposes of scientific measurement and predictability (particles, forces, fields, etc.) are now taken as underpinnings of the lifeworld and increasingly hypostasized as actual reality. The realm of everyday life experience is thus relegated to an illusion and true reality is what can be scientifically recorded by physics.

In their day, Galilei and Descartes already attempted to subvert the everyday belief in the truth of sensual experience and to make way for the new science of physics. Descartes thought of perception as being based on the physical movement of particles progressing from objects to the brain so that "we think we see the torch itself and hear the bell itself, rather than simply having the sensation of movements they have caused."[4] Scientific reductionism in this sense aims to divide the human subject from what it recognizes. To a certain extent, it cuts us off from the world. After all, the primary phenomenon of heat consists precisely in the *relationship* of our body to the surrounding world, such as the air or the sun. Color is produced through the eye's contact with an object; taste arises when the tongue touches food. Such relationships conveying to us the actual *qualities* of things are now cut off and reinterpreted as inner mental states. Actually, only particle movements, light waves, and chemical reactions exist. Thus, cleansing the world of all subjective, anthropomorphic properties leaves only the bare bones of nature that, as such, are far easier to dissect, manipulate, and technically control.

Step by step, all subjective and qualitative elements have been practically excluded from science's reinterpretation of the world. Even life itself could be reduced to biochemical molecular processes, albeit at a high cost: everything we associate with the existence of living organisms—sensation, feeling, self-movement, striving for something—was ruled out of research into living beings or equally displaced to an inner subjective world. With neurobiology as the new leading science, this program has reached a decisive point. Cleansing nature by the displacement of qualities into an inner subjective world is no longer enough. Subjective experience and indeed human consciousness itself must now be naturalized and reduced to physical processes. If this materialist explanation of brain functions was successful, the last citadel of subjective and qualitative being in the physical desert left by reductionism would finally be razed. "De-anthropomorphizing" nature would turn into the complete naturalization of the human being.

In this siege of the citadel, victory seems a prearranged destiny. The mission is almost complete, as large parts of the sanctuary are already under occupation;

---

[4] *The Passions of the Soul*, I, 23 (Descartes 2015, 205).

and while more and more squares and houses surrender, the hidden back streets are lit up by means of recent imaging techniques. Hardly any person doubts that the brain produces mental phenomena from a purely material grounding. In neuroscience and analytical philosophy of the mind, mind–body dualism is found equally outdated. However, the direct attack on the subject, supposedly the citadel's resident, has faltered for now. The victory was not won by the particular radical form of eliminativist materialism,[5] which declared subjective experience and "mentalistic" language as pre-scientific and naive institutions—thereby being destined to fade like the belief in ghosts, witches, aether, or phlogiston, and to make way for neurological language. Today, the majority of analytical philosophers and neuroscientists support a more moderate form of materialism that affords subjectivity a place, albeit only in identity with neuronal processes or as their accompaniment. In all instances there is no place for a causal role of subjectivity in the world. This explains the intense debates about free will: if consciousness cannot be denied, it shall at least remain the brain's product and therefore basically powerless. The human subject is granted continued dwelling in the citadel, but only on the condition that physicalism firmly and securely takes control.

Opening the citadel

It is not inconceivable for a surprising reversal of fortunes to redraw the battle lines around this last bastion of subjectivity. The brain could yet emerge as the real Achilles heel of the scientific world image. Firstly, the progressive elimination of the subjective dimension, which was hitherto so successful, now arrives at a methodological *impasse*. John Searle reasonably demonstrated how the segregation of the respective subjective component from the phenomena is no longer applicable when it comes to the reduction of subjectivity itself. There is simply no remaining space for the displacement of this sphere.[6] One possibility is to contest the existence of this sphere as a whole, but this is hardly convincing. Another option is to attempt to neutralize the sphere as an epiphenomenon of material processes, yet equally leaving the offence of subjectivity untouched.

At issue here is also that reductionism ends up in irresolvable epistemological aporias when called upon to analyze the brain. For according to the

---

[5] See, for example, Rorty 1970, Churchland 1988, Metzinger 2003.

[6] "Part of the point of the reductions was to carve off the subjective experiences and exclude them from the definition of the real phenomena, which are now defined in terms of those features that interest us most. But where the phenomena that interest us most are the subjective experiences themselves, there is no way to carve anything off" (Searle 1992, 122).

presuppositions, we recognize only what has already been processed by the neuronal machinery and thus appears as prefabricated subjective reality. Consequently, the brain as studied by the neuroscientist, like everything that he or she experiences, would be merely the product of his or her own brain. But how is the brain supposed to know itself? How should a physically describable and localized mechanism be in a position to bring forth the world of scientific experience in which it emerges at the same time? The conquered citadel would then be nothing more than the *fata morgana* imagined by the besieging forces, which never know for certain whether the citadel actually exists. Its presence could equally be a figment of imagination.

Any discourse about the brain clearly presupposes what the brain is alleged to bring forth: namely, conscious human persons who exist to communicate with each other. If this is indeed the case and neuroscience cannot escape its inherent dependence on subjectivity, intersubjectivity, and the lifeworld, then we are entitled to "turn things from the head to the feet again." Neurobiology and all other sciences emerge as a specialist form of human practice originating in the lifeworld, yet without ever gaining a position outside of it. The familiar world of everyday experience in which we coexist with others remains our primary and actual reality. It is neither merely the product of a different reality, which only science can comprehend, nor the illusionary figment or construct of the brain, but the foundation of all scientific knowledge. On the contrary, constructs are indeed the entities described by physics or neurobiology—electrons, atoms, molecules, action potentials, magnetic fields, or photon emissions. Their practical value is undisputedly great in developing explanations and prognoses of phenomena. But they can never serve to unmask the phenomena and experiences of the lifeworld as mere illusions.

On this condition, we have to suggest a novel interpretation of the brain: it does not give rise to our world like an invisible creator, and nor does it create or manipulate us like marionettes from some hidden headquarters. *Moreover, nowhere is the subject found in the brain.* Rather, the brain is the organ, which *mediates* our relationship towards the world, to other people, and to ourselves. The brain is the mediator making the world accessible to us, and the transformer connecting our perceptions and movements. But in isolation, the brain would be just a dead organ. It is only animated in connection with our senses, nerves, and muscles, with the internal organs, our skin, our environment, and in relation to other human beings. As soon as the *fata morgana* of the citadel dissolves and the lifeworld is restored to its legitimate position, the brain is no longer a last isolated sanctuary for the subject. On the contrary, it is a gregarious, animated trading venue, where messages and items of all kinds are exchanged, and which is connected with other places through an extensive network. In short, the brain is as an *organ of interrelations*.

Therefore, an adequate understanding of the human brain has to start from the phenomenology of our self-experience in the lifeworld, where we do not notice a mind–body division, but rather exist as embodied, animate, and mental beings, that is, as *human persons*. Only then can we go on to explore the brain's contribution to such unity on a biological level. First of all, I shall therefore develop the central thesis that all the brain's functions are dependent on *the human person's unity as a living organism* and may only be comprehensible on this basis. To do so, however, means to develop an adequate concept of life which is largely absent in current biomedical science. Here, recent concepts of embodied, embedded, and enactive cognition are of particular value, inasmuch as they contribute to an ecological theory of the living organism. My second thesis is that all higher brain functions presuppose the *human being's enactment of life in a shared social world*. This calls for a conception of human development as a continual anchorage of experiences within the mental and also cerebral structures of the human individual. It requires, in other words, a "cultural biology."

On the one hand, the dimension of life anchors the brain within the living organism and its natural environment. On the other hand, the sociocultural dimension anchors it within a shared, human world that shapes the brain throughout the course of a lifetime. Indeed, without the social world its specifically human functions would be incomprehensible. Both dimensions are united in a developmental and ecological view of the human brain as the organ of a "*zōon politikón*" or a living being that is formed by its social life to the very core of its biological structures. The brain then appears as an organ of mediation that enables the vegetative and sensorimotor relationships between the living organism and its environment. But it also transforms and intensifies these relationships to such a degree that it can become the medium for a human person's intentional relations towards the world. Hence, primary life processes are raised to emotional and intellectual enactments of life, opening up growing degrees of freedom. At the same time, the human brain is receptive to lifelong formation through intersubjective and cultural influences: it becomes a social, cultural, and historical organ—an organ of the human person.

Overview

In this volume, I shall turn to a detailed explanation of these matters only after developing a critique of the widespread influence of reductionist notions of the relationship of the brain and subjectivity. In the first part of the book, I develop this critique in two basic steps. In *Chapter 1*, I concentrate on an analysis of neuroconstructivist epistemology. Here, I oppose the concept of phenomenal reality as an internal modeling or representation via neuronal processes. This critique is mainly based (1) on a concept of perception as an active exploration of the environment, and (2) on the coextensivity of "lived body" and "physical body," or *Leib* and *Körper*. This refutes the notion of a non-extended inner mind and provides us with a space in the real world that we actually inhabit. In *Chapter 2*, I provide a critique of the notion of the "brain as a subject," demonstrating the non-reducibility of subjective, in particular, intentional and temporal, experience to neuronal processes. A critique of both the "mereological" and the "localizational fallacy" characterizing the attempts to identify subjective experience with brain processes completes this part.

In the second part of the book, I provide a step-by-step analysis of the brain as an organ of the human person, drawing on various approaches from phenomenology, philosophical biology, and embodied and enactive cognition. The crucial basis for this is developed in *Chapter 3*, namely an ontological concept of "embodied subjectivity," implying a person's *dual aspectivity* as a unity of "lived body" and "physical body." As a living being, the human person thus appears under two complementary aspects, the one experienced from a first- and second-person perspective, the other regarded from a third-person or observer perspective. The common reference of both aspects to the living organism and its life process replaces the dualism of mind and body: processes of life (*Leben*) and of experiencing life (*Erleben*) are inseparably linked, for experience is nothing else but the specifically integrated and intensified life process.

On this basis, I then develop an *ecological conception of the living organism* as an autopoietic system related to its environment. From a certain level on, the organism becomes a center of subjectivity, which should be regarded as a continuous integration of the life process itself. Moreover, this conception involves a closer examination of the specific causality of living systems which I describe as a connection of vertical (inner-organismic or part–whole) circular causality and horizontal (organism–environment) circular causality. This leads to a concept of *integral causality* by which living beings become the causes of their own conscious enactments of life.

Following on from this, in *Chapter 4*, I focus on a conception of the brain as an organ of a living being within its environment. First, the constant resonance between brain and organism is the basis of a background feeling of the body, a *"feeling of being alive"* that may be considered the foundation of all conscious experience. The relations of brain, organism, and environment are then, as a centerpiece of the investigation, portrayed in several related respects. The focus lies on the functional cycle of perception and movement, constituting a unity of brain, organism, and environment as a superordinate system. As a crucial consequence of this ecological model, consciousness is regarded as the *ongoing integral of the functional loops between organism and environment*. Replacing representationalist accounts of brain functioning, the concept of the brain as a *mediating* and *resonance organ* is developed in detail. The relation of conscious experience to the mediating processes is finally interpreted by Hegel's notion of "mediated immediacy."

In *Chapter 5*, emphasis is on the brain as a social, cultural, and biographical organ, including a broad view on recent research findings in developmental psychology and neuroscience. The impact of early intersubjectivity on brain development is considered, including the formation of implicit memory and attachment relations, as well as the social resonance systems connected to intercorporeality and primary empathy. Turning to secondary intersubjectivity, language acquisition is examined as the anchoring of an embodied interpersonal practice, connected with the neural resonance system of mirror neurons. After taking a look at the further development of perspective taking and reflective self-consciousness, the chapter closes with some fundamental considerations concerning brain and culture.

Chapter 6 offers an analysis of several outcomes of this ecological and dual-aspect approach for the mind–body problem. The unity of the living organism and its enactments of life provides an alternative to the separation of the mental and physical as well as to identity theories in philosophy of mind. The concept of integral causality is further differentiated in the light of emergence theories, emphasizing the primacy of holistic functions over their components. The role and function of consciousness as integral to the organism–environment interaction is discussed in detail. This gives rise to several conclusions regarding the intentional determination of neuronal processes, including in particular a concept of embodied freedom.

Chapter 7 examines the consequences of the conception for etiological and therapeutic approaches in psychiatry and psychological medicine. Contrary to current reductionist tendencies in neuropsychiatry, it develops a concept of mental illness as a circular process with a crucial impact on a person's self-experience and interpersonal relationships. The effects of somatic therapy and psychotherapy are then explained from the standpoint of dual aspectivity, providing an important example for the concrete applications of the concept in medical practice. In summary, an orientation towards embodied subjectivity is shown to be indispensable for psychological medicine.

Finally, *Chapter 8* summarizes fundamental concepts and insights of the book. The brain is presented as an organ of mediation, transformation, and resonance. Its functions are integrated by the living organism as a whole, or by the embodied person, respectively. In contrast to brain centrism, this leads to an integral, personalistic concept of the human being which has its basis in intercorporeality: it is in the concrete bodily encounter that we primarily recognize each other as embodied subjects or persons. All scientific endeavors to reveal the functioning of the brain are ultimately dependent on this fundament in the life-world.

Part 1

Criticism of neurobiological reductionism

Part

Chapter 1

Cosmos in the head?

> **Overview**
>
> Chapter 1 contains a criticism of the neuroconstructivist epistemology, according to which phenomenal reality is to be understood as an internal mirroring or a reconstruction of the outer world by means of neuronal processes. As it turns out, the idealistic theory of representation is still the basis of this conception (1.1). The criticism developed in the chapter emphasizes, in contrast, the enactive character of perception which is always connected with the operative capacities of the body. In order to prove that the subjective space of the body is not only virtual, its coextension with the space of the objective body or the entire organism is accounted for in detail (1.2). On this basis, in contrast to the conception of an interior phenomenal world, the objectifying achievement of perception, which brings us into direct connection with things by means of circular interactions, can be recognized (1.3). Finally, taking the example of colors, the claim of the mere virtuality of perceived qualities is rejected (1.4).

The assumption that everything that people experience is, in reality, a construction or even an illusion created by their brains is one of the common convictions of neuroscientists and neurophilosophers. In particular, neuroimaging results, due to their seemingly simple and suggestive presentation, have ignited the enthusiasm of researchers, lay people, and the media alike. Assuming that we can literally watch the brain thinking, perceiving, or feeling, there is hardly a phenomenon, from pain or anger to colors or music and even to love or faith, which is not accommodated somewhere in the brain. The almost taken-for-granted view that reality can be found in the head turns perception, so to speak, into a physiological illusion. Typical descriptions thus read as follows:

> What you see is not what is *really* there, it is what your brain *believes* is there. (Crick 1994, 31)
>
> Multimedia mind-show occurs constantly as the brain processes external and internal sensory events. (Damasio 1999b, 112)
>
> [T]he world around you, with its rich colours, textures, sounds, and scents is an illusion, a show put on for you by your brain [...] If you could perceive reality as it really is, you would be shocked by its colorless, odourless, tasteless silence. (Eagleman 2015, 37)

One of the most radical elaborations of such assumptions is found in neurophilosopher Metzinger's "Ego Tunnel":

> Conscious experience is like a tunnel. Modern neuroscience has demonstrated that the content of our conscious experience is not only an internal construct but also an extremely selective way of representing information […] First, our brains generate a world-simulation, so perfect that we do not recognise it as an image in our minds. Then, they generate an inner image of ourselves as a whole […] We are not in direct contact with outside reality or with ourselves […] We live our conscious lives in the Ego Tunnel. (Metzinger 2009, 6–7)[1]

According to this neuroconstructivist conception, the real world is dramatically different from the one that we experience. What we perceive are not the things themselves, rather the mere images that they evoke in us. We find ourselves in a dark room and look at a show projected on its walls by the tireless work of myriads of neuronal brownies. The real world is a rather bleak place of fields of energy and movements of particles, without any qualities whatsoever. The tree in front of me is actually not green, its blossoms are not fragrant, the bird in its branches does not sing melodically: all these are only useful illusory worlds, simulated realities, or models which the brain produces in place of bare, materially kinematic processes. In fact, we remain locked in the hollow of our skulls like Platonic cavemen. Metzinger himself points to this analogy (Metzinger 2009, 22); however, in the case of the brain, the wall of the cave

> is not a two-dimensional surface but the high-dimensional phenomenal state-space of human Technicolor phenomenology. Conscious experiences are full-blown mental models in the representational space opened up by the gigantic neural networks in our heads. (2009, 23)

Thus, we are enclosed in the skull as Plato's prisoners, yet with the cave being our mind itself, or rather a mental projection screen or "phenospace" (Metzinger 2009, 221). Indeed phenomenal experience is nothing else but "an *online* hallucination" (Metzinger 2003, 51).

Of course, even neuroscientists or neurophilosophers continue to live with this insight in the everyday world of "naive realism." And they are well advised to do so; for if the world of our experience were in fact only a virtual product of our brains, how could we ever find out anything about the actual world "out there?" If we were only in contact with a *reality simulation*, a world of mere appearances, how could neuroscientists make any statements about "real brains?" Already in terms of knowledge theory, such a position is obviously untenable.

---

[1] As Metzinger explains, the metaphor of the "tunnel" signifies that the world-simulation is not only present but also extended in time (2006, 23).

However, the result of the scientific reinterpretation is a creeping virtualization of perception—just as if we could not basically trust our senses, and only physics or neurobiology could enlighten us about the real nature of the world. In any case, we are told to give up our naive notion that in perception we are in contact with the things themselves.

1.1 The idealistic legacy of brain research

Where do such conceptions stem from? As we will see, the epistemology of neuroscience carries the burdensome legacy of its greatest opponent with it, that is, idealism.

In the "Introduction," it was already demonstrated how the reductionist program of the natural sciences has gradually eliminated all qualitative properties of nature. Color, heat, smell, taste, as well as categories such as purposiveness or goal-directedness of living organisms were assigned to the human subject as anthropomorphic constructions. Indeed, the atomism of Antiquity had already carried out this separation—in Democrit's words:

> Sweet exists by convention, bitter by convention, color by convention; but in reality atoms and the void alone exist. (See Soccio 2012, 72)

In modern times, Galilei took up this theory once again:

> To excite in us tastes, odors, and sounds I believe that nothing is required in external bodies except shapes, numbers, and slow or rapid movements. I think that if ears, tongues, and noses were removed, shapes and numbers and motions would remain, but not odors or tastes or sounds. (Galilei, *Il Saggiatore*, 1623; Morton 1997, 59)

John Locke canonized this viewpoint by distinguishing the primary and secondary qualities of perception: only the quantitative categories (volume, shape, number, and movement) are primary or "real," all qualitative characteristics (colors, smell, taste, sound) are secondary or anthropomorphic.

In parallel to this, the modern concept of *consciousness* emerged as that of a container, into which everything qualitative and subjective could be inserted. With the reinterpretation of life as a form of physical process, experience lost its embeddedness in life activity and was banished to its own sphere of the purely "mental." Conceived by Descartes as a refuge of the mind in the face of the sole reign of physics over the material world, consciousness was since then in danger of becoming a closed chamber, a windowless enclosure of the subject. Every possible object of Cartesian consciousness is, namely, an "*idea*"—a thought, a representation, or an image. Moreover, what we perceive are also images and not the things themselves. *Idealism* is the philosophy which, in the wake of Descartes, develops from the image-theory of perception. For Locke, Hume, and Kant, our perceptions are "impressions," "ideas," or "representations" from which we can only draw problematic conclusions regarding the reality, in which we believe we are living. The idealist sits in the enclosure of his consciousness and receives the "*ideae*" as the delegates and representatives of things which he never gets to see themselves—in Locke's words:

> For, methinks, the understanding is not much unlike a closet wholly shut from light, with only some little openings left, to let in external visible resemblances, or ideas of things without: would the pictures coming into such a dark room but stay there, and lie so orderly as to be found upon occasion, it would very much resemble the understanding of a man, in reference to all objects of sight, and the ideas of them. (Locke, *An Essay Concerning Human Understanding*, vol. I, ch. 11, §17; Locke 1813 151–152)

In Kantian epistemology too, the world is taken into this inner room: space and time are pure forms of intuition and thus produced by the mind. The world is recognizable, but only because we are not actually in it, rather *it is in us*.

> But appearances are only representations of things that exist without cognition of what they might be in themselves. As mere representations, however, they stand under no law of connection at all except that which the connecting faculty prescribes. (Kant, *Critique of Pure Reason*, B 164; Kant 1998, 263)

Whereas reason is given full authority to structure the world, this happens only within a closed jurisdiction. Goethe already argued against this with the unmistakable eye of the beholding naturalist that idealistic philosophy could never reach the object.[2]

The further development of idealism can only be hinted at here. In his "Wissenschaftslehre" (1794), Fichte seeks an answer to the question: "How do we come to assume that something external to us corresponds to the representations within us?" (Fichte 1992, 87). In his following deduction of how the world is, in principle, produced by the transcendental Ego, the notion of the "external world" (*Außenwelt*) has its first philosophical appearance (1992, 388). The way leads further on from Schopenhauer's "World as will and representation" (1819/1966) to Nietzsche's perspectivism and, finally, to the Radical Constructivism of the present. René Magritte's well-known picture *La condition humaine* (Figure 1.1) illustrates how much the idealistic conception of perception has molded enlightened consciousness in the twentieth century.

---

[2] Letter to Schultz, 18 September 1831; see Werke, *Hamburger Ausgabe*, Vol. 4, p. 450.

Figure 1.1 *La condition humaine.*
[*The Human Condition*], 1933 (oil on canvas), Magritte, René (1898–1967). National Gallery of Art, Washington DC, USA/Bridgeman Images. © ADAGP, Paris and DACS, London 2017.

The picture shows a painting with a landscape which is undistinguishable from the real landscape behind the painting (and in fact, both *are* paintings!). In a lecture in 1938, Magritte himself explained the picture as follows:

> The problem of the window resulted in *La Condition Humaine*. I placed in front of the window a canvas, which was to be seen from the interior of the room, and which represented precisely that piece of landscape which was hidden by the canvas. The tree which was represented on the canvas thus concealed the tree situated behind it outside of the room. For the viewer, it was thus placed inside the room, on the canvas, and at the same time, through the imagination (*pensée*), outside the room in the real landscape. That is

exactly how we see the world. We see it outside of ourselves and, nevertheless, we only have a representation of it in us.[3]

Here the doctrine of the "external world," with its strange duplication of reality, is indeed stylized as the *conditio humana* itself. The windows of our soul monads are closed and all that we receive from the outer world are representations—multicolored pictures which the painter of consciousness has created for us.

This idealistic epistemology—truly, under changed circumstances—has also made its way into brain research and the neurophilosophy related to it. For them, too, we only live in a subjective reality which is, however, now constructed or simulated by the brain. In the interior space of consciousness, the subject, the lonely prisoner in his own citadel, watches the pictures of the unreachable outer world. The only thing is that these pictures are no longer constructs of Kantian faculties of understanding, but rather of the underlying brain processes. What corresponds to the Cartesian *ideae* or images are the "neural representations"—specific excitation patterns through which the brain mirrors the structures of the outer world.

As can be seen, the idealistic chamber of consciousness and the neurobiological inner world of the brain match one another surprisingly well. Neuroconstructivism only makes the connection between the two traditions. Thus, materialism and subjective idealism paradoxically extend hands to each other as they ascertain the point they have in common: namely, that the subject has no part in the world. Admittedly, materialism can finally triumph because, with the reduction of the ability to recognize and act on the processes of the brain, the idealistic subject is no longer left even with the power over his own palace.

The picture of the world as an internal construct—this epistemological conception is to undergo a criticism in three steps. It will at heart consist in refuting the picture of a bodiless and worldless subject which underlies the idealistic theory of perception.

1.2 First criticism: embodied perception

1.2.1 Perception and motion

Let us return once again to the supposed "condition humaine." Is Magritte right, and do we, in reality, only see pictures? Of course, we could, in case of doubt, easily ascertain whether there are in fact meadows and trees outside of the window, in the so-called outer world, or whether it is a film set or another type of

---

[3] My own translation from D. Sylvester (Ed.), *René Magritte. Catalogue Raisonné II: Oil Paintings and Objects 1931–1948*. Antwerpen: Menil Foundations, Fonds Mercator 1993, p. 184.

illusion. We could simply go out and check it with our senses and movements. We never, indeed, perceive "from nowhere," but rather from our situated bodily position. The sight of the window "over there" already includes the possibility of moving to it. The perception of spatial depth itself only emerges in connection with the ability to measure its diameter and to grasp the objects from different aspects depending on our own movement. When we perceive, we are always situated in the same world as the things we perceive, that is, they are perceived *as available for our interaction* with them.

The underlying assumption of neuroconstructivism is that there is an external reality which is only given to us through representations in our mind. This fundamental assumption of an inner mind being separated from external reality is challenged by the current concepts of *embodied and enactive cognition* (Varela et al. 1991, O'Regan & Noe 2001, Thompson 2005, 2007, and others).[4] From an enactive point of view, reality is not something predetermined and external, but is continuously brought forth by a living being's *sensorimotor interaction* with its environment. Hence, the idealistic conception of perception ignores the fact that as embodied subjects we are not locked into our consciousness. Embodiment does not come as an external addition to perception, but, rather, it is constitutive for it. We must be physically in the world, be related to it, be able to move and act in order to perceive anything at all. It is only the dominance of an epistemology based on our visual sense and its metaphors (picture, perspective, representation, etc.) which makes us forget our embodiment. As a matter of fact, there is no "outer world" perceived by a passive, bodiless subject, as Magritte's picture suggests. This is also evidenced by the development of vision.

Half a century ago, Held and Hein (1963) carried out a classic experiment on newborn kittens who are blind at first. Two kittens in each case were placed in a cylinder marked with vertical stripes from which they got visual input (Figure 1.2). One kitten could walk around in the cylinder of its own accord, while the second kitten was riding in a gondola harnessed to the first and attached to the central axis. After some weeks of intermittent exposure to this procedure, the kittens of the first group were freed from their harness and they moved perfectly normally. In contrast, the other kittens who had remained passive in their gondola were incapable to orientate in space and to recognize objects, they stumbled and bumped helplessly against objects. In terms of visual input, they had received the same stimuli as the kittens of the first group but, nevertheless, remained blind to the structure and spatiality of their surrounds. This means that only a sensing and moving organism forms experienced space, namely from the coherently connected visual, motor, and vestibular patterns it receives.

---

[4] Concepts of enactivism will be more thoroughly dealt with in Chapters 3 and 4.

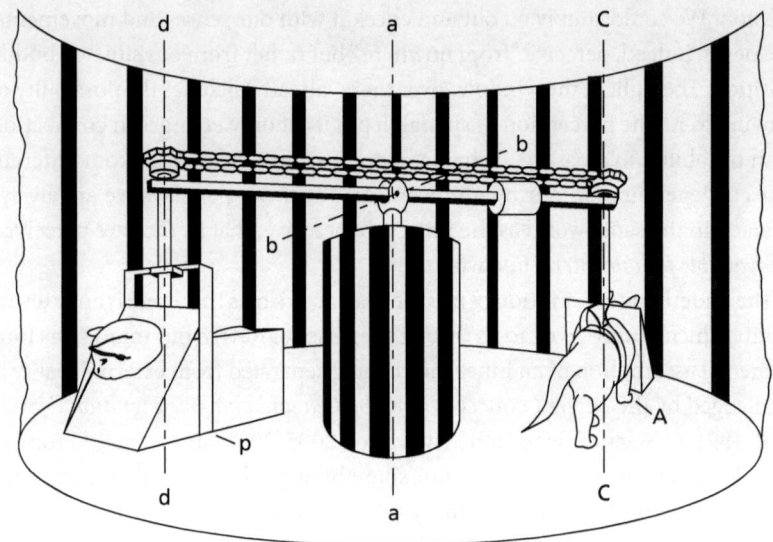

Figure 1.2 Spatial arrangement in Held and Hein's experiment (1963).
Reproduced from Richard Held and Alan Hein, Movement-produced stimulation in the development of visually guided behavior, *Journal of Comparative and Physiological Psychology*, 56 (5), pp. 872–876. http://dx.doi.org/10.1037/h0040546, Copyright © 1963, American Psychological Association.

Cataract surgery restoring vision in persons who had become blind early in their life led to comparable results: even though their retinas could now receive light, what they experienced was only a chaotic flickering of stimuli, no regular vision, and in particular no depth perception. Even after years of training, vision remained severely impaired in most cases, for beyond adolescence the brain could no longer adequately adapt to the unfamiliar input (von Senden 1960, Thinus-Blanc & Gaunet 1997).

From these observations it follows that something as basic as spatial perception is only possible for embodied and active beings.[5] If we were from birth unable to interact with the world in any way, we would never become able to

---

[5] This connection between visual perception and self-movement has been aptly analyzed by Hans Jonas (1966/2001, 152–156): "We may therefore say that the possession of a body in space, itself part of the space to be apprehended, and that body capable of self-motion in counterplay with other bodies, is the precondition for a vision of the world" (Jonas 2001, 156).—Moreover, in his ecological theory of perception, Gibson (1979) has demonstrated the dependence of perception on an organism acting within its environment: What we perceive is primarily what we can interact with, and what provides options or "affordances" for our action.

see. Vision is, like all other abilities to perceive, only an extension of the bodily basis of all experience. In perception, a living being is not in opposition to the world, but always already involved and entangled in it—as it is obvious from the very meaning of "perception" (from the Latin *capere* = to catch). Thus, our perceptual capacities develop in the course of our interaction with the world, implying the continuous circularity of perception and motion. We would not be able to recognize abstractly what the meaning of "long," "deep," "soft," "heavy," "hot," or other qualities is—we have to experience them as bodily beings. Likewise, the perception of doors and windows, meadows and trees, and humans and animals depends on our sensorimotor dealings with them. Perceiving has always meant taking part in the world, touching it and being touched by it. It is based on *embodied practice*.

Now, we may perhaps concede the embodied nature of perception—but is not the bodily subject itself only a construct? The spatial body schema, touch, proprioception, kinesthesia, and visceroception are these not all produced in certain well-known areas especially of the parietal cortex and projected into the virtual space construed by the brain? The phantom limb of amputated patients and related experiences of healthy persons, in which bodily feelings are localized outside the limits of the body, seem to prove amply that our subjective body is nothing more than a habitual phantom body, a simulation or construction of the brain. To demonstrate this, neuroscientist Ramachandran also points to the well-known rubber hand illusion (Botvinik & Cohen 1988, Ehrsson et al. 2004): if the concealed hand of a person is touched in the same rhythm as a rubber hand visibly lying before him or her on the table, then after some strokes the subject will feel the rubber hand actually being "touched" as if belonging to his own body. From this and similar body illusions, Ramachandran concludes: "*Your own body* is a phantom, one that your brain has temporarily constructed purely for convenience" (Ramachandran & Blakeslee 1998, 58).

The subjective body would thus also be a construct of the brain like the complete experienced reality. The result is a split between the organic body and the subjective body, as if these belonged to two different worlds—one to the physical world and the other to a virtual "inner world" of consciousness construed by the brain. This would apply to all bodily sensations:

> Pain itself is an illusion—constructed entirely in your brain like any other sensory experience. (Ramachandran & Blakeslee 1998, 58)

> You can reach out and touch the material of the physical world […] But this sense of touch is not a direct experience. Although it feels like the touch is happening in your fingers, in fact it's all happening in the mission control center of the brain. It's the same across all your sensory experiences […] your brain has never directly experienced the external world, and it never will. (Eagleman 2015, 40f)

Now certainly the brain has never experienced the external world, for it cannot in principle experience anything. But what about myself? Is my spatial sense of touch in my fingers or pain in my foot only an illusion? If perception is intended to convey more than a virtual world, apparently the alleged virtuality of bodily experience must be refuted. As we will see, the fundamental Cartesian division between subjective and objective body is indeed unable to withstand a closer analysis.

1.2.2 The coextension of lived body and physical body

Let us first envision the fact that we normally experience the subjective or lived body (*Leib*) and the organic body (*Körper*) as *coextensive*. The pain felt is located where the needle pierced the physical body. The potter feels the clay exactly where his hand, in fact, presses and forms it. Indeed, if a patient shows the doctor his painful foot, the latter will also look there for a cause. If the subjective experience of the lived body were only an illusion, he could ignore the statement of the patient and, instead of that, examine his brain. There is thus a spatial correspondence or *syntopy* of the lived body and the physical body. This syntopy was already analyzed by Husserl (1989), using the example of the hand feeling the touching of an object that simultaneously moves over the skin. In this "co-apprehension" of what is given in the subjective and the objective attitude, the body manifests itself as a unity:

> Thus there lies in the sensations an order which "co-incides" with the appearing extension [...] From the very outset, it [the hand] is apperceptively characterised as a hand with its field of sensation, with its constantly co-apprehended state of sensation which changes in consequence of the external actions on it, i.e. as a physical-aesthesiological unity. (Husserl 1989, 162–163)

Although the phenomenon of phantom pain shows us that the organism and brain can also induce a sensation of pain without the respective limb, this does not make the normal case any less astonishing. How is it actually possible that we feel the pain in reality where the matching wounded part of the body is situated, too, and not in the brain?

The coextension of the lived body and the organic body cannot be explained by a "projection" of bodily sensations into the space of the latter, for the objective space of the organic body would have no existence in a virtual subjective world. There cannot be a projection "towards the outside" if this outside is, according to the assumption, merely an interior world constructed by the brain. The projection concepts, which were rather common formerly, were thus largely replaced in the cognitive neurosciences by a unified virtual-phenomenal space, or a "phenospace" (Metzinger 2009, 221). Consequently, then, the perceived prick of the needle too, which causes the pain, must be declared a virtual construct or a simulation of the brain. We would then have absolutely no access to actual reality.

However, as soon as we enter an *intersubjective* situation as the patient already mentioned at a doctor's visit, it becomes immediately clear that the subjective experience and the objective situation, the sensation of pain and its observable physical cause, in no way belong to two separate worlds. The syntopy or the coincidence of the place of pain and injury now, indeed, involves the body perceived by both the doctor and the patient. Just there where the patient feels the pain and where he points to is where the doctor also finds its cause. *Both see the same foot which subjectively hurts and is objectively injured.* How is this possible?

Here we first have to ascertain that a reference to the respective "phenospaces" of doctor and patient is no longer possible—if talk about a reality of the body is to have any sense at all, then in the intersubjective situation. For in this context the subjective spaces of both persons coincide in a way *which cancels their mere subjectivity*. The argument goes as follows:

> Since, according to the neuroconstructivist premise, every brain only produces its own virtual space, there cannot be any "shared phenospace" of doctor and patient. For if perception could, without remainder, be described and explained as a physical process happening between an object and a brain, *then two persons could never observe one and the same object*. The two processes would run, starting from the object, in different directions and remain strictly separated from one another. Both persons would thus be locked in their particular world, all the more since they remained themselves only simulations for each other—in the end leading to a *neuro-solipsism*. To the extent, however, that the intersubjectively constituted space possesses objectivity—if it did not possess it, it would not be possible to agree on an understanding of mutually perceived objects, indeed not even on a simple exchange of goods as during shopping—it also shows that the particular subjectively experienced spaces, on the basis of which it is constituted, are *not only virtual*. The subjective view is thus, admittedly, an *individual, perspectival view*. It is, however, not "only subjective" in the sense as if what was seen was only "in the subject." When we see, we are always already in a space shared with other persons.

The body perceived by both doctor and patient in agreement can thus no longer be a merely subjective specter—rather, it is located in the shared, intersubjective and, as such, objective space. Now, the subjective place of the patient's pain concurs with the objective place of the body part. Hence, the subjective-bodily and the objective space *in reality* coincide and we must repeat the question: how is it possible that the patient feels the pain *there* and not, for example, in the brain?

The direction of the question admittedly shows that we in the Cartesian tradition are still used to categorically separating subjectivity from the living organism. It is completely different in evolutionary terms: originally the whole body was, in a sense, a sensing and feeling organ. Precisely at its surfaces which border on the environment, the organism is irritable, sensitive, and responsive. *Elementary sensitivity begins at the periphery of the body.*[6] The development of a central nervous system does not remove peripheral sensitivity, but *integrates* it by means of the peripheral nervous system spread over the whole body. The fact that bodily consciousness does remain coextensive with the organism shows that it does not spring up from it as a separate entity, like Athena from the head of Jupiter. Rather, it is, from the very beginning, an *embodied and extended consciousness*. It presents the *"integral" over the living organism altogether*, not a phenomenon encapsulated in the brain.

Seen in this way, the coextension of the subjective, lived body and the material organic body is no longer surprising. It is, however, functionally meaningful too: conscious experience is where the interactions with the environment take place—in the periphery, not in the brain. After all, the body is the actual "player on the field." That is why it is meaningful that its borders, positions, and movements in the environment are experienced in "analog" form, that is, in the space of the lived body, not only cognitively registered.

It would also be theoretically conceivable that pains would become conscious in a placeless manner, such as thoughts or memories. However, without the coincidence of the two spaces, we would only have our body as an external tool to be plied and would not be "incarnated" in it. Only because consciousness is *in the painful hand*, it is withdrawn involuntarily from the pricking needle.[7] Only

---

[6] This is in agreement with Antonio Damasio's opinion that perception in its evolutionary primal form consisted in experiencing "*the outside world in terms of the modification it causes in the body proper*" (Damasio 1995, 230). "In the beginning, there was no touching, or seeing, or hearing, or moving along by itself. There was, rather, *a feeling of the body* as it touched, or saw, or heard, or moved" (p. 232). The body is thus the mediating organ whose peripheral sensations, by becoming "transparent," enable to perceive the environment.

[7] This was even clearly recognized by Descartes in his "Meditations" (ch. VI). Granted, he writes, the stimulation of pain fibers in the foot only leads us to sense the pain "as though" they were in the foot. Nevertheless, this illusory local sensation is meaningful, for it lets us withdraw the foot or in another way remove the cause of the pain we are feeling there: "It is true that God could have constituted the nature of man in such a way that this same movement in the brain would have conveyed something quite different to the mind; for example, it might have produced consciousness of itself either in so far as it is in the brain, or as it is in the foot, or as it is in some other place between the foot and the brain; but none of all this would have contributed so well to the conservation of the body" (Descartes 1993, 99). Descartes only refuses to draw the necessary conclusion, namely to conceive of the subject of pain *as bodily and spatial*.

because the feeling of the potter is *in the touching hand* by which he feels the structure of the clay can he also mold it skillfully. A mere "central processing" in the brain could never achieve what the immediate presence of the subject in his hand makes possible, that is, the linking of perception, movement, and objects into *a common, intermodal action space*: "My body is wherever there is something to be done" (Merleau-Ponty 1962, 224). We can thus speak of a not merely embodied, but also an extended or "*ecological subjectivity*" (Bateson 1972, Neisser 1988).

Therefore, if I grope for something, I move *and I feel* not a virtual, but a real hand which, for its part, touches a real object. That becomes possible by the fact that the subjective bodily space *is embedded in the objective space of the organism in its environment*. This means: we are actually in the world as bodily beings (*leibhaftig*)—we are not beings who only have the illusory feeling of inhabiting their body.

Admittedly, the extension of the subjective, lived body is flexible—that is, corresponding to the particular functional requirements. It does not always square exactly with the limits of the objective body. That is why instruments can also be integrated in the subjective body schema: When groping with a walking cane, one does not feel the resistance of the surface being groped for in one's hand, but rather at the cane's tip.[8] The trained driver notices the quality of the street coating literally under the tires of his car. A person who has had a limb amputated learns to "incorporate" his prosthesis by adapting to it, so that it becomes a new body limb for him. In fact, even a rubber hand can temporarily connect to the felt body if it is included in the loops of sensation, perception, and movement in a coordinated manner—in just the same way as in ventriloquism the speaker's disguised voice is attributed to a dummy. In all these cases, far from being "merely illusions," the *optimal coherence* of the various sensory and kinesthetic modalities is established within the intermodal action-perception space of the body.[9]

---

[8] 'The blind man's stick has ceased to be an object for him, and is no longer perceived for itself; its point has become an area of sensitivity, extending the scope and active radius of touch' (Merleau-Ponty 1962, 127). See also Gallagher and Zahavi (2008, 141–148).

[9] On the one hand, one may call the rubber hand experience an illusion—after all, one's hand is actually being touched under the table. "But in another sense there's no illusion—or rather, the mechanisms at work in this illusion, if we want to call it that, are those of normal, successful perception" (Noë 2009, 74). Hence, such illusions do not prove perception *as such* to be illusory or merely a "veridical hallucination"; on the contrary, they point to the synthetic or gestalt-forming activity of perceiving which renders the environment available and viable for a moving and acting being. For experimental studies on this formation of intermodal coherence ("dynamic capture"), see Soto-Faraco et al. (2004).

Instead of being only a central construct, the subjective space of the lived body is thus modified depending on the particular border at which the real interaction with the environment takes place. This is, in turn, functionally appropriate: the physical contact with the actual resistance of the surrounds must feed into the person's subjective experience, so that adequate handling of objects and tools is made possible. The supposed illusions which arise from this are, in reality, highly useful extensions of our subjective body schema in flexible contact with the environment (Gallagher 2005, 142–146). As Merleau-Ponty remarked, the spatiality of the lived body is not "a *spatiality of position*, but a *spatiality of situation*" (1962, 100). This means that the objective space of the physical organism and the subjective space of bodily experience are intertwined and mutually modify one another.

Granted, the phenomenon of phantom limbs or phantom pain shows that the habitual body schema (anchored in the somatosensory cortex of the brain) is part of the subjective bodily space. As a consequence, the latter's extension may sometimes deviate from the objective or physical body to a surprising extent. However, just like the phenomena of extension in tool use mentioned before, such exceptions do not contradict the *basic syntopy*, that is, the coextensive spatiality of living and organic body—on the contrary, they even confirm it. If *Leib* and *Körper* would not be coextensive normally, a person with an amputated limb would not even notice his phantom limb as such, for there would not arise any discrepancy between both types of space. However, only the fundamental syntopy is at stake here, if we want to refute the illusion thesis or the idea of a mere "phantom body."

In order to make this central point for the further investigation quite clear, we ask once again: where is the pain now when my foot hurts me? According to common neuroscientific belief, it is where it is produced, that is, in the brain. Even John Searle, a prominent critic of neurobiological reductionism, is of this opinion:

> Common sense tells us that our pains are in physical space within our bodies, that for example, a pain in the foot is literally inside the area of the foot. But we now know that is false. The brain forms a body image, and pains, like all bodily sensations, are parts of the body image. The pain-in-the-foot is literally in the physical space of the brain. (Searle 1992, 63)

However, the brain does not feel pains, nor does it contain them. It does not produce a "body image" either, for the experienced body is not an image of a body, it is rather *the body itself as felt*. The only thing that can be found in the brain *when* somebody feels pain are neuronal activations in the somatosensory cortex and in the cingulate gyrus, and however much these may have to do with the pains—they *are not* the pains.[10]

The pain-in-the-foot is thus *neither* in the physical space of the foot, *nor* is it in the physical space of the brain, for pains are, after all, neither anatomical things such as sinews, bones, or neurons, nor are they physiological processes such as charge-transfers at neuronal cell membranes. Where is the pain then? It is in the "foot as a part of the living body," for this unified living body also produces—not least by means of the brain—a *spatially extended body subjectivity*. The fact that I can state meaningfully: "I have pains in my foot" and can also show the doctor the same foot presupposes that the subjective space of my pain and the objective space of my foot do not belong to two separate worlds which are only connected with one another in a causal way (namely via physiological processes in the brain). It presupposes *that the subjective and the objective space of my body syntopically coincide.*

This is certainly difficult to accept for a physicalist thinking. Is it not true that the "ghost in the machine" (Ryle 1949) is here being wakened again? Is it intended to allow the soul a secret readmission into the physically cleansed world? Indeed, it was a self-evident part of Aristotelian and Pre-Modernist belief that the soul was indivisible and, nevertheless, coextensive with the organic body.[11] Even Kant still wrote in his pre-critical period:

> I would, therefore, keep to common experience, and would say, provisionally, where I sense, there I am. I am just as immediately in the tips of my fingers, as in my head. It is myself who suffers in the heel and whose heart beats in affection. I feel the most painful impression when my corn torments me, not in a cerebral nerve, but at the end of my toes. No experience teaches me to believe some parts of my sensation to be removed from myself, to shut up my Ego into a microscopically small place in my brain from whence it may move the levers of my body-machine, and cause me to be thereby affected. Thus I should demand a strong proof to make inconsistent what the

---

[10] Of course, identity theory claims exactly this. Although the coextension of subjective and objective bodily space is incompatible with an identity of consciousness and brain processes, identity theory cannot be criticized extensively here (on this, see mainly sections 2.2.1 and 6.2). But even if one would assume an identity of neural processes with pain sensations (however this identity might be conceived), it would still not be possible to locate the pain *as pain* in the physical space of the brain. Searle's statement therefore commits a category error.

[11] See Aristotle, *De Anima* 411 b 24: "In each of the bodily parts there are present all the parts of the soul." Similar statements are found, for example, in Meister Eckhart: "The soul is one and indivisibly complete in the foot, and complete in the eye, and complete in every limb" (Meister Eckhart 1958, sermon 10, 161–165), or in Thomas Aquinas: "*Anima hominis est tota in toto corpore et tota in qualibet parte ipsius*" (Thomas Aquinas 1953, I q 93 a 3).

schoolmasters say: my soul is as a whole in my whole body, and wholly in each part. (Kant 1766/1900, 49)

If the phenomenal experience of lived body space is related to intersubjective and, hence, objective space, this is, in fact, linked to some extent to the doctrines of a coextensivity of "soul" and "body," admittedly with quite a different terminology. Descartes argued against this, saying the body was only a machine of parts and thus divisible like a corpse, whereas the soul represents an indivisible whole.[12] It is, however, not necessary to reanimate Descartes's independent soul substance, in order to reconcile the experience of our lived bodily being-in-the-world with an objective view of the physical body. The pre-condition is much rather *an adequate concept of life*: the organism itself represents, namely, a functional whole which is, as such, indivisible and, at the same time, extended in physical space—in parallel to the subjective body and its indivisible extension.[13] The fact that this whole living organism can become the bearer of a likewise spatially extended subjectivity does not add any new entity to the *purely physically* describable world, and thus does not contradict physical laws. However, it means a fundamental change for ourselves as living beings: we are no longer self-contained monads, to whom an image of the world is feigned, but rather *we inhabit our body and, by means of that, the world*. Phenomenology can thus put our primary experience in its rightful place again, namely to be in the world as incarnated beings.

Let us summarize for the time being: we started from the deliberation that perception does not mean recording images passively in an otherworldly consciousness. All perception is much rather embodied: it is based on sensorimotor interaction with things, on *concrete bodily practice*. It was further shown

---

[12] "[T]here is a great difference between mind and body, inasmuch as body is by nature always divisible, and the mind is entirely indivisible. [...] and although the whole mind seems to be united to the whole body, yet if a foot, or an arm, or some other part, is separated from my body, I am aware that nothing has been taken away from my mind" (Descartes, *Meditations VI*; 1993, 97). However, Descartes neglects that the *life* of the body does not consist of parts that could be severed ad libitum (*partes extra partes*).

[13] The organism as a functional whole is indivisible, for as Aristotle already remarked, an amputated hand is no longer a *hand* in the functional sense of the word (Aristotle, *Metaphysics*, Z 11, 1036 b 30ff). Here, the parts only exist as parts of a whole (on this, see 3.2.1). Likewise, the *subjective body* is indivisibly extended too, inasmuch as all spatially distributed bodily sensations nevertheless pertain to one and the same subject and are integrated in the body schema. In feeling one's own body parts with eyes closed this may easily be verified: one's felt arm and felt leg belong to an integrated whole of proprioception. Likewise spatial and yet indivisible are the feelings of vigor, tiredness, or sickness, which extend over the whole body.

that the subject of perception is extended over bodily space, and this is not in the form of a phantom entity, a mere construct of the brain, but rather as *an embodied subjectivity* incorporated in a living body, continuously growing out of, and coextensive with it. The somatosensory and motor structures in the brain are, admittedly, necessary requirements for this subject experience. However, that does not mean that the bodily subject could be localized in the brain like Descartes's soul in the pineal gland. The peripheral and autonomic nervous system, the senses, the skin, the muscles, the heart, the viscera—all these are carriers of subjectivity too. We belong to the world, with skin and hair—we are bodily, living, and thus more "organic" beings than neuroscientific cerebrocentrism would have us believe.

1.3 Second criticism: the objectivity of the phenomenal world

1.3.1 The space of perception

What was shown for the awareness of one's body is now to be extended to perception as a whole. Is the illusion-thesis true for this? Do we see in reality only pictures appropriately constructed and projected onto the mental screen of our consciousness by the *camera obscura* of the brain?

Of course, it works quite differently in phenomenal terms. When looking, as with every other sensory perception, we are not in our head but in the world, coexisting with the objects. Perception does not take place in a vessel called consciousness, into which sensory stimuli are imported from outside. I do not, in fact, perceive "visual sensations," pictures, or representations, but rather the writing desk, the window, or the sky. I do not hear "sound sensations," but voices or music. Perception establishes a direct relation between the perceiving person and the perceived object. Is this immediacy of our world experience really only an illusion?

The problem of how a phenomenal world arises at all, and what function it has, also occupies the neuroscientists. For example, neuropsychologist Prinz (1992) raised the question: why do I not perceive the visual stimuli of my retina, the action potentials of my optic nerve, or else my brain states directly if they are in reality the substratum of my perception? And why do I plan actions and not directly the corresponding neuro-muscular processes of my body? In other words: why are there "distal" and not "proximal representations" at all? The world of experience, as Prinz's answer goes, presents a "virtual space," in which the various sensory and motor "data formats" are made compatible with one another and integrated. In this space, we can perceive and act, without being burdened by knowledge about the physiological processes actually taking place.

Of course, Prinz's question as such displays a category error, namely a confusion of the causal and intentional level: we perceive light waves just as little as excitations in the nerves because they only represent physical carrier processes necessary for perception, and not perception itself. What forms the mediating substratum of perception can hardly become its object. Moreover, Prinz, with his answer, recognizes that it is precisely the phenomenal world which enables our orientation and action in the world. It only remains incomprehensible why he then terms it a "virtual space." After all, by integrating perception and movement *in the same space*, it allows us to jump over a ditch in such a way that our feet actually come down at the other side. The phenomenal world is formed by an intermodal integration, that is, a *sensus communis* or common framework for the various senses and movements. Thus the person whom I see, his voice which I hear, and his hand which I shake, are included in a unified space—and this unity is undoubtedly actually the case.[14] For an "illusory world," the world of experience disposes therefore of an amazing amount of objectivity. Let us look at this more closely.

1.3.2 The objectivizing achievement of perception

What we perceive are neither pictures nor models, rather things and people. This can initially not be taken for granted: if, for example, I perceive a house, I only actually see *one* aspect of it, perspectivally limited. Nevertheless, we certainly see the house itself, the full object. How does perception overcome its own restricted perspective?

As Husserl has shown, perception cancels its tie to one perspective by integrating further *possible* aspects of things (Husserl 1950, 91–94). Thus, we perceive the house not just by looking at its visible side, but also by implicitly co-perceiving its invisible aspects, which we anticipate to see once we move around the house. Likewise, we co-perceive its materiality, its solidity, as well as its "affordances" or possibilities for action, which would be available to our reaching, grasping, handling, etc. All this implicit content of perception is derived from earlier experiences that enable our skillful dealing with the object (e.g., moving towards the house, opening the door, going upstairs, and so on). Therefore, my experience of an object depends on a *horizon of possible further experiences of this object*—a horizon that is derived from my former dealings with it, but which is now implicitly given or "appresented."

---

[14] The problem of the *sensus communis*, which should integrate the different senses into one unified perception of objects, was first raised by Aristotle in *De Anima*. Today it is discussed under the terms of "intermodal perception," and of the "binding problem" in neuroscience.

In other words, it is my embodied engagement in the world, which enables me to see *the house itself* and not a mere sensory impression or a subjective picture.

However, there is still another level of objectivity which is characteristic of human perception. For in perceiving the house, we experience it not only as an object of our possible engagement or skilled coping, but also as *independent* of our present perception. The objects are not only there "for me," in the immanence of my subjectivity, they are given *as such*. How is this independence possible? Husserl's later answer referred to the intersubjectivity of perception: The house that I see is also *a possible object for others* who could see it simultaneously from other sides. Thus, the object gains its actual objectivity, that is, its independence from my own perspective, through the *implicit presence of a plurality of other perspectives*. Husserl also speaks of an "apperceptive horizon of possible experiences, my own and those of others" which turns the mere subjectivity of my experience into an "*open intersubjectivity*" (Husserl 1973b, 107, 289; see also Zahavi 1996). Thus, there is again a horizon of perception, but one that is shared with others. The plurality of possible subjects corresponds to the plurality of aspects that the objects afford. In perceiving, we always enact and inhabit a space that we share with others.

As we can see, the perspectivity of perception does not mean mere subjectivity or even virtuality. On the contrary, through interacting with the objects and through our interactions with others, we are able to overcome our primary subjectivity. Gestalt psychology, moreover, has shown that perception completes fragments to wholes (e.g., missing letters are added to a word) and produces color- or size-consistency even where the field of perception is discontinuous or distorted (a square looked at from an angle shows not as a rhombus, rather still as a square). Indeed, even the illusions of perception are based on its inherent tendency to compensate for expectable aberrations, which serves the purpose to provide a constant and viable experience of the environment. Neuroconstructivists are usually happy to point out such illusions in order to prove the virtuality of perception. Actually it works the other way around: it is precisely the gestalt-oriented, actively shaping, and intentional structure of our perception that enables us to recognize *real things* instead of receiving one-to-one mappings of mere stimuli.

It is not the physical elementary events taking place between the objects, our sensory receptors, and the brain that are the "actually real" events in perception. This complete cascade of physical and physiological processes is only its material basis. There would be no world of meadows, trees, cats, or human beings for us and, of course, no adequate action in this world if perception had not long since integrated the elementary processes or individual stimuli into meaningful forms and gestalts. Perception means an action-directed openness to the world, not a photograph. We do not perceive stimuli or images, rather gestalt units, meanings and affordances. Perception avails itself of the mediating processes, in order to establish a direct relation to things—in other words, a *mediated immediacy*.[15]

I must thus be happy not to be able to perceive my brain states because they themselves could not give me the faintest inkling about reality—just as radio waves themselves could not allow the music they transmit to be heard. That is why neuronal processes are not in any way "more real" than the perceptions of things which they convey. This becomes irrefutable no later than *when it is connected with my fellow human beings*. If physical reality were the "actual one," they would ultimately only be forms of matter- and energy-states. My integrating, gestalt-forming perception displays other people for me as *what they are in reality, too*—or should we still speak about "constructs," "images," and "simulations" when we, as human beings, look at one another? Here, too, neuroconstructivism can only be maintained as long as one ignores interpersonal relationships. Perception is, of course, not a pure copy of stimuli constellations, for it selects, shapes, and integrates what is to be perceived. However, it is therefore not a mere construct—it rather presents things and people to us *as themselves*, and *in their relation to us*.

Here a decisive quality of perception becomes clear which cannot be found on the physical or physiological level: it produces a *coexistence* between the perceiving person and the perceived thing. If I see the other human being, I also see him in relation to me, within a shared space. Indeed, only to the extent that the phenomenal world presents itself as accessible, comprehensible, and relevant to us, can it become reality for us at all. That becomes above all possible by our having always been part of the world as embodied subjects. The supposedly only subjective or virtual space of phenomenal experience is thus anything but an inner world to be localized in the skull, or a brain-generated "out-of-the-brain-illusion" (Revonsuo 2003). On the contrary, it is the space of our *being-in-the-world*—the space of our relationship to everything that gains relevance and importance for us.

---

[15] This important notion was introduced by Hegel in his "Science of Logic" (*Wissenschaft der Logik*, vol. 1). The German philosopher Helmuth Plessner regarded this as a fundamental structure of life processes, in particular of perception (Plessner 1975, 48, 168, 321–325). I will come back to this in section 4.2.6.

1.4 Third criticism: the reality of colors

But, finally, what about the qualities which we experience when perceiving—the colors, sounds, and smells of our world? All those things which make the world familiar and inhabitable, are these only internal constructs that have no existence outside of our brain or consciousness? At least, this is what neuroconstructivism suggests:

> It is unsettling to discover that there are no colours out there in front of your eyes. The apricot-pink of the setting sun is not a property of the evening sky; it is a property of the internal model of the evening sky, a model created by your brain. The evening sky is colourless. […] It is just as your physics teacher in high school told you: Out there, in front of your eyes, there is just an ocean of electromagnetic radiation, a wild and raging mixture of different wavelengths. (Metzinger 2009, 20)

I have already mentioned the aim of the natural-scientific program to cleanse nature of all non-mathematical properties by shifting qualities into the perceiving or feeling subject. Colors—I choose these qualities as an example—no longer appear in this reduced world. Let us assume that a person sees a green tree in front of her in the meadow: even a comprehensive physical examination and description of all of what happens outside and inside the person's body would not allow any statement about her perception of color as long as she remained silent about it. Without our experience of color science would have no reason to even suspect its existence. Although we could learn from the person that she did, in fact, see a green tree, the physical description could not contribute in the least to the explanation of this perception, for, according to the data, the person could equally see any other color or none at all. Color cannot be physically explained or reduced—that is why, from this viewpoint, it suggests itself that they be eliminated from the inventory of the real world.

Moreover, a neuroscientist too can only ascertain that during the act of perceiving the color green, the light of a certain wavelength falls on the retina and triggers off a cascade of neuronal processes, arriving at area V4 of the occipital lobe, which he knows is necessary for color perception (Zeki 1992). But however carefully he follows the neural signals from the retina along the optic nerve and across the brain, nowhere will he come across anything like a color itself, or anything that explains color perception *as such*—as little as the observations of the physicist outside of the body. Undoubtedly, it needs the light waves which, reflected by an object, stimulate the retina and optic tract, *in order for us to see* colors, or the sound waves which cause our eardrum to vibrate, *so that we hear* tones. However, we do not see any light waves and do not hear any sound waves, rather colors and tones. Should we regard this as an illusion created by the brain?

Of course, the existence of sensory qualities in the perceived environment cannot simply be refuted. But colors are obviously of a different type than, for example, the size or mass of an object. The green of the tree fades in the evening light and vanishes at night, whereas his height remains the same. Hence, already on the physical level, colors are dependent on light, that is, on the respective interaction of object and illumination. But even then the wavelength of the emitted light may be only approximately correlated with the perceived color. The same wavelength, for example, will give rise to quite different colors if the surrounding environment is different—color consistency or color illusions demonstrate this well enough. Obviously an *interaction of object, light, and perceiving organism* is required in order for a certain color to appear in the world. Yet from a physical or neurobiological point of view, only *conditions or correlates* of color perception may ever be ascertained, which may neither explain nor predict it *as such* (Stroud 2000).

Now the physicist need not worry if the tree, apart from its structure of particles, is green in addition or not. This question simply no longer arises for his measurements and the formation of his theories. The denial of qualities in the world thus emerges not from a scientific necessity. In fact, it rather emerges from a *physicalist world view* which hypostasizes to "actual" reality the quantifiable extracts of reality, originally chosen by science for certain aims, and the theoretical constructs (atoms, photons, electromagnetic fields, etc.) derived from it. Physical descriptions, explanations, and constructs are now alleged to be valid for all areas of the lifeworld. Then the green tree is now only a big stack of molecules, the nightingale's song in its branches is only an irregular sequence of air pressure variations, and the joy of the wanderer who listens to it only a certain neuronal excitation pattern in his brain.

However, this scientistic worldview is by no means inescapable. The fact that light waves are not colored or sound waves not loud themselves is no reason to refute the reality of colors and tones. After all, there is a host of other characteristics of reality which likewise fall through the coarse grid of physical descriptions—for example, the fruitfulness of fruit trees, the brood care behavior of gray geese, a debate in the British Parliament, or the German trade surplus in the year 2017. Should all that not be described as reality just because physics has nothing to say about it? Physicalism claims that everything we can state about the world is reducible to physical facts (see Quine 1960). Of course, at least this does certainly *not* apply to this statement itself: for the knowledge of *what a physical fact is in the first place* cannot be contained in the set of all physical facts. But the phenomena to be reduced—perceived colors, sounds, or smells—cannot be contents of physical statements either, for they belong to the realm of psychological facts. The physicalist reductionist is faced with phenomena which he *cannot even describe* in what he presumes to be the only valid language, let alone is able to reduce.

Of course, in a purely physical world, there would be no sense in speaking about colors and tones. However, such a world is only a conceived abstraction of the world, which we as living beings dwell in and experience. In this world our organism makes qualitative distinctions which cannot be found on the physical level, and thus structures the environment into the meaningful and the relevant in order to sustain itself in it. In this way, it becomes possible that things and living beings *show themselves* and *enter into a relation with us*, that is to say, in colors, sounds, and odors. To that extent, the sense qualities are the results of the relation of a living being to its environment. But this relation has a world-disclosing quality and, insofar, a thoroughly objective character. Even the so-called primary qualities of physics can only become accessible to us via the secondary ones.

Is the tree in fact green then? It depends on whether we look at it as a part of our shared lifeworld—then we can agree on its color, and it is thus not "only subjective"—or whether we descend in a physical construct world, in which, according to its premise, none of the lifeworld qualities can be found any longer. Color is neither an objective characteristic of the material world ("naive realism"), nor is it a mere product of an inner world (neuroconstructivism). Colors and other sensory qualities are rather the expression of a *complementarity* of living beings and their environment. They emerge in the interaction of an organism's perceptual capacities and the characteristics of objects. Thus, it can be shown that the development of color patterns in flowering plants took place in constant interaction with the development of color vision in insects. The property and its perception arose in various species co-evolutionarily in the context of a comprehensive ecological system (Ehrlich & Raven 1964).

Similarly, it is true for life as a whole: the world also changed with its development; new, systemic relations and corresponding relational properties emerged. Living beings produce qualitative differences from quantitative ones and, by this means, they transform the world, for the specific relationship between color property and color perception now *belongs as such to its objective features*. This transformed world is our lifeworld. As long as we are not blind, we all see colors, and we can compare our perception of them with that of others. We dress in certain tones of color, in order to appeal to others, and painters design the canvas with colors, in order to invoke certain impressions in us. That is why our world contains colors and other qualities just as necessary as it contains fruits, trees, animals, and human beings—we cannot deny the one and let the other exist.

1.5 Summary

We started from the neuroconstructivist thesis that the ontological status of experienced reality is that of a subjective image or a virtual model which is constructed by the brain. This thesis is ultimately based on a still dualistic division of the world into a bodiless and worldless subjectivity on the one hand, and a physically reduced material world on the other hand. Subjectivity is conceived of idealistically—though in the new robe of constructivism—while it is, at the same time, ascribed as a construct to purely material processes in the brain. The result is a peculiarly hybrid doctrine, composed of a disembodied mind and a disembodied brain, which could rightly be called "Cartesian materialism" (Rockwell 2005). It is connected with the assumption that there is an external physical reality as such, which is only given to us through representations in our mind.

In the criticism I have shown that perception is not to be understood as an internal representation, model, or construct, but rather as *the active relation of an embodied subject to its environment*. When we perceive, we are not enclosed in the skull to see pictures from the world outside. On the contrary, we interact with the world as embodied beings, coexisting with things and other persons in a shared space. Human perception is thus based on interaction in a twofold sense:

1. According to the enactive approach, living beings generally do not passively receive information from their environment which they then translate into internal representations. Rather, they constitute or *enact* their world through a process of *sense-making* (Varela et al. 1991, Thompson 2007, Di Paolo 2009): by actively searching and probing the environment for relevant cues—moving their head and eyes, touching a surface, walking towards a goal, grasping a fruit, etc.—they make sense of their surroundings. In other words, they constitute their experienced world or *Umwelt* (von Uexküll 1920) through their ongoing sensorimotor interaction and embodied coping with the environment.

2. In addition, through their *social interactions* and implicit relation to others, human beings are able to transcend their primary perspective and gain access to a shared, objective reality. From early childhood on, experiences of joint attention, pointing towards objects, shared reference to situations, mutual understanding, and cooperative practice result in a *participatory sense-making* (De Jaegher & Di Paolo 2007). In this way, a shared reality is constituted, which becomes an implicit part of our relation to the world. This is why we perceive a given experiential object as transcending its momentary appearance: it could also be seen by others. The objects are not

only there "for me": even Robinson Crusoe on his island always perceived his surroundings "with others' eyes," already before Friday appeared on the scene. This is also what enables us to see things *as such*, objectively, or in independence from our momentary perception. For objectivity ultimately indicates that the objects are experienced as intersubjectively accessible, in the co-presence of possible other subjects, or "as actually there for everyone" (Husserl 1960, 91). Human reality is therefore always *co-constituted* or, as we might say, "*interenacted*."

Hence, human perception is anything but a parade of images in a disembodied, worldless, and solitary mind. Rather, it is an activity that transcends the boundaries of the body and the centrality of the subjective perspective through interactions on two levels:

1. On the first level, the *sensorimotor interaction* of the mobile body with the environment implies a constant changing of perspectives that relativizes the momentary relation of organism and environment: each perception is enriched by a history of former experiences and a horizon of possible further interactions with the object.

2. On the second level, the *social interaction* with others implies a shared reference to objects as well as a contrast and alignment of perspectives which helps to overcome a merely subject-centered worldview through participatory sense-making. The capacity to share one's perceptions with others in principle results in an *increased distance* of the subject from the object of perception, that is, in an objectification.

Thus, we live in a world of real objects, because we are involved in its constitution through our sensorimotor engagement. And we live in a shared objective reality because we continuously "interenact" it through our joint activities and participatory sense-making.

The acid test of every epistemology is, when all is said and done, the intersubjective relationship. When it is a matter of another person, we cannot simply withdraw to a radical constructivist viewpoint. The reason for this is that we would not only declare their very presence a virtual one; we would rather also remove the necessary *limitation* which the other represents for our own self-being. The other is real for me—and it is only through him that I gain reality myself. I can no longer be a solipsistic or constructed being. At the same time, it is the *consensually* apprehended reality which guarantees the reality of my perceptions and embeds my subjective bodily space into an objective one—into the shared space of "open intersubjectivity" (Husserl 1973b).

On this condition we were able to restore its objectivity to perception, as well as its qualities to nature, without thus falling back into a "naive realism."

Certainly, perception is not a one-to-one mapping of the physically described world. But we have seen that it is precisely the active, gestalt-like, and intentional characteristics of perception which allow us to see the *things as such*. Physicalism, with its extremely reduced data basis, eliminates all qualitative and gestalt-like perceptions from the definition of the real. The primacy of the lifeworld counters that: it is only in the latter that things, living beings, and persons show themselves *as what they really are*. Perception thus presents us truly more than the mere stimulus configurations contained in the perceptual field. Nevertheless, it does not thus present any constructs, rather the real world—of course, not as "world in itself," rather *as the world in its relation to us*, the perceiving persons.

We may finally ask why this conflict is actually so important. Would it be so bad if we acknowledged subjective reality as a construct of the brain from a natural scientific perspective—as long as we, nevertheless, in our everyday life continue to work from the adequacy of our perception? The answer is: whatever we declare to be a mere semblance, we also gradually look on as no longer relevant and meaningful. It is given a subordinate and derivative existence and is devalued in its meaning. Moreover, our own judgement and trust in the world is fundamentally undermined: after all, we are trapped in a cave watching shadows on the wall while reality is outside, beyond our grasp, and even knowledge. We all live in error and need the authority of science to enlighten us about what reality is really like. It feels as if we are being blamed, or condescended to, for not perceiving things as science thinks we ought to. If we thus declare our shared reality to a virtual construct, we rob ourselves of the basis of our autonomy and self-confidence. In the last analysis, the question of what is "really real"—physical matter instead of animated bodies, brains instead of selves, neural computation instead of conscious experience—is an ethical question.

Chapter 2

The brain as the subject's heir?

Overview

Chapter 2 critiques the claims according to which subjectivity is to be regarded as a construct or epiphenomenon of neuronal processes and thus one's experience of agency and freedom of choice should be seen as an illusion. First it is shown that the subjectivity of "experiential facts" cannot be reduced to objective or physical facts about brain processes. Likewise, the reduction of the intentionality of consciousness to relations of representation is refuted (2.1). Moreover, the identification of the subject with the brain leads to fundamental category mistakes which will be examined as the "mereological fallacy" and the "localization fallacy" (2.2). On this basis, a critique of the thesis of the powerlessness of the subject is developed (2.3). Finally, the summary analyzes the basic "naturalistic fallacy" of an objectifying account of consciousness which believes it can remove itself from its rootedness in the lifeworld (2.4).

Just as the world we experience, the experiencing and acting subject also becomes a product of brain processes from a reductionist standpoint. If the physical world is deemed actual reality, then the subject can, of course, only be allotted an epiphenomenal status. A rising choral song of materialist neurophilosophy heralds the message that our subjective experience is nothing else but the colored user interface of a neurocomputer, which even creates the illusion of the user itself.[1] Our experience of being the authors of our thoughts and actions is only part of this "grand illusion." Actual reality consists of the computational processes of the neuronal machinery running in the background:

> Our thoughts and our dreams, our memories and our experiences all arise from this strange neural material. Who we are is found within its intricate firing patterns of electrochemical impulses. (Eagleman 2015, 5)

---

[1] See Slaby (2011). Daniel Dennett was probably the first to claim that consciousness is "the brain's user illusion of itself" (Dennett 1991).

> The neurobiology of consciousness faces two problems: the problem of how the movie-in-the-brain is generated, and the problem of how the brain also generates the sense that there is an owner and observer for that movie. In effect, the second problem is that of generating the *appearance* of an owner and observer of the movie *within the movie*. (Damasio 1999a, 11)

The complex, even paradoxical structure of human self-consciousness, which is most difficult to grasp philosophically, is here subsumed with a flick of the wrist under the general neuroconstructivist thesis. If consciousness is only an "out-of-the-brain illusion," why not also the subject who has this conscious experience? One just needs to add a "meta-representation" to the inner representations (thoughts about thoughts, images of images), and self-consciousness is explained: we are only dream subjects within a dream. The brain is a "world simulator" and, at the same time, a "self-simulator." In Metzinger's "self-model" theory, the subject is consequently conceived as analogous to a pilot who believes that he experiences reality, while he is in fact placed into a flight simulator—and who is indeed himself only a product of this simulator, or a virtual self-model:

> The human brain can be compared to a modern flight simulator in several respects. Like a flight simulator, it constructs and continuously updates an internal model of external reality [...]. However, there is a difference. [...] there is no user, no pilot who controls it. The brain is like a *total flight simulator*, a self-modeling airplane that, rather than being flown by a pilot, generates a complex internal image of itself within its own internal flight simulator [...]. Operating under the condition of a naive-realistic self-misunderstanding, the system interprets the control element in this image as a non-physical object: the "pilot" is born into a virtual reality with no opportunity to discover this fact. (Metzinger 2009, 107–108)

Of course, this comparison suggests the Cartesian picture of a pilot steering the body-plane—only to then refute this picture as a naive or dualistic self-deception. However, no serious philosopher nowadays claims that the subject or self should be regarded as some type of "non-physical object," a *thing* or an *entity* that could be distinguished from the person as a whole. And to be fair, even Descartes himself explicitly declared "that I am not lodged in my body as a pilot in a vessel, but that I am very closely united to it" (Descartes 1993, 93). Only if one would assume "the Ego" or "the Self" (writ large, as it were) to be a pilot or homunculus somewhere within the body, would it indeed be justified to speak of a "self-misunderstanding." But why should it be an error or an illusion if the airplane-system—or rather, the whole living being or the embodied person—is simply aware of itself *as itself*? There is nothing self-contradictory or illusionary involved, nor a "myth of the self" which Metzinger has avowedly set out "to shatter" (2009, 1). Hence, if he boldly claims

> that to the best of our current knowledge there is no thing, no indivisible entity that is *us*, neither in the brain, nor in some metaphysical realm beyond this world (2009, 1)

then the simple reply is that Metzinger searches for "us" in the wrong places. The indivisible entity that we are, is indeed neither in the brain nor in an otherworld, but it is quite visibly *our bodily being*—a living, functionally indivisible and self-aware organism. We do not exist a second time within our bodies or somewhere else. The "myth of the self" is just that: a myth. Like many neuroscientists and neurophilosophers, however, Metzinger prefers to keep on fighting against this "ghost in the machine" (Ryle 1949), for this lends more clout to his thesis of the "grand illusion":

> We are Ego-Machines, but we do not have selves. We cannot leave the Ego Tunnel because there is nobody who could leave. […] Ultimately, subjective experience is a biological data format, a highly specific mode of presenting information about the world, and the Ego is merely a complex physical event—an activation pattern in your central nervous system. (Metzinger 2009, 208)

Of course, the question immediately arises how Thomas Metzinger could become aware of living in the Ego Tunnel, if there is no escape from it—indeed, if there is even "nobody who could leave." A dreamer who becomes aware of dreaming can no longer be *only* a dream (this was already Descartes's bastion against an assumed "malign genius" who could deceive me of everything—except that I am the one who doubts). But be that as it may, let us return to the question of the self: granted, we may not "*have*" a self, but why should we not *be ourselves*, only because our self-awareness as living beings requires, as one necessary condition, an integrating activity of the brain? Later in his text, Metzinger himself concedes that it might also be possible to term the *organism*—as a self-organizing and self-sustaining system—a "self." In this case, the self, as he continues, would not be "a thing but a process":

> As long as the life process—the ongoing process of self-stabilisation and self-sustainment—is reflected in a conscious Ego Tunnel we are indeed selves. Or rather, we are "selfing" organisms: At the very moment we wake up in the morning […], [a] new chain of conscious events begins; once again, on a higher level of complexity, the life process *comes to itself*. (2009, 208)

From the point of view of embodied subjectivity which I will develop in this book, this seems a fairly acceptable position—provided only that we replace the provocative catchword "Ego Tunnel" by the more appropriate term "self-awareness." As I will argue further below, there is indeed an inherent continuity of life and awareness, or *Leben* and *Erleben*. Hence, in self-awareness, the life process of the organism in fact comes to itself, for it has always been a *self*-organizing process. But Metzinger does not seem really satisfied with this option. After all, he has already stated in his introduction that

> No such things as selves exist in the world. A biological organism, as such, is not a self. (2009, 8)

Therefore, he is now eager to assure that:

> True, upon your awakening from deep sleep, the conscious experience of selfhood emerges. [...] But there is no one doing the waking up, no one behind the scenes pushing the Reboot button, no transcendental technician of subjectivity. [...] Strictly speaking, there is no essence within us that stays the same across time, nothing that could not in principle be divided into parts, no substantial self that could exist independently of the body [...] We must face this fact: We are *self-less* Ego-Machines. (2009, 208)

As we can easily see, in order to push the unwanted option aside, Metzinger needs to revive the Cartesian strawman once more: the reader should believe that in order to speak of a self, it must be a "transcendental technician" that steers the life process just as a user operates his computer. And instead of assuming the continuity of the living organism as the basic continuity of ourselves, there should be an "essence within us" which stays the same across time.[2] But this indivisible and bodiless Cartesian entity sadly does not exist (we must face it ...), and Metzinger is glad to renew his illusion thesis.

Our first exposition of the reductionist claim has already pointed out one of its central weaknesses: the imputation of a Cartesian "Self" which no one actually supports, and the corresponding lack of a concept of the living being. Nevertheless, the concepts of an epiphenomenal or illusory subjectivity will now be critiqued in more detail in three steps. First, it will be demonstrated that subjectivity and intentionality cannot be reduced to physical descriptions of brain processes. The second step will examine the false conclusions and aporias to which the identification of the subject with the brain necessarily leads. In the third step, the claim of the subject's ineffectiveness and impotence will be refuted.

2.1 **First criticism: the irreducibility of subjectivity**

2.1.1 **Phenomenal consciousness**

The notion of a self-model implies that subjectivity or phenomenal consciousness is only an image or representation of the neuronal processes constructing it. However, the catchy term "model" only conceals the crucial problem: how could a physically implemented structure possibly give rise to consciousness of the world and of itself? After all, it is the same case with consciousness as with color: without our experience of it, science would not have any reason to even suspect its existence. To put it more pointedly: in a purely physically described

---

[2] On the continuity of the *embodied* self, see my publication entitled "Self across time: The diachronic unity of bodily existence" (Fuchs 2017a).

world, however complex its processes, something like consciousness simply does not show up, just like colors. In contrast to the brain, consciousness is not an object in the world—on the contrary, it is the presence of the world for a subject.

In his famous essay: "What is it like to be a bat?," Thomas Nagel has defended the resistance of subjective experience against its complete objectivization: even if we could fully describe the processes and behavior of a bat neurophysiologically, we would not have the slightest idea what it experiences or how it feels pain or ultrasound, in other words, what "it feels like to be a bat" (Nagel 1974). Therefore, there is basically, according to Nagel, an epistemological boundary for the neurosciences: subjective or experiential facts which are each only accessible from a unique perspective cannot be transferred completely into objective facts which can be observed by various individuals. The subject is the center of a world, and such centers cannot be found in a purely physical world, including neuronal processes.

It has become common to express this contrast in terms of the phenomenal or *first-person perspective* and the naturalistic, objectifying or *third-person perspective*. However, the source of the notion of perspective from an optical point of view should not allow us to forget that, in the case of the first-person perspective, more is at stake than just a particular angle, namely precisely "how it is" or "what it feels like" to be in a certain mental state, that is, an *elementary affective self-experience* before any self-reflection.[3] Subjectivity in this basic sense does not mean a perspectival view on contents or objects, connected to a conscious ego-experience; we are rather dealing with a primary bodily-affective self-feeling as the core of all conscious processes. Even before every perspective and cognition, there is a form of immediate, pre-reflective self-presence, an affectively colored familiarity of consciousness with itself, which may, according to Michel Henry (1963), also be designated "auto-affection."[4]

This self-affection may be taken to ground the *first-personal givenness* of every experience, which Zahavi (1999, 2005) has elaborated. Thus, any sensation, any

---

[3] For this, the term "what-is-it-likeness" has also come to use since Nagel's argument.

[4] This is in contrast to common higher-order or representational theories of consciousness (e.g., Carruthers 2005, Rosenthal 2005); on their critique, see mainly Zahavi (1999). In this respect, the analyses of Michel Henry are also comparable with concepts of the "Heidelberg School" (Henrichs 1970, Frank 1986, 1991) who assume a pre-reflective self-familiarity of consciousness as the basis of all higher-order reflective self-recognition. "Familiarity" (*Vertrautheit*) implies an affective element which is not explicitly thematized by Henrich and Frank, however. For an overview on phenomenological accounts of pre-reflective self-consciousness, see also Thompson and Zahavi (2007).

perception or action directed towards an object implies a tacit self-awareness without requiring introspection; it is given immediately, non-inferentially as mine:

> This first-personal experiential givenness is manifest in the very having of the experience. It is a givenness that obtains even when we are not explicitly aware of it [...]. A conscious mental state is not merely conscious of something, its object; it is simultaneously self-disclosing or self-revealing. (Zahavi 2017, 198)[5]

Further, the basic affective self-awareness grounds the existence of *subjective or experiential facts*—for example, the fact that *I* experience pain, feel hunger, am happy or sad.[6] Thus, it is also the basis for everything existentially meaningful, for what constitutes my personal concerns and cannot be replaced by taking a general or scientific point of view.[7] Can such subjective facts be reduced to objective ones, for example, facts which can be described in neurobiological terms? Is it possible to describe the fact that I am now feeling pain as a certain neuronal activity pattern without its losing its significance? No, because even the seemingly unproblematic re-formulation "Thomas Fuchs feels pain at this moment" no longer expresses the fact that it is *my* pain and that it is I *myself* who suffers from it.[8] Even if this statement from the third-person perspective were reliably true in all cases (e.g., on the basis of the simultaneous observation of my brain processes), it lacks the decisive feature of subjectivity, namely that *I myself* am that T.F., about whom this statement is made. This would be all the more true for an exact description of the physical processes in the brain of T.F.—nowhere in it could the *mineness* of the pain be found. Between both manners of stating this, there is an ontological leap. The reality of my pain is of *a basically*

---

[5] Zahavi's concept of pre-reflective self-awareness, also termed "minimal self," does not emphasize its affective aspect which is highlighted in my account, yet certainly does not exclude it.

[6] These are not primarily *propositional* facts, that is, facts that are expressed in propositional terms ("I feel pain," "I am sad"). The feeling of pain or sadness is before any verbalization, in which I could also be mistaken (e.g., because I feel not precisely "sadness" but some related emotion, say, disappointment). Primary experiential facts, as such, are "immune to error through misidentification" (Shoemaker 1968).

[7] "We feel that even if all possible scientific questions be answered, the problems of life have still not been touched at all," writes Wittgenstein (Tractatus Logico-Philosophicus, 6.52; Wittgenstein 1961).

[8] Here I draw from Hermann Schmitz's analysis of subjective facts: "A fact [...] is *subjective*, if at most *one*, and only on his own behalf, can state it, while others may well speak about it with unequivocal labelling, but never ever can state what is meant" (Schmitz 1995, 6; own translation).

different kind to the reality of objective physiological facts—and nevertheless it is no less "real" than these.

Facts of self-experience cannot be transferred into objective facts without a decisive loss. And indeed not so much because of their special qualities or "qualia,"[9] but rather, above all, because of their subjectivity itself: it constitutes an *absolute epistemic asymmetry of facts*. The natural scientific reduction is based, as already presented in the "Introduction," on stripping subjectivity from the experienced facts and reducing the remainder to elementary physical processes. It thus transforms what is subjectively experienced into objective statements, which is connected with a loss and alienation, but which is practical and successful for the purposes of explaining and predicting nature. The reduction fails, however, when subjectivity as such is at stake. Even if it could be proved that subjective experience is always produced by certain neuronal activities (we will see whether this is in fact the case), the explanation would still remain incomplete—the radically new ontological characteristic of the subjective itself could only be accepted and not be explained further from the physical processes.

The principal asymmetry between subjective and objective facts also manifests itself in the performative function of certain speech acts. The statement "I promise to visit you tomorrow" is obviously not equivalent to the statement "Somebody promises to visit you tomorrow, and the person who promises that is Thomas Fuchs." The act of promising as a performative self-commitment can only be expressed in the first person; the report about the promise of a third person, even if it is completely correct, does not include this commitment.[10] It becomes clear that the I-statement of a speaker cannot be transformed into the report about a third person without crucial semantic loss. For the fact that the promise concerns *me* and my affectively colored experience of self-congruency and self-commitment, which I put in the balance, is eliminated from the objectifying description. The performative effect of certain speech acts thus marks a subject as the irreducible center of self-related meanings and of

---

[9] The problem of the "explanatory gap" (Levine 1983) in philosophy of mind is usually explained on the basis of "qualia": even if we were certain that phenomenally conscious states are identical with brain processes, we would not have a scientific explanation for the fact that these processes are experienced in the special qualitative way of pain, color, sadness, and the like. However, the qualia problem only concerns a partial aspect of subjectivity; in my view, it does not constitute the decisive explanatory gap which is rather based on the "mineness" of any conscious awareness as such.

[10] On this, see Ricoeur's analysis of the subjectivity of performative speech acts (Ricoeur 1992, 42–43).

being affectively concerned. In other words: based on the absolute epistemic asymmetry of *facts*, there is also an absolute performative asymmetry of *certain actions*.

2.1.2 Intentionality

Whereas our line of argument at first applied to subjective experience, as it manifests itself in conditions of pain, hunger, sadness, or the like, our last considerations went beyond that. Subjectivity is not merely state-like, it is moreover essentially oriented *to what it is not itself*: it is open to the world, related to objects, and directed to contents and meanings. Experiential states which are of such types that they are directed to something, that is, perceptions, thoughts, wishes, ideas, or memories, possess the characteristic of *intentionality*. That is to say, they have an intrinsic content to which they relate and which can be expressed by a that-clause (e.g., promising "that I will come tomorrow"; believing "that Monica is wrong"; wishing "that the rain stops"; etc.). In other words, intentionality opens the dimension of *sense and meaning*.

It is obvious that the intentionality of consciousness represents a serious problem for a physicalist reduction—more serious, in fact, than its subjectivity—because experiences with missing or weak intentional content, such as pain or moods, could still be objectified as "mental states" and thus possibly be equated with neuronal processes. Intentionality can, however, no longer be adequately defined as a mere mental state; for what is meant or intended by them belongs to the definition of intentional acts. The mental state of the intention to buy a book does not exist independently of the book, the way to the bookshop, the purchasing process, and so on. In other words, it presupposes its *embedding in a situational and meaningful context*.[11] A definition of intentional acts, independent of object and context, would, however, be the precondition for its description as states of the brain. Physical processes, such as the activations of neurons, can, as such, not be aimed at a context, and the imaging of brain activities during intentional acts cannot basically capture the direction of their sense.

2.1.2.1 Intentionality and phenomenal consciousness

Nonetheless, in the analytical philosophy of mind, the naturalization of intentionality is attempted in two steps. First, the phenomenal characteristics of

---

[11] On this, see also Searle: "semantic contents, that is, meanings, cannot be entirely in our head, because what is in our heads is insufficient to determine how language relates to reality. [...] If the meaning of the sentence 'Water is wet' cannot be explained in terms of what is inside the head of speakers of English, then the belief that water is wet is not a matter solely of what is in their heads either" (Searle 1992, 49).

consciousness and subjectivity, as so-called qualia, must be separated from the intentional characteristics. Intentional meanings, as Chalmers (1996) argues, for example, could then be construed in terms of a functionalist theory which would explain them as neuronal representations. Certain neuronal system states are, through the previous history of the brain, functionally connected with certain configurations of the environment. That is why, in each case, they produce the suitable output for a certain input and make the intentionality of consciousness dispensable. Functionalism thus seems the suitable strategy for reducing intentionality. The qualia problem would then be left as the only "hard problem of consciousness" (Chalmers 1996)—but this problem could as well be ignored as negligible for the overall course of the world. A functional definition, for example, of pain would consist of the connection of physical input (tissue damage or trauma) and behavioral output (aversive or avoidant behavior). The *feeling* of pain is irrelevant for this connection.

It has already been shown that, with the problem of subjectivity, there is indeed more at stake than certain individual qualities, such as "red" or "warm," that is to say, subjective experience as such. Is it possible to separate intentionality from subjectivity? Is the *experience* of meanings in principle a dispensable addition? The claimed separability of subjectivity from meaning presupposes a reductionist re-definition: "meaning" would then consist only of the two-place assignment of sign and signified, or *representatum* (the representing internal state) and *representandum* (the represented part of reality), and this assignment would be purely functionally realized by the regular connection between the input and appropriate output of the brain. However, Galen Strawson has emphasized that meanings only exist *for someone:* "Meaning is always a matter of something meaning something to someone. In this sense, nothing means anything in an experienceless world. There is no possible meaning, hence no possible intention, hence no possible intentionality, on an experienceless planet" (Strawson 1994, 208–209).

Intentionality is thus a three-place relation: *something* means *something for somebody.* "I believe that Monica will come" puts (1) Monica in relation to (2) an act of supposition, which can only be attributed to me (3) as a conscious person. Intentional acts and attitudes are something whose meaning is *experienced* and which, thus, necessarily belongs to a phenomenal consciousness. Wishing or wanting something, remembering or recognizing something, understanding words—all these possess a certain quality of "what it is like" to experience this state. Seeing an apple is different to imagining an apple (Zahavi 2003). Each is connected with a particular way of experiencing and self-experiencing—just like experiencing pain, hunger, or sadness. Thus, intentionality and subjectivity cannot be separated from one another.

2.1.2.2 Intentionality and representation

The decisive notion, which is, nevertheless, intended to achieve the naturalization of intentionality, is the concept of *representation*. Let us therefore consider this central concept of cognitive neurosciences or neuroinformatics more closely.[12] *Neuronal representations* should depict an external fact or a set of facts in a neuronal system in such a way that they can represent ("mean") this in the cognitive operations of the system. All information about the fact is mirrored in representing patterns of neuronal activity and can, as such, be further processed. They are usually regarded as the basis of *"mental representations"*—the contents of consciousness. Renewed pictures of the neuronal representations on a higher level, in other words, meta-representations, would then be the basis for reflective processes. Thus, the intentional contents of consciousness would be physically realized and as such could have effects on the output of the system, that is, on the behavior, without the phenomenal intentionality of a subject being required for that.

Searle has shown that in reality only an "as-if" intentionality is constructed in this functionalist account (Searle 1992, 78–84). For a meaningful connection cannot be ascribed to functional, rule-consistent procedures without there being someone who *understands* this connection. In order to illustrate this, Searle has developed the thought experiment of the "Chinese Room," which has become commonly known but shall nevertheless be briefly described here (Searle 1980):

> Imagine that someone who does not understand a word of Chinese is locked in a room, in which there is a program with all the rules for answering questions in Chinese. The person now receives questions that are passed into the room which are written in Chinese symbols ("input" into the system) and works out completely correct answers with the help of the program, which he then returns ("output" of the system)—of course, these are purely rule-consistent and he does not understand any of them. Let us presume that the program is so perfect and the answers are so good that even a Chinese person outside the room would not notice the deception. Nevertheless, one could not say about the man in the room that he *understands Chinese*. The semantic content or meaning of the language thus contains more than its mere grammar and syntax.

Searle's "Chinese Room" is, of course, the image for an information processing machine in which a central processor works according to the algorithms

---

[12] Main proponents of representationalism in philosophy of mind are, for example, Dretske (1995), Tye (1995), Lycan (1987), and Metzinger (2003). However, the concept is common in most neurocognitive theories as well as in accounts of empirical studies.

of a program ("If you get input X in the context Y, then give output Z"). The machine functions completely adequately as a system, but, nevertheless, it lacks the decisive characteristic of intentionality, namely the semantic content—*experienced meaning* or *comprehension*. Hence, our understanding cannot be reduced to program procedures or information processing in the brain.

This can be transferred to all technical cybernetic systems: a torpedo is programmed so as to detect a moving target and pursue it. We can also say that the object is "represented" in its steering system. However, this representational function only exists *for us*, namely on the basis of our previous construction and programming, which places the torpedo in a regular connection with a target object. The steering mechanism allows the torpedo to make corrections in movement, by means of which it finally reaches its target. Nonetheless, it would, of course, be nonsensical to say that the torpedo "seeks its target," that is, in fact, it has an intentional and time-spanning relation to its target object. Every correction only serves the internal set-point regulation of the mechanism and occurs purely momentarily without relating in any way to a target anticipated *as such*. For this goal itself, the mechanism remains blind and deaf. If it reaches it, the program is simply over—its purpose is, however, only "fulfilled" from our point of view.[13] The "representation" of external facts in a system is, thus, completely different to the intentional directedness to these facts.

The notion of representation is meant to eliminate this experienced significance—that is why it is so cherished in neurophilosophy. In fact, however, it is *only we ourselves* who can ascertain the representation of one fact or event by another fact; it does not exist *as such*. As a rule, contexts of representation are created by us. The map of a country which we produce represents a landscape; a portrait, a human being; and a sentence, a set of facts. In an improper sense, representations may also be ascribed to objects of nature as the result of causal connections—in this sense, a track in the snow "represents" an animal, smoke "represents" fire, and the rings in a tree trunk's cross-section "represent" the life years of the tree.[14] In all these cases, however, the representation exists *only for us* who can establish the context of meaning, insofar as we dispose of intentionality. For nothing prevents us from attributing representations not only to the smoke or the growth rings but to all effects traced back to a certain cause: the warmth of the earth at night "represents" the daily solar radiation, the

---

[13] On this crucial difference, see also Jonas's critique of "cybernetics and purpose" (Jonas 2001, 108–127).

[14] Both Dretske (1995) and Tye (1995) take the growth rings as an example of a "natural" representational relationship on the grounds that the number of rings causally co-varies with the number of years. For a poignant critique, see also Bennett and Hacker (2003, 142).

tides "represent" the moon's gravitation, and the stomach mucosa "represents" the incoming food by producing a regular output, namely gastric acid. So if representations existed "as such," in the subjectless nature, it is obvious that they would exist everywhere as well as nowhere.

Each semiotic relation is three-place too: *something presents a sign of something for somebody.*[15] That is why in a computer *as such* nothing more takes place than transitions from one electrical state to another. Only the programmer or user can interpret these processes as symbol manipulations or information processing, thus *lending* them *meaning*. Briefly: in a world without subjective experience there are no longer signs, nor symbols or information, representations or meta-representations, meaning or sense. "Reading" representations "into" a purely objective causal connection of natural processes is, in this respect, a conceptually unsound manner of speaking, intended to give the neuronal processes an appearance of intentionality.

One can, admittedly, attempt to define representation in terms of a three-place relation *without* a subject, as Metzinger does:

> Mental representation is a process whose function *for* the system consists in representing actual physical reality [...] [An] internal state X represents a part of the world Y *for* system S. (Metzinger 2003, 26)

Certain neuronal processes, as representata, thus depict an external state for the system, by which Metzinger means an information processing system such as a human organism or its brain (2003, 24–25). But this seemingly three-place relation cannot be maintained. The preposition "for" indicates either the reference to an intended goal or purpose ("what is this good *for*?") or to a subjective point of view ("*for* me it is clear that …"). Both kinds of relation cannot apply here, for a subjectless system neither pursues goals (like the torpedo, it only passes through regulations and adaptations, but is indifferent to its state), nor does it have a point of view. A goal could only be ascribed to it by its engineer or designer, but this external view would not solve the problem. Nonetheless, Metzinger speaks of the "for" relation as a "teleological criterion" and regards mental representations "as internal tools, which are currently used by certain systems in order to achieve certain goals" (2003, 26–27). Granted, at present these can only be biological systems:

> Artificial systems—as we knew them in the last century—do no possess any interests. Their internal states do not fulfil a function for the system itself, but only for the larger unit of the man-machine system. (2003, 27)

---

[15] Peirce's definition of the sign is in accordance with this: "A sign, or representamen, is something which stands to somebody for something in some respect or capacity" (Peirce 1932, 228).

However, neither an artificial nor a biological system, *taken only as a cybernetic system*, has an "interest" in "achieving certain goals." Granted, it may be in the position to fulfill certain functions—be it for human purposes or for its own preservation. But these functions may only be ascertained from the outside. As long as nothing is *at stake for* the system and it does not have *concerns* or goals, his functionality does not imply any teleology. It is not "too cold" in the room for a thermostat, nor for a brain, and a torpedo does not "experience failure" when it misses the ship.

In contrast to machines, a biological system admittedly perishes if its "representations" are not functionally adequate. They have thus a function for the preservation of the system—a function which may be traced back to a causal history of evolutionary selection.[16] However, one can still not talk about interests and goals which the biological system pursues, rather only about a natural causal history which produced systems of a kind that their internal processes may be described *from our point of view* as "functional" in the sense of self- or species-preservation. *For the systems as such*, it does not matter at all whether they perish or not (of course, as long as they do not have *subjectivity*, and thus concerns and interests—but this is not implied in Metzinger's definition). With this, however, the precondition for a three-place concept of representation, which could refer to a subjectless system, is lost. Metzinger's definition can then imply no more than that the neural system produces certain activation patterns or "data formats" which *we* can interpret as "representations" and as tools for self-sustainment. Whichever way you look at it, the representational relation—something *stands for, points towards*, or *means* something else—cannot be reinterpreted as a functional–causal connection, without there being subjects *for whom this is functional*.

A neuroscientist may nevertheless continue to speak about "representations" or "maps" in the brain in the sense that certain neuronal activation patterns are causally connected and correlated with a perceived object, an imagined object, or the like. He may also use such observed correlations to make inferences about the present perception or imagination of the owner of the brain. However, these patterns *as such* are not therefore *symbols* of objects, *they do not refer to them, do not mean them, and do not represent them*—no more than a tree presents its years of age in its

---

[16] This is the strategy of "teleofunctionalism," to which also Metzinger consents (2003, 27); on this, see Block (1978), Lycan (1987), and Millikan (1984). According to Millikan, the project of teleofunctionalism is to derive functions (and accordingly malfunctions or misrepresentations) from a causal natural history, or in her own words, "to let Darwinian natural purposes set the standards against which failures, untruths, incorrectness, etc., are measured" (Millikan 1991, 151). The concept and the critique it has received cannot be dealt here in more detail.

growth rings. There are no representations of the outer world in the brain, either in the semantic or the iconic sense of the word.

Should one not at least speak of traces of *memory* as representations of what is experienced in the brain? Without them, the person could surely not remember their knowledge of, say, World War I. Well, remembering something realizes an *ability*, such as the ability to recite a poem or to play a sonata by Schubert on the piano. When learning a poem, the brain undoubtedly develops the preconditions for a person being able to remember it later, for example, certain synaptic connections and dispositions for neuronal excitation. The poem is, however, not "stored" in the brain as a "representation," no more than their memory of the dates of World War I or of their voyage to Morocco, for the brain contains neither sentences nor pictures. Sentences in books represent facts *for us*, pictures in photo albums represent memories *for us*. However, there is no homunculus in the brain who would be able to grasp neuronal patterns of activity *as* representations, to see them *as* pictures or to read them as traces of memory. Neither rings in the tree, nor tracks in the snow, nor neuronal activity patterns in the brain are, *as such*, "representations" of past events.[17]

Hence, a valid concept of representation in the cognitive neurosciences would have to include the point of view of the observer. Representative connections can only be ascertained from the perspective of researcher subjects, who are, in addition, dependent on the statements of their test subjects in the first-person perspective, if they wish to arrive at correlations with subjective experiences. Talk about functions or functional connections is, for its part, *necessarily teleological*: in order to be able to determine the function of certain processes within

---

[17] Again, see Bennett and Hacker (2003, 154–171). Similarly Edelman and Tononi reject a representationalist account of memory: "Representation implies symbolic activity, an activity that is certainly at the center of our syntactic and semantical language skills. It is no wonder that in thinking about how the brain can repeat a performance—that it can, for example, call up what may appear to be an image already experienced—we are tempted to say that the brain represents. The flaws in yielding to this temptation, however, are obvious: There is no precoded message in the signal, no structures capable of the high precision storage of a code, no judge in nature to provide decisions on alternative patterns, and no homunculus in the head to read a message. For these reasons, memory in the brain cannot be representational in the same way as it is in our devices" (Edelman & Tononi 2000, 94). Instead, memory should be regarded as a "system property," which enables the brain to dynamically react to current situations and, on the basis of established neuronal dispositions, to activate varying response patterns not in a replicative, but in a creative way. In short, memory is never based on fixed "engrams," "copies," or "representations," but always recreates *similar* images or actions.

a system, I must, as an observer, presuppose the sustainment of the system as a purpose. Hence, if the notion of representation should serve to eliminate subjective experience or to identify subjective with brain states, the neuroscientist loses sight of the prerequisite for his research: his own subjectivity. However, since the neurocognitive notion of representation may hardly be purged from its semblance of objective givenness any more, it seems more reasonable to replace it generally, for example, by the term *pattern* and *pattern resonance* (on this, see section 4.2)

Let us sum up: ascribing intentionality to certain (not all) processes of consciousness identifies its inherent directedness to objects. However, intentionality cannot exist without subjectivity. Although the *performance* of intentional acts is linked with certain organic processes of a living being, its content, namely, "grasping something *as* something," does not tally with any physical or physiological description. There is, in fact, no meaning, no sense without subjects. The concept of representation is intended to indicate a two- or three-place relation, which could be described purely functionally. Nevertheless, each relationship of representation only exists for a person, who recognizes and interprets it as such. A picture is not a picture without someone who grasps it *as* a picture; a sign means nothing unless there is someone who understands it as a sign; a track refers to nothing without a tracker: the concept of representation cannot replace subject-dependent intentionality.

2.2 Second criticism: category mistakes

2.2.1 The mereological fallacy

Let us now examine the category mistakes and fallacies which result from the identification of the subject with the brain. These include, first and foremost, the neuroscientific practice of personalizing the brain and ascribing to it the most varied human activities. Brains can then, for example, "recognize faces" (Caharel et al. 2009), "perceive taste with all senses,"[18] but also "perceive alcohol" (Hodge et al. 2006). The inferotemporal cortex "identifies objects,"[19] the brain "decides when to work and when to rest" (Meyniel et al. 2014), and it even "recognizes itself as the subject of recognition" (Northoff 2004a, 17). If one reads neuroscientific literature, one can almost come to the conclusion that the brain genuinely calculates, believes, interprets, construes hypotheses,

---

[18] Science Daily 2016 (https://www.sciencedaily.com/releases/2016/08/160831133706.htm).

[19] MIT News 2015 (http://news.mit.edu/2015/how-brain-recognises-objects-1005).

recognizes, and decides. The category mistake occurs so often that Bennett and Hacker (2003) have given it a name of its own, namely, that of the "mereological fallacy."[20] A part of the organism, the brain, or one of its subsystems, thus has psychological and personal activities ascribed to it, which, in fact, only belong to the person as a whole. Examples abound, but I give only one more of them here:

> This simple fact makes it clear that you are your brain. The neurons interconnecting in its vast network, discharging in certain patterns modulated by certain chemicals, controlled by thousands of feedback networks—that is you. And in order to be you, all of those systems have to work properly. (Gazzaniga 2005, 31)

Well, of course I am not my brain—for my brain is certainly not married, not a psychiatrist, and it has no children. Even worse, it does not see nor hear anything, it cannot read or write, it cannot dance or play the piano, and so on. Thus, I am rather glad not to *be* my brain, but to only *have* it.

However, the personalizing language is not only meant figuratively or metaphorically, as the defense of this position is often articulated—on the contrary, it is precisely a successful naturalization which requires infiltrating intentional vocabulary into the description of subpersonal processes. For what could be explained about man if one only described monotonous, electrochemical processes on his neuronal membranes? The dissection of the live whole into micro-processes must, at least verbally, be undone, in order to reach the level of perceptions, motives, and actions again. The neurosciences, for that reason, attempt to insert a "hybrid" level in between which blends the physical and intentional descriptions, thus, to a certain extent, implanting personality in the brain.

That seems less problematic the more one goes over from actions to "pure" cognitions. Does the brain write? Does it hear, does it see?—Hardly. But does it think perhaps?—That may well seem so. Nevertheless, what could we make of a sentence such as this: "Peter's brain intensely deliberated about what it should do. When it could not find a solution, it decided first of all to wait and see." If thinking, feeling, and deciding were, in fact, activities of the brain, this would not be a ridiculous sentence, rather a quite meaningful one. Yet we rightly ascribe such activities to Peter, and not to his brain, because they are simply not "cognitions" or "mental states" in which Peter is, rather they are *life acts* which can only be ascribed to Peter as an embodied and conscious being. Reflecting, feeling, wanting, and deciding—none of these can be found at the physiological

---

[20] Mereology means the relation of parts and whole (from the Greek *méros* = part).

level of description *because these concepts do not exist there at all.* It is not wrong for empirical reasons to speak of the thinking, feeling, or perceiving brain—it is much rather conceptual non-sense. Erwin Straus formulated this insight briefly and appropriately: "Man thinks, not the brain" (Straus 1956, 112).

In their critique of the mereological fallacy, Bennett and Hacker (2003, 71–72) show that, behind the "as-if" subjectivity of the brain, there is again a latent Cartesianism: the "I" or "Ego" is thought of by neuroscientists as a substantialized, supposedly autonomous, freely acting center of decision, which is then declared to be non-existent: "It is not the Ego, but my brain, which has decided." This still assumes that there could be something like a Cartesian "Ego" making decisions. This Ego, the non-material soul is thus toppled and in its place comes the brain, only to immediately do the same as the Ego in Descartes, namely, to putatively imagine, to perceive, and to decide. Nevertheless, brains think or decide just as little as bodiless Egos—in both cases, one part is put in the place of the whole. This does not change if the Ego is replaced by "consciousness" or the "mind," as long as these concepts are, for their part, understood in the sense of a bodiless inner world. However, consciousness is a characteristic of living beings or, more precisely, an *enactment of life*. It manifests itself in life utterances and activities which are experienced by the living being as a whole and can be recognized by others in its behavior: being frightened, afraid, or happy, reflecting, speaking, writing a letter, or playing football.

That seems to be just a matter of course, which it is not, however. Even for John Searle, mental states are "simply higher-level features of the brain" and consciousness is "an emergent property of the brain" (Searle 1992, 14). On the other hand, shortly afterwards, he emphasizes that "the ontology of the mental is essentially a first-person ontology. Mental states are always somebody's mental states" (p. 20). But *somebody*, that is a person, therefore not "an Ego," "a consciousness," not to speak of a brain; it is rather a complete human being of flesh and blood. Can we, nevertheless, ascribe somebody's mental states to his brain? No, this is where Searle is contradictory: consciousness is a feature of human beings, that is, of organisms, not of brains. A neuroscientist may well be able to ascertain indications of a person being conscious in her brain—however, in order to find out whether she actually is conscious, he must observe her embodied behavior or engage in interaction with her. The brain may well be the central place for physiological processes which are necessary for her being conscious, but *it is not aware*, it does not perceive, it does not move, it does not get angry or feel happy—all of those are the activities of *living beings who are conscious.*

The basic problem of neurobiological research into consciousness consists, when all is said and done, in the *reification of consciousness itself*. It then no longer appears as an activity of living organisms, no longer as a relationship between subject and world which transcends the boundaries of the body. It is rather transferred into the objective world, as if it were an object in spatiotemporal reality which could be physically described or, at least, made indirectly visible by physical means. This leads us to a further fallacy.

2.2.2 The localization fallacy

A category mistake connected with the mereological fallacy consists in localizing single phenomena of experience in specific brain areas—we can speak of the "localization fallacy." According to it, visual perceptions are produced in the visual association cortex, fear in the amygdala, or memories in the temporal lobes. Constantly, new areas are found for all types of mental phenomena—pain, sadness, racist prejudice, deliberate deception, self-criticism, taking another's perspective, empathy, indeed even personality traits.[21] This research program is, first and foremost, based on imaging techniques which reflect the specific brain activities in vivo and seem to suggest that mental functions should be located in certain areas of the brain.

> The confrontation between localizational and holistic paradigms in brain physiology goes back as far as the eighteenth century. For a long time, localization theory was discredited by the "phrenology" of Franz Josef Gall (1758–1828), who speculatively related features of character, such as love of children, domesticity, or superstition, with certain areas of the cerebral cortex and corresponding protrusions of the skull. Albrecht Haller (1708–1777) and later Pierre Flourens (1794–1867) proposed a contrasting, holistic theory of the function of the brain, the so-called equipotential theory, according to which the complete brain always takes part in mental functions (Hagner 1997, 89–92, 248–50, Karenberg 2009). By means of the discovery of brain areas, whose failures are responsible for motor and sensory aphasias, Broca (1861) and Wernicke (1874), however, contributed greatly to the rehabilitation of the localization project, which enjoys particular success today. Accordingly, theories of the *modularity* of the mind (Fodor 1983, Pinker 1997), implying the construction of consciousness from separable single functions, are still preferred in cognitive science.

---

[21] See, for example, Phelps et al. 2000, Vogeley et al. 2001, Langleben et al. 2002, Etkin et al. 2004, Eisenberger et al. 2005, or Singer and Lamm 2009.

Undoubtedly, the localization theory has its own justification. The brain is regionally specialized; various neuronal areas and centers fulfill different functions. For this reason, it is also possible to connect certain *features or components* of conscious processes with local activities. Thus it is possible, by means of brain imaging and other procedures (single neuron recording, electroencephalography (EEG)), to ascertain with high probability whether someone is speaking silently to himself, imagining different categories of visual objects, adding or subtracting numbers, paying attention to a vertical or horizontal patterns of stripes, is preparing to press the right or the left button before him, and also whether a person is feeling pain, fear, or happiness (Edelman et al. 1998, Cox & Savoy 2003, Kamitani & Tong 2005, Soon et al. 2008, 2013). This is, however, only possible if corresponding correlations have been established by imaging beforehand, namely according to the information given by the test persons. Such advances are based on the functional specialization of the regions of the brain.

On the other hand, none of these regions is per se capable of producing the complex achievements of integration which are the basis of processes of consciousness. In fact, widely distributed brain areas and centers outside the cortex also contribute to this, so that a dynamically changing network of neuronal assemblies and activity patterns spread over the whole brain is involved in a special subjective experience.[22] Last but not least, the unsolved "binding problem"—the question of how the scattered activities and processing paths are reintegrated, as, for example, in unified intermodal perceptions (see 1.3.1)—points to the limitations of the localization paradigm (Uttal 2001).

According to the classic cognitivist or modular view, the brain implements encapsulated mechanisms for cognizing (perceiving, planning, evaluating, decision-making, etc.). Each module is believed to be responsible for computing an independent cognitive function, largely unaffected by the working of other modules and disconnected from bodily and environmental processes. This conception still fuels experimental cognitive research, not least because of its suitability for isolated study designs. However, it has now come under

---

[22] Edelman and Tononi (2000, 139–142) have proposed the "dynamic core hypothesis," according to which conscious states emerge from an ever-changing functional cluster of networks, characterized by strong interactions and "reentry" feedback mechanisms, and situated mainly within the thalamocortical system. "A dynamic core is therefore a process, not a thing or a place, and it is defined in terms of neural interactions, rather than in terms of specific neural locations, connectivity or activity" (2000, 144). What is neglected in this theory, however, is the role of body–brainstem interactions for the emergence of consciousness; this will be investigated in section 4.1.

growing criticism as being inadequate for the distributed functioning of the central nervous system, multitasking at every level and highly dependent on contextual variables (Van Orden et al. 2001, Hardcastle & Stewart 2002, Gibbs & Van Orden 2010). Therefore, the modular model is increasingly replaced by thinking in overarching functional systems and highly flexible brain connectivity patterns, where the same cortical or subcortical area may be co-opted into different functions depending on which of its interconnected networks is activated (Friston et al. 2003, Sporns et al. 2005; for an overview, see Cosmelli et al. 2007).

This also corresponds to the complexity of experience itself: all terms for special functions, such as seeing, hearing, thinking, feeling, wishing, and so on, single out components of consciousness, whereas factually subjective states of experience always remain holistic. Thus, all perceptions are not only embedded in a bodily background experience, but are also connected with feelings, memories, and linguistic concepts. There is no "pure" pain, no "plain" seeing or hearing. Conscious experience is not put together from components at all; it is, conversely, a *primary unified process* or a *"stream of consciousness,"* which differentiates into specific activities and achievements according to the particular demands of the situation. Hence, brain functions may best be conceptualized along two polar organizational principles: functional segregation *and* functional integration. Their interplay is enabled by connectivity and distributed neuronal assemblies that transiently oscillate at the same frequency (Friston 1994, Cosmelli et al. 2007).

For that reason, however, talking about circumscribed "neuronal correlates of consciousness" is not appropriate. It implies that phenomena such as perceptions, feelings, or thinking processes could be isolated from the holistic activity of consciousness. These phenomena, however, are not states which can be isolated; they rather presuppose a *subject* that perceives, feels, thinks, and so forth. However, what kind of "correlate" subjectivity has, how far its organic base extends, and whether it does not include the complete organism, is still unexplained up to now. As long as this is not the case, the search for correlates of consciousness still remains at a speculative stage (Cosmelli et al. 2007).

> Noë and Thompson (2004) have pointed out that even in the best studied subsystem of the brain, namely the visual cortex areas V1–V5, it is not possible to unambiguously attribute visual content to certain neural assemblies. The reason is that even with regard to the same object, the activity of these neurons depends on the living being's body posture, behavior, state of attention, and the relevance of the object for current tasks, in short: on the overall state of the organism in relation to its environmental context.

Moreover, each perception of a moving object contains not only the object itself, but also its motion dynamics, the background of the visual field, the eye, head, and body movements by which one follows the object, one's proprioceptive body awareness, and so on. Thus, perception is not a momentary snapshot of a stimulus configuration, but rather a dynamic, intentional, and attention-directed process which ultimately includes the whole system of brain, body, and environment. The search for neural correlates of consciousness can therefore only grasp certain partial components, not perception as a situated, bodily, and spatial process.

If attempts toward localization of consciousness or conscious functions lead to impasses, one may ask what misleads neuroscientists to localization fallacies time and again?—Above all, three kinds of observations contribute to this:

1. To begin with, it is specific *function failures* as a result of local lesions in the brain which seem to pinpoint the "seat" of the function in the relevant area. Because of the high plasticity of the brain, however, lost functions can in many cases be taken over by other brain areas. But even apart from that, the failure as a result of a lesion allows at best for the conclusion that the area represents the *necessary*, but not the *sufficient* condition for a function. There are always other areas and connections required within the complete neuronal system, as we have seen earlier in the case of perception. Hence, it is not functions that may be strictly localized, but only disturbances of functions.

2. The new *imaging techniques* seem to establish the place of the function in vivo. In a world of pictorial media, neuroscience has developed its public power of persuasion not least by means of its colorful staging. That is why it is all the more important to know the methodical limitations of these techniques.

    First, imaging techniques do not in any way measure neuronal activity as such, rather indirect parameters, as, for instance the BOLD signal (the blood oxygen level-dependent signal, i.e., increased blood flow and oxygen use in certain brain areas, from which the neuronal activation can be inferred) in functional magnetic resonance imaging (fMRI). In order to create a sufficient contrast, the basic activity of the brain is determined in advance and then "subtracted" so that the locally increased activations emerge. Thus we are not dealing with "images of the brain," rather of the visualizations of statistic calculations, that is, scientific constructs produced in an intricate manner. Further, mean values are formed from greater samples of test persons since no significant results can be individually gained as a result of the extremely limited differences in local activity. Not surprisingly, the validity of the achieved correlations has also been strongly questioned, for example in affective neuroscience (Vul et al. 2009).

Moreover, it is not in any way clear whether the experiential phenomena investigated correspond to the most colorful flashing structures. In the case of pathological phenomena, local increases in activity can also correspond to secondary, compensatory reactions to the actual functional disturbances at another place. In any case, all other brain regions, in which nothing appears to happen in the image, are active at the same time and in various ways involved in the experience and the function. Thus, the resting state of the brain, a basic activity spread over the cortex and known as the "default mode" (Raichle et al. 2001), seems to represent the basis of a background experience, on which specific activities of consciousness can only develop. Finally, Anderson and Pessoa (2001), in a meta-analysis of 2603 fMRI studies in 11 task domains (e.g., vision, audition, attention, emotion, language, memory, action execution, etc.), found that in fact most regions of the brain are involved in supporting multiple tasks and can perform different operations under different circumstances, again pointing to the limits of the localization paradigm.

What the images actually show and what really happens in the brain thus require careful interpretation. Moreover, imaging occurs in laboratory situations, where the relation of conscious processes to the environmental context remains largely excluded, as does their prehistory and their temporal course. These aspects are, however, essential features of consciousness. The technique of imaging thus, as it were, freezes the stream of consciousness and isolates it from its context. If one takes all these methodical limitations together, data on the local metabolic activity of the brain can to a certain degree reflect its functional specialization, but it can only offer limited *indications* of ongoing mental processes.

3. The localizability of mental functions seems to be impressively shown by the fact that certain conscious phenomena can be evoked by direct electrical stimulation of the brain (see Selimbeyoglu & Parvizi 2010 for an overview). Thus, in the 1960s, the neurosurgeon Wilder Penfield succeeded in triggering, by means of targeted stimulation during brain surgery, the kind of experiences in conscious patients that are known as epileptic auras (Penfield & Perot 1963). Among these experiences were changes in perception (distortions of sounds or visual objects, experiencing déjà vu), feelings of pain, fear, sadness, or disgust, as well as memory flashbacks, voices of familiar persons, well-known melodies, or fragments of experienced scenes. Prior to the neuroimaging era, such brain stimulation experiments provided the most direct evidence for a possible localization of functions.

But what actually follows from Penfield's and similar experiments? It is tempting to infer localization from causal production or even to identify experiences with circumscribed brain processes, but it is, however, misleading. For even the stimulation of my foot by a needle produces a sensation of pain—nevertheless, this would not cause any brain researcher to localize the pain in the pain receptors of the skin. Pain sensation is the *integral reaction* of the living being to a peripheral stimulus, for which, undoubtedly, the activation of certain neuronal networks is also necessary.

It is in principle possible that that the same pain could be produced by the direct stimulation of the somatosensory cerebral cortex or the insula (Selimbeyoglu & Parvizi 2010). This does not, however, change the fact that the pains, in both cases, represent expressions of life, that is, *reactions of the whole organism*. The pain is experienced as suffering, it is accompanied by tensing in the body, defense movements of the foot, and an expression of pain in the face, as well as with an activation of the sympathicus system, that is, a stress reaction of the organism—all that is the pain. If it is thus not situated in the skin receptors, what speaks for localizing it in certain centers of the brain?

One possible argument is that the sensation of pain in the periphery can be suppressed by a blockade of nerve conduction, so that it is no longer felt, and can thus not represent the correlate to the sensation of pain. However, the same applies to any region of the brain. If a sufficient number of its neuronal connections are severed, its stimulation can no longer produce any sensation. Hence, a certain, sufficiently extended totality of brain activities in connection with the organism is necessary so that we can experience pains.[23] That is why those experiences cannot be localized at their triggering point and are not "identical" with certain neuronal processes. The temporal lobe does not contain any memories or sensations of smell, nor does the insula have any pain sensations, even if they can be provoked there by an electrode. Only the living being as a whole has memories and sensations.

This leads to the following conclusion, in agreement with Rockwell's (2005) account: a pain in the foot is not caused by an unconscious signal that travels up the leg and transports "information" about the event into the brain. Instead, the pain should rather be regarded as " … a network property that arises out of the relationship between the nerves in the foot, the spinal cord, and the various

---

[23] Selimbeyoglu and Parvizi (2010, 9) come to a similar conclusion: "Today, the phrenological notion is outdated […] perceptual and behavioral phenomena induced by electrical charge delivery to a brain region are most likely due to change of activity in a network of brain areas (including subcortical regions) rather than the excitation or inhibition of a blob of cortical grey matter per se" (Selimbeyoglu & Parvizi 2010).

neuronal ensembles in the cranium" (Rockwell 2005, 32). Whether the stimulation occurs in the foot or in the brain, in each case it means a reconfiguration of the *whole nervous system* which embodies the pain. Similarly, as we will see further later in this book, consciousness is not in the head, but spread over the whole body, and it is only modified, not "brought forth" by the local stimulation.

We see that increasing research into the functional specialization of the brain is not suited to supporting a localization of consciousness as such. The decisive reason for this is that it represents an integral activity of the organism, which, as we will still see more closely, requires continuous embedding in an environmental context. Granted, partial functions of consciousness can to a certain extent be assigned to certain specialized regions, damage to which then also results in the failure of the function. However, every theory which views consciousness as being assembled from localizable individual functions or modules incurs the problem of how these individual functions are to be integrated into a united activity—a question which is mirrored in different variations of the "binding problem." The entire project of the spatialization and materialization of consciousness all too easily loses sight of its object because of looking too closely at it, thus ending up with only fragments. Hence, what Georg Christoph Lichtenberg wrote at the end of the eighteenth century about the attempt of the anatomist von Soemmerring to localize the soul in the ventricles of the brain is still valid today:

> If I, when viewing the setting sun, take a step towards it, I come closer to it, little and all as it may be. In the case of the organ of the soul, this is quite different. Indeed, it would be possible, by means of coming exaggeratedly close, such as with the microscope, to once again distance oneself from what one can approach. (Lichtenberg 1973, 852; own translation)

How far we must step back to set eyes upon the locus of consciousness still remains to be investigated.

2.3 Third criticism: the powerless subject?

2.3.1 The unity of action

In the first step of the criticism (2.1), it was explained why subjectivity and intentionality cannot be completely reduced to physical descriptions. In a further step (2.2), we investigated the mereological and localization fallacies, to which an identification of the subject with the brain leads. A third question remains to be addressed. A reductionist neurobiologist could argue: "Sure enough, consciousness is real and possibly not completely reducible. However, it is certainly *produced by* the brain. That is also why the brain possesses reality to a greater extent than consciousness. It is the *actual* reality. And because this reality is of a physical nature and, as a result, subject to physical principles, subjectivity itself cannot have its own effectiveness in the world." We may well believe that we ourselves direct our thoughts and actions, in reality, however, they are designed by neuronal systems, and they surface in consciousness like film scenes which a projector on our back casts on the screen.

By this means, we arrive at the discourse about free will, which has been debated for years. Indeed, it is surprising that, of all things, the human brain is called as the crown witness of determinism. For it is precisely the brain that is the organ whose growing complexity in the course of evolution has relaxed the rigid stimulus–response mechanism, thus enabling organisms to attain increasing degrees of freedom—seen from that point of view, it is the organ of freedom. We talk, for example, in psychiatry about a lack of freedom, above all in the various impairments or dysfunctions of the brain. Patients with frontal brain injuries suffer from aimlessness and a lack of initiative; they can no longer maintain a directed intentional arc, spanning longer stages. Patients with Tourette syndrome are compelled to make spasmodic movements or to express swear words, and are unable to restrain themselves. People with compulsions cannot help doing things which they themselves find meaningless, or think what they do not want to think. Schizophrenic patients even experience their actions as being directed by foreign powers. In all these cases, it is rather *disturbances* of brain functions that restrict the patients' freedom or dictate to them what they must do.

It is, however, precisely this, according to the opinion of some neurobiologists, that applies to us all: brain processes work deterministically, and we cannot do otherwise to what our brain determines. In fact, decisions are ultimately directed by unconscious emotional processes in the limbic system, and the actions are then triggered by the premotor areas of brain, before the person becomes conscious of this. Thus, the brain only deludes us into believing we are acting and responsible persons, whereas we can in fact only ratify its decisions in hindsight.

> [O]ur actions are clearly the result of a causal chain of neuronal activity in premotor and motor areas of the brain. […] although we may experience that our conscious decisions and thoughts cause our actions, these experiences are in fact based on readouts of brain activity in a network of brain areas that control voluntary action. (Haggard 2011, 404)
>
> [O]ur brains have to function as efficient, unconscious computers that nevertheless make rational decisions. (Swaab 2014, 331)

Although published over 30 years ago, Benjamin Libet's demonstration of a preceding readiness potential in the brain, in the case of subjectively experienced arbitrary movements, still functions as an *experimentum crucis* for the neuroscience of voluntary action (Libet et al. 1983, Libet 1985). In this study, test persons were asked to wait for the impulse or "urge" to move a certain finger, and then, to state the point of time of this impulse, with the help of a rotating clock hand. EEG activity was measured at the same time, showing the emergence of the so-called readiness potential over the supplementary motor cortex 1 second or more before the actual movement, and about 500 milliseconds before the stated impulse to move. This seemed to demonstrate that action is prepared and triggered by the brain even "before you know it," at least challenging any versions of free will where intention occurs at the beginning of the decision process.

The deterministic interpretation of this experiment has frequently been criticized, above all, because it isolates human action experimentally from its intentional context and restricts it to the level of accidental movements.[24] It seems, to say the least, adventurous that the denial of free will should be based on an experiment which certainly depends on the voluntary participation of test persons, who would never have moved their finger without their consent. This preceding component, that is, the actual process of deliberation and decision is not included in this experiment at all. It thus disassembles the temporal and meaningful unity of forming one's will and acting on one's will, with the result that a final, artificial "moment of decision," a "tug of will" is created. Similarly, all further experiments on brain and volition have so far only dealt with decisions made in time frames of seconds and on extremely simple actions such as moving a finger.

> Moreover, an experiment carried out by Herrmann et al. (2008) rather suggests that the readiness potential may reflect an unspecific anticipatory stance. In this study, test persons carried out a *choice reaction* task: depending on geometrical figures presented to them at the last moment, they had to choose between pressing either one of two buttons. This was preceded by readiness potentials too, however, *before* the presentation of the respective picture, thus at a time at which the choice between the buttons could not have begun in the brain. Thus it seems likely that the readiness potential serves the general preparation of expected movements, corresponding to what Jeannerod (1997) has termed "motor imagery," but does not yet determine the final action.
>
> Libet's paradigm has meanwhile been further developed into *action prediction* by applying massive computational technology to whole-brain fMRI scans. Also using a choice task, Haynes and his group were able to predict with 60% accuracy whether subjects would press a button with their left or

---

[24] See Gallagher 2005, 237–240, for a critique.

right hand up to 10 seconds before they became aware of their choice (Soon et al. 2008). This seemed to question the idea of conscious decision-making. However, a more recent study by the same group confirmed Libet's initial assumption that a conscious *veto* is still possible even in the last fraction of a second: while the computer tried to predict their actions from brain activity, test subjects were able to stop their already initiated action until up to 200 milliseconds before the actual movement (Schultze-Kraft et al. 2016).

We have already pointed out the implicit dualistic preconditions of the neurobiological position (see 1.5). This also applies to the arguments against free will: they are based on the fiction of a Cartesian ego, separated from its body, its feelings, and its enactment of life, which reaches a decision in unlimited arbitrariness and then imposes its execution on the body. The effectiveness of this fictitious ego is then declared refuted by referring to the closed causal chain of bodily processes. Consciousness always comes too late compared with its neuronal construction mechanisms. The physical world leaves no scope for the causality of the subject. Consequently, decisions and actions ought to be ascribed to the brain.

Such argumentations are basically subject to the criticism regarding the mereological fallacy. Brains decide just as little as they are in the position to act. Indeed, attributing decisions to brains also negates the concept of decision itself (Fuchs 2007a): a computational, neuronal process as such, regardless of whether it proceeds in a strictly deterministic, probabilistic, or indeterministic way, is incapable of grasping alternative possibilities *as possibilities*. Indeed, it is even unable *to grasp the future*. That is why it is no more a process of decision-making than a cube falling or the function of a random generator.

The term "readiness potential" does not mean that the brain or the motor cortex could actually be "ready" or "prepared" for something to happen. This readiness can only emerge with conscious life, for only consciousness is able to integrate time into a span that includes the immediate past, present, and future. This integration has been famously described by William James (1890) as extended or "specious present," by Henri Bergson (1950) as "duration," and by Husserl (1991) as "inner time consciousness." To explain it briefly: the mere succession of conscious moments, as such, could not establish the experience of continuity. It is only when these moments mutually relate to each other in a forward and backward directed intention that the sequence of experiences is integrated into a unified process. Husserl conceived this as the synthesis of *protention* (indeterminate anticipation of what is yet to come), *presentation* (primal or momentary impression), and *retention* (retaining what has just been experienced as it slips away). This can be illustrated by a melody or a spoken sentence: we hear the current tone (presentation), but are at the same time still aware of the tones just heard (retention), and vaguely expect the continuation of the melody (protention). Consequently, what is perceived is not a sequence of single moments but a dynamic, self-organizing process, which integrates the tones heard into a melody, or words into a sentence.[25] From this follows that being ready or prepared for something, or anticipating the next-to-come, is only possible for a conscious living being. Indeed, to anticipate the not-yet and to retain the no-longer is one of the most fundamental functions of consciousness.

A fortiori, the anticipation of possibilities *as possibilities* is only available to a human being who finds herself in *future-oriented life conduct*, who disposes of *embodied capacities of action* and who can counterfactually also *imagine the not-being*—"to do or not to do?" is the question at every decision. Comprehending the alternatives *as alternatives* (left button or right button?) in the first place is even the precondition for all so-called decisions in the above-mentioned experiments. If, however, this subjective perspective is eliminated as illusory, then there are no alternative pathways of events; the world runs as it runs, and, consequently, brains decide nothing. Apart from this, psychology has always been aware that not only conscious and rational considerations are included in the subjective decision-making process, but rather also unconscious or partially conscious motives, dispositions, and tendencies. This does not change the fact that every decision needs anticipation and thus, consciousness.

The same applies for the concept of action. We can only speak of actions (in contrast to events) if there is a person acting, and this is the complete human being. Monica goes to school—not her Ego, her brain, or her legs. If Monica moves her legs for this, they usually do that by themselves, and there is no need for a willed decision (it suffices that she wants *to go to school*). Should Monica have the idea of moving her legs intentionally and in a targeted manner, as the Libet experiment requires it of the test persons, her legs will certainly obey her. Nevertheless, this particular instrumental relationship, which the human being can have to her body, does not produce a bodiless "Ego" or an ominous

---

[25] Perceptual experiments on the so-called flash-lag effect also demonstrate that we are slightly ahead of the present: if subjects are watching a continuously moving object, and a sudden flash is presented at the exact location of the object on its trajectory, the subjects erroneously see the object as having already moved past this point (Changizi et al. 2008, Nijhawan 2008).

"will" which gets the body moving from outside. Monica would not know at all how she should do that—"to move her leg," like she would move a plate out of the cupboard. She remains, also with intentional movements, an embodied being which *moves itself*—and does not transport its own legs, like two pieces of wood, from here to there.[26]

Now if specific motor readiness potentials emerge in Monica's brain shortly before she sets off, she, of course, does not become an automatic machine or a marionette of her brain. Monica could, for example, have come to the conclusion rather to play truant and to go swimming. As soon as she turns this decision into action, however, *precisely the same* readiness potentials would appear in her motor cortex. These brain activities are therefore necessary, and at a very late stage also sufficient, conditions for Monica's *muscular movements*, but are not sufficient for her *future-directed action*. For the action of going to school is, undoubtedly, a completely different action than playing truant, although they both use the same muscles and motion sequences. What the neurobiological description explains is therefore, at best, a body movement in the sense of a physiological event. In other words, it explains the *proximate* or subordinate causes of the action. To explain the movement *as action*, however, a knowledge of Monica's motives, thoughts, wishes, and aims is required—that is, thus, a quite different, namely, psychological, teleological, or intentional description. Physiological causes are completely irrelevant for the question of the *meaning* of an action. Of course, too, these subjective phenomena do not exist in a transcendental world of the mind; they are, rather, just like Monica's ability to go, manifestations of her embodied subjectivity. Hence, if one wishes to give the cause for the action *as* action, it can, therefore, neither lie in an Ego or will, nor in the brain, but rather in the complete human being with all his or her mental and bodily capacities.

2.3.2 The role of consciousness

Of course, one can further radicalize reductionism and can award subjectivity a merely epiphenomenal status also in the processes of consideration, evaluating, and deciding. The question is therefore whether the process of the subjective assessment of possibilities *co-determines the result*, or whether it is only a powerless mirroring of physical processes. If subjective experience in fact remained without consequences for the course of the world itself, this would indeed strike at the heart of the idea of personal freedom and agency. Is it then crucial that I seriously consult with myself about what I should do in a certain situation? Does it make a difference in the world? Would we really be able to act otherwise?

---

[26] This corresponds evidently to the conception of Aristotle who spoke of living beings as "self-moved."

If it is true that we do not find possibilities, evaluations, reasons, and, finally, decisions in the physical world, then it does make a difference in fact. For it means that the processes of deliberating, evaluating, preferring, and deciding cannot completely be reduced to physical-chemical laws. That brain processes are not solely determined by such laws can easily be seen, as the brain is essentially shaped by cultural, ideational, and symbolically mediated influences. For example, what counts as a logically valid inference or what the result of "$x=\sqrt{16}$" is, is not determined by natural laws of physics. So if we find "$x=\pm 4$" as the solution of the equation, its correctness does not result from physical or neurophysiological but from mathematical laws. The brain is only a highly malleable carrier medium, which is capable to adopt such general laws. Such shaping of neural dispositions, however, is crucially mediated by subjective experience; we will come back to this in Chapter 6.

Now, the shaping of the brain by means of language, ideas, and culture is commonly also conceded by neuroscientists. This, however, is assumed not to change anything about our being completely physically determined: in that case, it is argued, functional equivalents of meanings and cultural programs become part of the neural algorithms, for instance, equivalents of mathematical, logical, or moral rules. But it is still the brain that carries out these programs, calculates, thinks, and "decides," since it was programmed in this and not another manner. Subjectivity and conscious experience, however, are assumed not to have an influence on the process of deliberation:

> The *sense* of will is an invention of the brain. Like so much of what the brain does, the feeling of choice is a mental model—a plausible account of how we act, which tells us no more about how decisions are really taken in the brain than our perception of the world tells us about the computations involved in deriving it. (Blakemore 1988, 272)

A central argument against such a position is based on the theory of evolution: why should subjectivity and consciousness have evolved at all? What is the point of investing such developmental efforts and energy into a phenomenon without any significance and consequence, a systematic self-deception of billions of living creatures?[27] If the brain functions perfectly well without an ancillary support of consciousness, then there seems to be no causal role for conscious processes that could improve the odds of a living being's survival.

In his account of consciousness, neuroscientist Edelman explicitly poses the question whether phenomenal consciousness has causal efficacy and thus an

---

[27] This kind of objection against epiphenomenalism was already put forward by Puccetti (1974) and Popper and Eccles (1977).

adaptive function (Edelman 2004, 76–88). Granted, he argues, with certain neural processes, the simultaneous property of consciousness is given in a no further deducible manner—a "phenomenal transform" of the "dynamic core" (see 2.2.2), including "what-it-is-likeness" and qualia. However, the causal closure of the physical world demands that it is not the phenomenal experiences C, but only their carrier processes C′ which can cause physical effects. These processes were selected for by evolution in order to enable efficient planning and acting, and it is they that realize causal links. The phenomenological transform only serves as a "reliable indicator of the underlying causal C′ events" for the individual (2004, 79).

Now Edelman himself does not seem entirely sure what purpose this indicator might serve if the conscious individual is nothing but a powerless accompaniment of their neurons and, for this reason, he adds another function: consciousness, at least, enables higher animals to communicate to others the states of their C′ brain regions:

> Animals so evolved would communicate efficacious C′ states in terms of C. C, after all, is the only information available that reflects C′ states to each animal and to others. (2004, 81)

Of course, Edelman has to concede that the dynamic core as a carrier of consciousness will already have developed in species "without extensive communicative abilities" (p. 81). Therefore, the only option left is to conceive of C as an "epiphenomenon" (p. 85) that is necessarily linked with C′ processes, without itself having a function. Nonetheless, Edelman finally states that "the phenomenal transform is an elegant means of conveying the integrated states of C′ on a first-person basis" (p. 86). But which function does this elegance fulfill? The claim remains tautological, for "conveying C′ states on a first-person basis," in the final analysis, means nothing else than transforming them into phenomenal experience. So in that case, phenomenal experience is good for phenomenal experience.

Here we encounter once again the basic dilemma of neurobiological approaches: the more complete the alleged physiological description of the neural foundation of consciousness, the more precarious the question of the function of consciousness itself becomes. As Hans Jonas has pointed out, it becomes "a dead-end alley off of the highway of causality, past which the traffic of cause and effect rolls as if it were not there at all" (Jonas 1966/2001, 128). More so, it becomes one of the properties that natural science wanted to eradicate from its world, namely a "*qualitas occulta*," a hidden, unprovable property of certain material processes that is manifested in no effect. Hence, there is no way around the insight that if we do not want to buy into the ontological as well as the biological absurdity of an inconsequential subjectivity, we have to conceive of the brain in a manner that it cannot only be shaped by social and cultural influences, but also be *currently* integrated into the superordinate conscious enactment of a human being's life.

We have already been able to ascertain in various ways which fundamentally novel phenomena appear in the world with the emergence of consciousness. I summarize its most important dimensions as follows:

- The integration of the living being's sensorimotor interactions with the environment into an *intermodal action space* ("*sensus communis*"), allowing for skilled coping with environmental affordances and opening up possibilities for action.
- The *intentional and affective directedness* of a living being towards relevances and meaningful situations in its environment; that is to say, consciousness is teleological, oriented towards goals and purposes.
- The *integration of experience over time*, in the sense of being directed towards the immediate future and its possibilities (protention) as well as retaining past experiences (retention)—in other words, the temporal coherence of consciousness.
- The awareness of *alternatives of action* offering themselves in a given situation, in human beings also including counterfactual imagination ("as if").
- Last not least, the self-experience of the living being in relation to the environment, that is, a basic sense of *self-awareness and self-affection*, integrating the organism's current overall state with regard to its own self-preservation. This integration also manifests itself in the spatially extended and yet indivisible unity of the subject-body (see 1.2.2).

All these phenomena and properties are nowhere to be found in the physical world: neither a unified action space filled with qualitative affordances, nor an intentional and affective directedness, nor an integration of time, nor finally the dimension of self-awareness, which turns higher animals into centers of their own world. *Unlike physical mechanisms, consciousness is not analyzable into distinct spatiotemporal components; it covers space, time, and the body.*

> To demonstrate this with regard to temporal integration: physical processes, including neural processes in the brain, are always only present, irrespective of how complex they may be. They are never more than *linear sequences of events*, at any time restricted to the current moment, without any anticipation of a future (physiological control loops and even "feedforward" mechanisms cannot actually "anticipate" anything), or a memory of the past. It is only the overarching temporal continuity of consciousness (see 2.2.1) that allows higher animals to grasp the possible future, in particular to anticipate possible action.

Although recent neurocognitive theories posit the brain as a "predictive organ" or "prediction machine" (e.g., Downing 2009, Clark 2013, Hohwy 2013), this should not blind us to the fact that brains are neither in the condition to advance hypotheses about possible events nor to make inferences about remote objects or predictions about the future—simply because they are not "ahead of themselves" and therefore *unable to anticipate what is yet-to-come, even less to grasp the future as such*. There may well occur an alignment of predisposed excitation patterns and incoming stimuli in the dynamical state space of the brain, in the sense that "forward models" are either matched by the input or not. But this is not principally different from correction mechanisms in "target-seeking" missiles; it means neither a "confirmation" nor a "disconfirmation" of hypotheses or anticipations. No matter how important stochastic (Bayesian) adjustment processes may be in the brain's processing of incoming stimuli, a "predictive brain" as such does not exist.

Given the irreducible integrative properties of consciousness, it seems nearly absurd to assume that this multidimensional integration, and with it, the appearance of a fundamentally novel phenomenon in the world, should have remained without consequence for the behavior and the adaptation of living beings which dispose of such a function. On the contrary, over the course of evolution, the brain has developed as an organ whose complexity enabled the emergence of feeling, emotion, thought, and volition, and which became the crucial (though not sufficient) basis of integrative conscious experience. In this way, the developing brain allowed for ever greater degrees of freedom of living beings and multiplied their scope of choice and action—up to the possibility of free deliberation and decision in human beings.

Thus, the brain is rather an organ of freedom than of necessity. There are neural processes that can function, so to speak, as a "matrix" for motives, considerations, imaginations, and evaluations, no less than for mathematical or logical laws. Neural conditions of consciousness do not exclude freedom, but are its conditions of possibility—though it is only consciousness itself which is able of envisaging possibilities as such. Hence, the alleged causal closure of the physical world should not blind us to the particular possibilities of emergence and "downward causation" that made their appearance with living beings, and which may also enable a consistent account of embodied human freedom. We will return to this issue in Chapters 3 and 6.

2.4 Summary: the primacy of the lifeworld

In this chapter, the idea that subjectivity could be reduced to the description of neuronal processes was criticized and refuted. The characteristics of phenomenal consciousness, especially the subjectivity of experiential facts, the phenomenon of intentionality, and the integration of time, cannot be sufficiently explained by the description of correlated physiological events. Moreover, the attempts at reduction run into category mistakes which were analyzed as the mereological and localization fallacies. Finally, the claim that processes of consciousness only possess an illusory efficacy leads to the aporia that their appearance and function in evolution become a riddle. In contrast, it was shown that consciousness enables an integration of space, time, and self that is not found in the physical world and multiplies the possibilities of living beings to cope with the environment and to preserve themselves.

Following on from the "Introduction," I would now like, at the end of this first part of the book, to grasp the problems posed by neurosciences at their root and will additionally use a culturalist approach, as it was developed by Janich (1996) and Hartmann (1996, 1998). My thesis reads as follows: the problems of the relationship between brain and mind, as they present themselves today, emerge from a *short circuit* between the level of natural scientific, in this case, especially neurobiological constructs, and the level of intersubjective, lifeworld experience, from which the neurobiological special practice has developed and with which it remains always bound.

The basic paradigm which directs the cognitive neurosciences is, in the last analysis, a *metaphysical realism*: there is an objective, material world "out there" which is independent of our process of observation and of our anchoring in the lifeworld, and of which there must, in principle, be a complete, and, in fact, *physical* description (even if this description has to use certain constructs and we can only approximate completeness). If we had this complete description, it would include everything that happens in the world, that is, also *our experience and observation of the world itself*. In other words, it would have to include all that could be known about consciousness and its contents. Otherwise consciousness would be an additional, non-natural property of the world, which would contradict the precondition.

The basic problem of this approach lies in its manifest, though mostly not comprehended, circularity. It is based on the assumption that there could be a position of observation and recognition beyond our lifeworld experience which is, however, always presupposed with the observation. Independently of this experience, physical objects cannot be identified at all. What makes up a human being, a brain, neurons, molecules, or atoms can only be gathered from our common prior understanding or from conventional agreement. Metaphysical realism or physicalism is thus incoherent insofar as it overlooks its own dependence on the intersubjectively constituted lifeworld. This lifeworld is based on the basic relationship structure "We–It"; that is, as members of a community of interaction and communication, we are jointly directed to objects in our environment. The *perspective of the participant*, that is, the "we"-perspective of the first person plural is the primary and permanent basis for the scientific observational or third-person perspective. It follows from this that a nature regarded purely physically, in which no subjects occur, must always remain a theoretical construct, from which consciousness and intersubjectivity cannot be deduced.[28]

Neurobiology is primarily a highly specialized form of common practice arising from the lifeworld. "The lifeworld includes everything we can speak about in pre-scientific terms: fellow humans, cats, sunflowers, stones, weapons, cathedrals, but also sounds, afterimages, thoughts, memories, hunger, happiness and fear" (Hartmann 1998, 322; own translation). However, initially it does not contain any constructs such as atoms, molecules, or action potentials. Within the lifeworld, human beings form cultural, linguistic, and action communities, among them also special practice forms such as the natural sciences, which raise the perspective of the observer to its methodological ruling principle. In that way, they cut out certain quantifiable and objectifiable areas from the phenomenal lifeworld, in the way described in the "Introduction." In order to describe the structures of the section of reality they choose, they develop certain terminologies, and, in due course, certain constructs (atoms, electrons, waves, potentials, fields, etc.), which serve to explain the processes observed and which, in connection with certain laws, are of high prognostic, and thus also practical value for the community. In this way, methodical norms, such as the causal principle, which were initially only research directives, gain increasing undisputed, indeed metaphysical status (such as "universal determinism").

The "second naturalistic fallacy"[29] consists, according to Hartmann, in the fact that the structures and processes postulated on the construct level are now

---

[28] This is in line with Merleau-Ponty's argument: "For what precisely is meant by saying that the world existed before any human consciousness? An example of what is meant is that the earth originally issued from a primitive nebula from which the combination of conditions necessary to life was absent. But every one of these words, like every equation in physics, presupposes *our* pre-scientific experience of the world, and this reference to the world in which we *live* goes to make up the proposition's valid meaning. Nothing will ever bring home to my comprehension what a nebula that no one sees could possibly be. Laplace's nebula [or today, the big bang, T. F.] is not behind us, at our remote beginnings, but in front of us in the cultural world" (Merleau-Ponty 1962, 385).

[29] The "second," because the term "naturalistic fallacy" is already used to describe the deduction of an "ought" from an "is," that means, drawing ethical conclusions from natural facts.

increasingly pushed *underneath the lifeworld experience* and, in the long run, hypostasized as actual reality:

> A knife consists of a blade and a handle, the material of the blade is an alloy which consists of molecules which are a combination of atoms, which, in turn, consist of even more minute particles—all just a matter of looking "ever more closely." It is overlooked here that the construct objects, in contrast to the objects on the phenomenal level, are not accessible independent of the theories in which they arise. (Hartmann 1998, 326)

This gradual substitution of the phenomena by quantifiable constructs remains unproblematic for the primary, that is, inorganic and mechanical objects of the natural sciences. It already becomes, however, reductionist for the phenomena of life as these presuppose complex or holistically structured and, thus, *macroscopic* bodies; they disappear from sight in the course of ever progressing division. This approach must all the more remain reductionist in the face of the phenomena of experience and consciousness because these per se evade the objectifying perspective. According to the fallacy of the ontological hypostasizing of the constructs, physical description shall now apply universally, that is, capture all conceivable aspects of reality. The lifeworld must thus be reconstructed from the constructs: a dog barking happily then consists of certain collections of organic molecules, and his barking can be explained from genetic programs. The performance of Mozart's "Requiem" consists of transitory fluctuations in air pressure in the surroundings of human beings and the heard melody is explained from the firing of neurons in the brain of the listener.[30]

This naturalistic fallacy is also the basis of all mereological and localization fallacies in the neurosciences. Their belief in an ultimately valid material reality and its lawfulness, existing independent from any observer, is drawn from physicalism. According to it, the subjective worlds must be grasped as constructs which are produced by the physics of the brain. The general, naturalistic short circuit between the level of physical-chemical substructures and the level of the lifeworld then becomes the short circuit between brain and mind, or brain and subject.

Of course, quantum physics has long since shown that it is no longer possible to exclude the point of view of the observer, particularly in exploring the

---

[30] "The physicist's atoms will always appear more real than the historical and qualitative face of the world, the physico-chemical processes more real than the organic forms [...] as long as the attempt is made to construct the shape of the world (life, perception, mind) instead of recognizing, as the next source and as the ultimate court of appeal in our knowledge of these things, our *experience* of them" (Merleau-Ponty 1962, 20; translation slightly modified according to the French original, T. F.).

elementary processes, whereby the allegedly solid ground for reductionism becomes shaky. The physicist is left with neither fixed "building blocks," nor completely objectifiable "facts," from which the world could be assembled as from a construction kit. The idea of matter in the sense of interacting pieces such as billiard balls is long since outdated. The processes of the material world are no more directly given than other aspects of reality. Since, consequently, the neurosciences are also dependent on the observer, they cannot explain observation itself as a product of their object.

The basic thesis of physicalism that all areas of reality can be described either by physical concepts and laws or that their own local theories can be reduced to physical theories is untenable as well. The practice of empirical sciences, such as biology, psychology, or sociology, more than underlines that their explanations of the phenomena in their particular branch have nothing at all to do with physical theories. The prerequisite that their explanations *do not contradict* basic physical principles (thus, e.g., no non-physical natural powers are introduced) suffices for them. However, the description and explanation of phenomena in accordance with physical laws does not mean that the explanation itself can be a physical one. The happy barking of the dog cannot be satisfactorily elucidated either by the biochemical analysis of motor endplate activation in his vocal muscles or by a physical description of the atomic or subatomic processes in his brain. Physical or physiological descriptions cannot explain the Russian Revolution, just because the people and things involved in it consisted of matter and cells. Admittedly, the Communist Program did not exist without material carrier substances, for example, in the form of black lettering on newspaper pages, or in the form of certain excitation patterns in Lenin's brain. Nevertheless, it can at best be neurobiologically sufficiently explained why Lenin was no longer able to pursue his program in his last years of life—namely because of several strokes he suffered.

The basic naturalistic fallacy on which the search for the substrata of consciousness in the neurosciences is based has, as of now, not been worked out. Even if the concept of "social cognitive neuroscience" (Cacioppo et al. 2002, Decety & Ickes 2011, Cozolino 2014, and many others) is meanwhile firmly established—the neuro- and cognitive sciences can only become *social* neurosciences when they incorporate not only the observer perspective, but also the participant perspective in their concepts and research. The latter is, in contrast to the observer perspective, the actual social perspective in which people recognize one another as persons and, as such, communicate with one another. Their experiencing, perceiving, feeling, and acting can only be captured from this perspective and then, with certain restrictions, also be correlated with neuroscientific findings. If someone does not know what "seeing" is, and if they cannot communicate with other seeing persons, they cannot perform any neurophysiology of visual perception. The very constitution of his objects demands that the neuroscientist takes the perspective of the participant. Moreover, scientific discourse, too, presupposes that the persons involved recognize one another as judicious and capable of freely reaching agreement. Hereby, they do not relate to a construct level of physical descriptions, rather they relate to a common lifeworld as their meaningful context and horizon, which is represented by cultural patterns of interpretation, handed down by tradition. "Without intersubjectivity of understanding, there can be no objectivity of knowledge" (Habermas 2004, 885).

Thus the lifeworld experience gains a weight which puts the complete burden of proof on its denial. The special practice of brain research is justified as long as it does not lead to hyperbolic conclusions, intended to highlight lifeworld experience in its entirety as secondary or even illusory. Whoever would wish to undermine this experience by physiological constructs or brain-generated self-models, cannot invoke scientific doctrines such as the complete physical reducibility and causal determination of all phenomena. In fact, it is rather the other way around: the models of brain research, as soon as they transgress the level of pure anatomic and physiological research and touch the field of subjectivity and consciousness, must orient themselves primarily to plausibility for our experience—thus, for example, stating which neuronal conditions exist for this experience—and not to a physicalist world view, in which colors, tones, feelings, actions, and, above all, subjects do not occur a priori anymore.

It follows that a theoretical model which is suitable for an adequate interpretation of the neurobiological data and insights must start from the perspective of the first and second person, that is, from the self-experience of living persons and must return to it, without losing it on the way. On this assumption, in what follows I shall develop a view of the brain compatible with lifeworld experience.

Part 2

Body, person, and the brain

> It is neither the soul which thinks and senses […], nor the brain; for the brain is a *physiological abstraction*—an organ removed from the totality, separated from the skull, face, and body as a whole, and fixed within itself. However, the brain is only an organ of thought as long as it is connected with a human head and body.
>
> Ludwig Feuerbach (1985a, 177)

In Part 1, we concentrated on a critique of the dominant paradigm in cognitive neuroscience. This paradigm views the brain as a *constructor*, asking how the neuronal machinery produces the experienced world and the experiencing subject. Consciousness thus appears not as the relation of a living being to the world, but becomes an internal representation of the external world inside the head. In this conception, the brain is considered as a system in itself, in opposition to the remainder of the body as well as to the surrounding world. The body remains a physiological carrier mechanism for the brain, which supposedly even as a bodiless brain-in-a-vat could bring forth consciousness, as a "cosmos inside the head."

This approach may be successful in deciphering an increasing number of circumscribed neuronal mechanisms. However, it neglects the *reciprocal relationships and circular processes* in which the brain is embedded. This would be analogous to an attempt to understand the heart without considering circulation, or the lungs without observing the breathing cycle. In Part 2, our focus will be on the development of an ecological theory of the brain, based on the premise that it can only be adequately understood *as an organ of a living being in its environment*. In this sense, the brain is, on the one hand, connected to the living organism, and on the other hand, embedded in the natural and social environment by means of the organism's manifold, and in particular, sensorimotor interactions. In other words, the *body* always constitutes the connecting link for these interactions. This continual mediation is overlooked if one attempts to find a direct relationship between brain and environment. In this case, only a mirroring or representational relationship can be assumed between the two separate systems, as is generally assumed in the cognitive sciences. But seen in isolation, the brain is only a "physiological abstraction." The dynamic and continually modified relationship between the brain and the environment is only possible by means of the body or the organism as a whole. In these interactions, the brain functions—as we shall see—primarily as an organ of *mediation* and *transformation*, for example, transforming perceptual into motor activity, or experiences into memory, thereby expanding possibilities for future interaction with the natural and social environment. The brain functions as an *organ of interrelations* and only as such does it become an *organ of the person*.

This theory can only be devised in several stages and with reference to several distinct conceptual approaches. The basis will be an initial study in Chapter 3 of a phenomenological concept of "embodied subjectivity," subsequently leading to an idea of a *dual aspect* of the person as a unity of the subjective or "lived body" and the "physical body." I then turn to an ecological conception of the living organism that especially involves close analysis of the specific causality of living systems. On this basis, the following chapters move towards an investigation of the brain itself, first, as an organ of the living being and thereafter as an organ of the person.

Chapter 3

Foundations: Subjectivity and life

> **Overview**
>
> Chapter 3 develops the concept of embodied subjectivity, initially grounded in the phenomenology of bodily existence. Subsequently, a central concept for the following investigation will be the *dual aspect of the living person* as a dialectical unity of the "subjective body" and the "physical body." The mind–brain problem is therefore described as the "subjective body–physical body problem" (*Leib–Körper* problem) (3.1). Consequently, an ecological conception of the living organism is developed. This focuses, on the one hand, on a living being's self-organization and subjectivity, and on the other hand, on its relationship to the environment with reference to metabolism and the sensorimotor cycle (3.2). The theoretical conception concludes with an analysis of the specific, *circular causality* of living systems. Essentially, this incorporates the concept of *capacity* as a living being's holistic, dispositional property, by means of which it becomes the cause of its own enactments of life (3.2).

3.1 Embodied subjectivity

The results of the previous chapters showed that consciousness cannot be envisaged as an invisible chamber that is literally contained in the head and concealed behind the sensory organs. Indeed, it is not contained at all "in the physical body," but rather is *embodied*: conscious acts are particular, integral activities of a living, self-sustaining, sensory-receptive, and mobile organism. Therefore, the primary dimension of consciousness is the reciprocal, homeostatic, sensorimotor, and active–receptive relationship of the living organism and the environment. Granted, it is characteristic for human consciousness that it is able to decouple itself from the present interaction to a certain extent and to intentionally direct itself on "re-presentations" of absent, virtual, or possible objects, such as recollections, imaginations, fantasies, or future projects. Moreover, human beings have the peculiar capacity to take a stance towards themselves, to observe their own experience, and to reflect on themselves as subjects of experience, thus seemingly becoming an inner world of their own.

However, the possibility of representation and self-distance that human persons have does not suspend their primary, embodied being-in-the-world. It does not turn all their experience into representation or into a reduplication of the world in the mind as a separate container. We are no conscious monads for which an image of the world is merely projected. On the contrary, we are living beings, inhabiting our living body as the means of experiencing the world. This grounding relationship was described in phenomenology, especially in Merleau-Ponty's philosophy (1962), as *bodily subjectivity* or as "being-towards-the-world" (*être-au-monde*) through the medium of the lived body. Life and the lived body are the grounding source of our enactment of life, including our conscious activities. We must therefore consider this grounding of consciousness in greater detail, before we enquire further into its organic basis.

3.1.1 The body as subject

If we aim to describe or define life externally, as is the case in today's "life sciences," we consider it from the outset as something objective and determinable. But if we take our self-experience of life as the starting point, the peculiar thing is that this experience lies precisely in a *self-withdrawal*.[1] Our enactment of life is removed from immediate self-observation and always precedes any act of reflective determination.

To feel hungry is not yet to be *conscious of* hunger, nor is the sensation of a feeling to be *conscious of* that feeling. For in order to realize that we are hungry, thirsty, tired, happy, or sad, we must have already *become* hungry, thirsty, tired, happy, or sad. It is impossible to say what this hunger, thirst, or tiredness was prior to our becoming aware of it—similar to the way in which sometimes we only notice a latent repetitive sound once it has ceased, and silence returns. An experience only becomes conscious as such when it acquires a certain degree of intensity or contrast. Nonetheless, it was *our* experience in the first place.[2]

Life is therefore what has already happened to us and affected us before we clearly notice it. And as regards our bodily affections, such as hunger,

---

[1] See Waldenfels 2002, 412.

[2] Thus, consciousness in the extended sense of the term starts with pre-reflective consciousness or being affected: "Affect is there before being there for me in full consciousness: I am affected before knowing that I am affected. It is in that sense that affect can be said to be primordial" (Depraz 1994, 75; own translation).

thirst, pain, vital energy, and fatigue, we learn that we are never totally in control of ourselves. Indeed, what essentially constitutes us is something that we can neither cause nor control. The spontaneous and autonomous activity of life springs from an elementary drive, an impulse or motive for something. In experiencing a drive, such as hunger, thirst, or lust, we are presented with certain impulses of our lived body, which is autonomously motivated to pursue what it lacks, whether or not we follow these drives. Using one of Spinoza's terms, we may call this energetic source of our lives *conatus* or *conation*.[3]

It is even true of intentional acts of thought and deed that we do not fully "have them in hand," but rather "allow them to happen." Merleau-Ponty therefore speaks of the "passivity of our activity": "it is not I who makes myself think any more than it is I who makes my heart beat" (Merleau-Ponty 1968, 221). The movements of my thought and my arm are *self-movements*, which I cannot make, but at best, release and direct. This is even truer of involuntary enactments of life such as breathing, falling asleep, walking, crying, or even feelings of joy or rage: they occur spontaneously *of their own accord*, and any deliberate act of will rather tends to disrupt them. Therefore, within ourselves we experience a source of becoming, an *origin of spontaneity and movement* that we cannot take possession of, and that is removed from our determination and precise definition. Equally, our individual life story commences with an unconscious pre-history of the self. And nowhere is there any obvious point at which a purely biological development would suddenly turn into conscious life.

It follows that life can neither be attributed to a pure consciousness nor be reduced to the physical organism. Life manifests itself in a basic, bodily movement and subjectivity: it is the grounding principle, yet not the object of experience. Moreover, life always precedes the act of becoming conscious; the self only experiences itself in the mode of its self-withdrawal. Whatever we plan or do consciously, we live on the basis of an unconscious, bodily background which we are never able to fully reveal to ourselves. This background permeates all acts of perception, thought, and deed, insofar as they require a *medium* by which they are accomplished, and which itself remains transparent. This medium is the *subjective body*.

---

[3] From the Latin *conatus* = endeavor, effort, drive, urge. The concept dates back to Stoic philosophy and was later used by Hobbes and Spinoza in particular to denote the living being's striving for self-preservation (*conatus sese conservandi*), in close connection with affective-volitional life. For further reading, see, for example Lin (2004) and Fuchs (2012b).

All conscious experience is therefore not only dependent on the physiological body, but emerges from the subjective body.[4] At the most basic level, it is the body as the locus of a vague background feeling, for example, of well-being or discomfort, of vitality, energy, or fatigue, which may be summarized as the auto-affection of life (Henry 1963, 1975), or the *feeling of being alive* (Fuchs 2012b). This is closely connected to the body as a source of drive, spontaneity, and activity, which was earlier described as *conation*. One could refer to this dimension of vitality and conation as the "deep body." Furthermore, the lived body functions as a "resonance board" for all kinds of moods and emotions which we may sense.[5] Finally, the lived body is the center and, simultaneously, the medium for all perceptions, movements, and actions. Indeed, even the alleged "pure thinking" cannot be detached from bodily consciousness. For, even if my thoughts are freely able to move in time and space as regards their intentional content, as conscious activities they certainly remain bound to my bodily self-experience and to my sense of presence.

Thus, the lived body carries all our enactments of life, even and especially when it functions as an unheeded, transparent medium of our being-towards-the-world, as is the case in all skilled operations such as walking, cycling, speaking, writing, and so on. We can never entirely apprehend our lived body nor bring it fully into the scope of our conscious attention. Indeed, a part of it always remains "behind" our perception, as the source and center of our enactment of life. Therefore, all feeling, perception, imagination, thought, and action are completed on the basis of a *bodily background* or in other words: these activities always have a bodily subject. In this sense, Merleau-Ponty described the lived body as the "*natural subject*," which is the precedent and foundation for all conscious and reflective acts: pure consciousness without a subjective body is a dualistic abstraction. Nowhere does this exist in experience.[6]

---

[4] The literature on the phenomenology of the body is not easily summarizable. I only refer to the major texts by Husserl (1952), Merleau-Ponty (1962), Leder (1990), Schmitz (1995), Waldenfels (2000), Fuchs (2000a), and Taipale (2014).

[5] William James already described the body and its organs as a "sort of sounding-board, which every change of our consciousness, however slight, may make reverberate" (James 1884, 191). This will be dealt with in detail in Chapter 4.

[6] "There is, therefore, another subject beneath me, for whom a world exists before I am here, and who marks out my place in it. This captive or natural spirit is my body" (Merleau-Ponty 1962, 228).—"I am my body, at least wholly to the extent that I possess experience, and yet at the same time my body is as it were a 'natural' subject, a provisional sketch of my total being" (p. 178).—"[T]he body is a natural self and, as it were, the subject of perception" (p. 184).

By the same token, the subjective body is the *ensemble of all skills and capacities* at our disposal. As "habitual body" (Merleau-Ponty 1962, 71), it contains the preliminary drafts of our enactments of life and thus conveys the founding experience of "I can" (Husserl 1989, 266). I can dance a waltz because my lived body moves of its own accord to the rhythm of the music and completes these movements. And yet, I myself am the one who can dance and who moves, not a mind giving orders to its body for programmed movements. Similarly, I can recognize a familiar face because my capacity for perception already contains the preforms (*Vorgestalten*) of possible sensory objects. That is to say, my subjective body connects to suitable conditions and counterparts in the surrounding world, endowing them with affordances and meaning. Thus, they become, for instance, the suitable rhythm and space for the body's dancing, the objects of its grasping or throwing, the goals of its desire, and so on.[7] In this way, the subjective body forms a superordinate system of the organism and its environment, which manifests itself in the bodily subject's being-towards-the-world, and in our fundamental familiarity with the world. My subjective body is therefore initially not the physical body that I see, touch, or sense; rather, it is my *capacity* to see, touch, and sense. It is not an object in the world, but the medium, the field, or the capacity that reveals the world to me.

If we start from these primary experiences of the lifeworld, then dualism—already in Descartes's works as well as in the "Cartesian materialism" of his modern-day successors—is based on a twofold "disembodiment." On the one hand, the subjective body is objectivized as a mere physical thing; on the other hand, the bodily subject is hypostasized as a pure ego-consciousness. Neurobiological reductionism then necessarily emerges from the endeavor to reassemble this abstracted consciousness with the objectivized physical body, or rather with the brain as its *pars pro toto*. Indeed, this short circuit of mind and brain occurs *de facto* only in the form of correlations: "if, in brain area XY an activation occurs, then the person concerned will feel this or that." However, the correlations are then described as causations, so that the disembodied subject is relegated to an epiphenomenon. But this reductionist consequence does not necessarily follow as a result of the study of the physical body or the brain. Indeed, it can be avoided if we consider human beings as unified living organisms, and yet at the same time under a *dual aspect*—both as a subjective and physical body.

---

[7] "The movements of his [i.e., the subject's, T.F.] body are naturally invested with a certain perceptual significance, and form, with the external phenomena, […] a well-articulated system" (Merleau-Ponty 1962, 42).

3.1.2 The dual aspect of subjective and physical body

A person is a lived body (*Leib*) inasmuch as his or her subjective states, experiences, and actions are bound to the medium of the body. However, persons are also lived bodies *for others*, who directly perceive them "in the flesh" through their expression, attitudes, and acts—thus, not as a combination of pure physical body and hidden psyche, but as a unified entity. If someone greets me by extending a hand, this does not represent for me an inner, "mental act" involving a movement of the physical body as an outward symbol. Rather, this person is present for me by virtue of his greeting, in his offered hand.

In contrast, the body reveals itself as a *physical body* (*Körper*) as the result of an objectification. This already occurs in everyday experience where the body gains an exteriority like other objects that can be seen or touched. However, as Husserl remarked, it is never constituted as a full object that would be in opposition to us. Moreover, it shows the peculiar phenomenon of "double sensations" (*Doppelempfindungen*; Husserl 1973a, 378). For example, when touching one's left hand with the right, this yields both a sense of touching the left hand (like an external object) and a sense of the left hand *feeling* the touch; the same applies for the right hand. Thus, double sensation marks a *point of conversion* ("*Umschlagspunkt*," in Husserl's term) of subjectivity into objectivity, and vice versa. Hence, the lived body is never only subject and never only object, it rather is a "subject-object" (Husserl 1952, 195) or it is both *Leib* and *Körper*.[8]

Frequently, an objectification of the body also occurs during *disruptions* to the customary enactment of life as, for example, in a clumsy movement or involuntary fall, in a state of exhaustion, when feeling unwell, in an injury, or in physical illness. In such experiences, our body is to a certain extent alienated from us and shows itself as an obstinate, fragile, or vulnerable body. In a sense, this alienation culminates in witnessing another's *dead* body, the corpse. It is not least these experiences that have essentially motivated medical research, and ultimately form the basis of all scientific objectification of the body. In that sense, an individual is an objective or *physical body* as the entirety of material-anatomical structures and physiological processes that may especially be objectivized from a medical third-person perspective.[9]

---

[8] As Taipale (2014, 49) rightly remarks, this phenomenon of "self-palpation" is quite pervasive, for example when crossing one's legs, touching one's torso with the arm, or one's palate with the tongue.

[9] "There is my arm seen as sustaining familiar acts, my body as giving rise to determinate action having a field or scope known to me in advance [...]—and there is, furthermore, my arm as a mechanism of muscles and bones, as a contrivance for bending and stretching, as

Husserl (1989, 183–188) traced each of these aspects back to two different attitudes that human beings can adopt in relation to each other. As lived bodies, we relate to each other by means of a primary *"personalistic attitude"* which always grounds our common lifeworld and experience of life. As physical bodies, we only appear to each other in the *"naturalistic attitude,"* making the subjective body a measurable object of nature, which researchers can study in any conceivable degree of detail. On a daily basis, a doctor undergoes this change in attitudes, for instance, when greeting a patient and seeing her (friendly, anxious or similar) gaze, yet shortly afterwards taking hold of the ophthalmoscope to examine the patient's eyes as physical organs: at this point, looking at them from too close a distance, her gaze has vanished. In other words, the lived body may only be perceived in a holistic way.

Nevertheless, each of these attitudes is directed to *the same entity*, that is, the living organism or living person. The lived and subjective body as the location of sensations and affections (fatigue, pain, hunger, etc.), the body as the medium of the enactment of life or of contact with others—none of these emerges as a construct in the brain, mysteriously projected into external space. Rather, this lived body *is* the organism itself under the aspect of its holistic aliveness that is manifested both subjectively as well as intersubjectively. Conversely, the objectivized physical body is no *mere* object. Instead, it is still *somebody's* physical body—otherwise, no neuroscientist could connect his observations of the brain with a person's experience. We can thus consider the same entity in much the same way as a reversible figure, like Necker's Cube, in two distinct and non-transferable ways—as the lived and as the physical body. To this extent, the body, as already mentioned, is the *"point of conversion"* where the subject itself is revealed as embodied. Similarly, Merleau-Ponty refers to the "ambiguity" of the lived body that "forms between the pure subject and the object a third genus of being" and thus undermines the dualism of inside and outside. It is neither mere *consciousness of* the body nor objective *physical body* (Merleau-Ponty 1962, 314).

Plessner (1928/1975) defined the anthropological basis for both attitudes, and thus for the dual aspect of lived and physical body in terms of his concept of man's "eccentric positionality." In contrast to an animal's "centric" position in its environment, "eccentricity" means the capacity of human beings to relate to themselves and their bodily existence, to perceive themselves "from outside," that is, from the possible perspective of others, and finally, to confront

---

an articulated object [...]. It is never our objective body that we move, but our phenomenal body" (Merleau-Ponty 1962, 91–92).

themselves through reflection. This establishes a fundamentally ambivalent relationship of man to his bodily existence: on the one hand, the lived body is a medium and unobserved center of the enactment of life; on the other hand, the body becomes a consciously used instrument or even an obstacle. At one time, the body is a state of being and existing, at another time, it again becomes an object. The human person is a bodily being and yet "at the same time, extrabodily in character and living in a state of tension with his physical existence, while being utterly bound to it" (Plessner 1970, 39, own translation).

In the first chapter, we already described the ambivalent unity of embodied subjectivity as a *spatially* dual aspect under which the body appears: subjective (lived) bodily space and objective (physical) bodily space normally coincide syntopically (see 1.2.2). Although they are not identical, they are basically *coextensive*. A pinprick to the hand feels painful at this visible point—similar to the case of double sensations. What we experience, as Husserl remarked, are not two different hands—one hand being a physical, visible, and tangible body, and the other hand being the locus of pain. Rather, the hand is given as a "physical-aesthesiological unity" (Husserl 1989, 163). This unity also makes it reasonable for the doctor to search for the cause of a pain at the very point that emits it. However, the pain is not localized in the physical-anatomical hand, but only in the lived experience of the hand. The inner and outer perspective may coincide spatially, yet they are distinct from each other in principle. The subjective bodily space is *embedded* in the space of the living organism, yet without being identical to it.

This spatial dual aspect may also be described as the difference between *relative* and *absolute* spatiality.[10] Considered as an object or as a physical thing, I occupy a place, which changes relative to other things within physical space. As a living bodily subject, however, I always occupy my own unmistakable place and cannot remove myself from it. My lived body forms the "zero-point" of all orientation; it is an *absolute here* that can never become a "there."[11] Of course, this "here" now coincides with a place in objective space, relative to other things—from my position, for instance, I can leap over a ditch and land in the correct place. In other words, absolute space and relative space overlap. Nevertheless, my absolute "here" is clearly not identical with the relative place that I currently occupy, since this "here" always moves with me.

---

[10] For this distinction see Schmitz 1985, 117–118, and Fuchs 2000a, 97–98.

[11] As Husserl notes, the lived body is "entirely unique by virtue of the fact that it always 'bears within it' the zero-point, the absolute Here, in relation to which every other object is a There" (Husserl 2001, 584).

Finally, the dual aspect that the body displays also becomes evident in *intersubjective relationships*. The other person primarily appears to me as an animate and bodily unity. In other words, he is always present "in the flesh." His body is not merely the anatomical carrier of an "inner life," which remains inaccessible to me, and which I can only assume to exist on the basis of relevant knowledge.[12] Embodiment is rather the basis of intersubjectivity, insofar as we do not assign abstract inner states to other persons in our bodily interactions, but experience their facial expressions, gestures, and behavior in the context of the situation *as the direct expression* of their sensations, feelings, and intentions. Thus, the primary perception of others is not based on a *"Theory of Mind"*—not on hypothetical assumptions or inferences about an invisible inner world beyond the body. Rather, its basis is *intercorporeality*: the bodily communication and reciprocal empathy between embodied subjects (Fuchs 2017d). Only on a secondary level are we able to proceed to an objectivizing attitude and, on this basis, also to a scientific examination of the physical body of others.

Consequently, subjectivity is essentially embodied: the body is not merely the content or object of consciousness, but as the lived body becomes the constitutive basis of the subject itself. We experience all our feelings, thoughts, perceptions, and actions as subjective bodily beings and, at the same time, as physical beings.[13] The question, which emerges at this point, is therefore: *what is the nature of a physical body that reveals this type of dual aspect?* How can a body be a complex composition made up of physical matter, and yet at the same time function as the carrier of unified, conscious acts of life?

3.1.3 The dual aspect of life

An initial response to this question is as follows: this body must be a *living organism*, that is, an indivisible entity, functioning as a composite whole, and yet also extended in space. We will see how this unity of the living organism is to be conceived later in this chapter. This unified organism must further reveal a *dual character* in its manifestations of life, which means that these, on the one

---

[12] "Therefore the body is not an object. […] Whether it is a question of another's body or my own, I have no means of knowing the human body other than that of living it" (Merleau-Ponty 1962, 178).

[13] "Physical" (*physisch*) does not mean here "to be explained by physics" (*physikalisch*), as from the point of view of physical science, the body is neither a living nor a lived body, but an assemblage of particles and energy fields. In the following, I use the term "physical" as denoting the body's material aspect including emergent higher-level phenomena not to be explained by physics. The German language accordingly permits a distinction between "*physisch*" (in the sense of "material" or "somatic") and "*physikalisch*" (= pertaining to an object of physical science).

hand, represent specific configurations of physiological (including neuronal) processes and, on the other hand, constitute experiences and activities of the individual as a complete, living entity. In other words, the organism must also be a *subject*.

Accordingly, the living organism is in one sense a *composite entity of physiological processes* and in another sense a *subject of life acts*, whether these are passive *impulses or affects* such as lust, pain, hunger, or fear, or whether they are *activities* such as thinking, perceiving, moving, acting, or the like. All propositions, with which we refer to acts of life, that is, to sensations, perceptions, or actions of living beings, also require the grammatical insertion of a living being in the subject position.[14] This is not only grammatically, but also ontologically founded. For even though such acts of life are carried by single physiological processes, they are nevertheless generated by the living being overall, insofar as the latter exists as a whole and continues its life through its acts. Human beings move, perceive, or think, not their brains.

This conception of life under a dual aspect is fundamentally distinct from the framework of the conventional mind–body problem and the various attempts to identify solutions. Emotional, cognitive, or volitional acts of life are not transferred here to a pure "mental" sphere, but always *also* remain embodied, physical events. However, it is precisely because of this that they do not represent single processes in specific regions of the body describable by physics. Instead, they appear as acts and experiences of the *entire living being* as a unified physical organism. Therefore, acts of life cannot be divided into a purely mental and a physiological part. Psychic or mental conditions are always conditions of a living organism, that is, inseparable aspects of life acts. We may also express it like this: all *experience* (*Erleben*) is a form of *living* (*Leben*).[15]

This can be illustrated with reference to the following schemas (see Figure 3.1 and Figure 3.2). Most contemporary theories in philosophy of mind rest on the assumption of two fundamentally distinct entities, that is, "body" and "mind," or *physical* and *mental* processes. The former are accessible from an observer or third-person perspective, whereas the latter are allegedly only accessible from the inner or first-person perspective. These essentially distinctly positioned domains must now be reconnected by means of specific theoretical constructions. Usually,

---

[14] Here I follow Buchheim (2006, 39).

[15] In the German language, *Er-leben* literally means "experiencing life" and precisely denotes that with it, the life process "comes to itself." Of course, this does not yet include reflective consciousness (becoming conscious of one's experience as such), as it is possible for human beings. On the other hand, pre-reflective experience emerges only in higher forms of animals; the question of its presuppositions is dealt with in section 3.2.3.

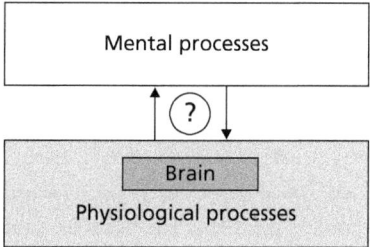

Figure 3.1 Mind and body in contemporary philosophy of mind.

the physiological basis of the mental sphere is reduced to specific brain processes (Figure 3.1). Mental processes must then be regarded as either *identical with, epiphenomenal of, supervenient on,* or *emergent from* the neuronal processes, or equally as independent in a *dualist* sense. However, regardless of which solution is chosen, in any case the *living being or person as a whole* no longer appears in these theories as a fundamental entity. Therefore, mental and brain processes can only be directly related to each other or "short circuited."

In contrast to this dichotomy, the conception of a dual aspect put forward earlier in this chapter considers *the living being itself* as the primary entity (Figure 3.2). Its manifestations of life may be regarded, on the one hand, as integral (bodily, emotional, intellectual) acts, which are experienced subjectively and, on the other hand, as physiological processes in any degree of detail. Thus,

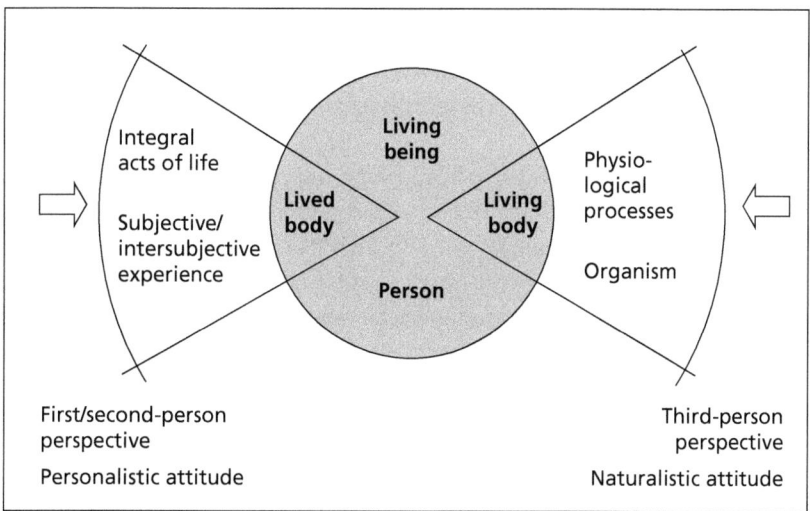

Figure 3.2 Dual aspect of the living being.

the living being displays a dual aspect, both as a *lived body* which is experienced subjectively (first-person perspective) and perceived by others in a personalistic attitude (second-person perspective), and as a *living body* which corresponds to the composite unity of the organism, including all of its physiological structures and processes as observable from a third-person perspective. One might also use the terms "body-as-subject" "body-as-object" as suggested by Legrand (2006), if it remains clear that the body-as-subject is also perceivable for others from a second-person point of view.

Epistemologically, these aspects are *complementary* to each other, that is, their respective descriptions are non-transferable and only exhibit certain correlations and structural similarities.[16] Nevertheless, these are aspects of the living being as an ontological unity, as manifested, among other things, in the fundamental coextensivity of subjective body and physical body. Since the living organism and its life process form the foundation of each aspect, the dual aspect does not signify a new ontological "dualism," but a mediated monism, or in Hegelian terms, a dialectical unity of unity and diversity: the aspects are *objectively distinct* characteristics of *one and the same living being*.

> I will avoid the obvious term of "aspect dualism," since it is nowadays used for the psychophysical aspect dualism traced back to Spinoza and Ernst Mach and revived in identity theory. According to this conception, mental and material processes or mind and brain are two aspects of the same, yet not independently identifiable event, without considering an underlying unity of the living organism (one could pointedly say "two sides of a coin without the coin itself"). To differentiate my vocabulary from this alternative conception, I will refer in the following to the "dual aspect" or to "aspect-duality" of the living organism or person.

The general aspect-duality of a living being is specifically defined in the case of human beings as a *"personal dual aspect."* A person is characterized as a living being in a position of taking a stance vis-à-vis its primary bodily existence. In this way, persons can appear to themselves and to others both as a subjective *and* a physical body. The *eccentric position* achieved by human beings (see 3.1.2) is therefore the basis for the duality of perspectives that we can take on ourselves. It also leads to the distinction of the personalistic and the naturalistic attitude.

With some restrictions, the complementary nature of these aspects may be compared to two sides of a coin—only one side is constantly visible to the exclusion of the other. Thus, each side is neither identical to the other, nor do they

---

[16] I will return to these in Chapter 6.

overlap; they can only refer to each other.[17] Therefore, it is erroneous to ask which side is "actual reality" and "brings forth" the other. Even if everyday or out-of-the-ordinary observations appear to approach this kind of "bringing forth"—a pinprick "produces" pain, a stimulation of the temporal lobe "produces" remembered images—in both cases what occurs are only *stimuli* that the living being or person in question responds to as *a composite whole* with an utterance of pain or a reminiscence. Physical events (= described by physics) can only give rise to, and are not the cause of, integral manifestations of life.

It should further be noted that *life acts* (the first aspect), unlike the "mental" sphere in philosophy of the mind, are not merely subjective. They can be experienced both from the inner perspective of the first person as well as perceived from the outer perspective of the *second person*. We can also describe the latter as the participant or "we"-perspective, where we perceive each other not in an objectivizing manner, but as living persons. Typical examples of such life acts might be, for instance, "laughing," "suffering pain," "playing tennis," "speaking," or "greeting." These predicates can be applied both from the perspective of the actor or experiencing person as well as from the participant's perspective. Nevertheless, they denote the *same* given fact of the lifeworld that may not be subdivided into "mental" and "physical" parts.

> Strawson (1959) described these statements as "P predicates" (personal predicates), which can be attributed to a person both from an inner and an outer perspective, and are to be regarded as "logically primary." The distinction of these perspectives remains secondary, as the person cannot be composed from physical and mental parts or of "behavior" and "experience." However, such dual meaning of specific predicates is not only valid for human persons. Insofar as we are related to higher animals, we can also perceive acts of life in them such as "enduring pain," "being startled," "seeking prey," and so forth. Also, Strawson does not clearly distinguish between the perspective of the second and third person, that is, the participant and observer perspective. The one-sided opposition of the first- and third-person perspective, while neglecting the second-person perspective is, however, one of the most significant roots of the brain–mind problem and its aporias.

It should be noted that the existence of personal predicates does not entirely eliminate the difference between inner and outer perspective. Thus, the uniqueness of subjective facts is preserved (see 2.1.1). Indeed,

---

[17] All current hybrid terms such as "neurobiology of consciousness," "neurophysiology of emotions," "biochemistry of pain," etc., already suggest a commensurability which neglects this irrevocable double character of the living person.

this is the pivotal point in our perception of the physical body as a person's subjective body and thus as the *center of a world*. However, this exclusivity of subjectivity is restricted to the extent that a particular act of life is by no means sufficiently characterized by the inner perspective. Personhood denotes a primary unit of inner and outer, which always includes the private, merely subjective sphere as only one component. Since being a person always entails being in relationship to others, we can never adequately and completely recognize ourselves from the first-person perspective.

A person's life acts therefore both exist in an inner and outer sense—they encompass lived experience *and* expressive behavior. By contrast, the *living or objective body* with its physiological processes (the second aspect) can only be perceived from an observer or third-person perspective, in the naturalistic attitude. Here, however, we can still differentiate two possible views: on the one hand, the physical body can be described from a viewpoint of *physics* as a material entity, which is, for example, 60 kg in weight, comprised of different material parts and processes, and so on. However, from a *biological-systemic* viewpoint, it can also be considered as an *organism*, that is, an integral and functioning system in exchange with the environment. What comes to the forefront here is an integrative viewpoint also within the second aspect. Nevertheless, this may not be readily equated with the first aspect of integral life acts. Rather, the integral aspect of life acts *corresponds* to the systemic-integrative viewpoint of the living organism. All research into dynamical systems and organisms remains bound to an objectivizing attitude and never converts to the perception of life acts. "Life can only be known by life" (Jonas 2001, 91). In order to perceive and investigate a living being *as such*, one has to participate in its life; this applies even more to human beings.

Furthermore, it should be noted that the *physical* dimension in this conception is not contrasted with the psychic or mental, but emerges in both aspects, that is, as a lived and as a living body (Figure 3.3). The "mind–body problem,"

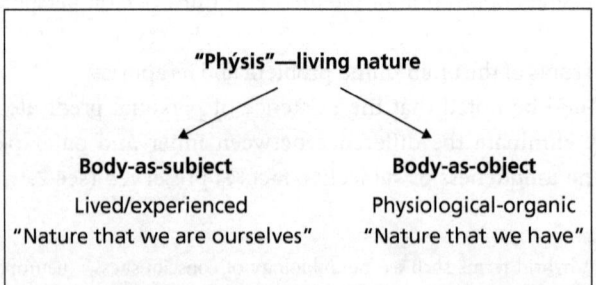

Figure 3.3 Dual aspect of "phýsis."

which is already dualistically pre-structured in the concept and nowadays generally abbreviated as the "mind–brain" problem, is thus reinterpreted under the dual aspect as the *"lived body–living body problem"* (*Leib–Körper* problem)—as such it has also been envisaged by Thompson (2007, 235–237). In all its integral, and also emotional–intellectual acts the living organism is a "physical" being, in other words, it is precisely *one* material and living substance, an *individual*. The problem is that we have no suitable concept for this special form of integral materiality. One might think of the term "*psycho*physical"; however, this expression has already been coined on a dualist basis and no longer reconstructs the unity of the living being. On the other hand, the original, in particular, Aristotelian meaning of *phýsis* as "living nature" gives an accurate summary of this dimension: it means a natural entity, which like the lived body is its own source of motion, and which we regard in analogy to our own self-experience as living beings, that is, as a unity of subjectivity and objectivity.[18]

On the one hand, therefore, "physical" refers to the subjective or *lived body* and, on the other hand, to the organic or *living body*. To apply the concept of *phýsis* as nature in this extended sense, we can also characterize the subjective body as "nature that we are ourselves" (Böhme 1992, 77), and the physical body as "nature that we have." This would mean that we connect the phenomenology of bodily existence with the Aristotelian tradition: on the one hand, a person's body is the phenomenal lived body—subjectively experienced, yet also perceived as animate *phýsis* by others; and on the other hand, it is the physiological–organic body. Under both aspects, it is a unified physical, lived, and living substance.

This concludes the basic description of our conception from which we derive the following analyses: only if we understand human experience and action primarily as *acts of a living being* will it become possible to overcome the dualism of "brain" and "mind" and to comprehend the brain as an organ of the human person. Our next task is to describe the living organism, and the brain contained within it, so that conscious experiences become tangible and can be grasped as acts of a living being. In Chapters 4 and 6, I shall return to the basic conception, which I have developed here, in order to explore issues in greater detail.

3.2 Ecological and enactive biology

The brain's function can only be adequately understood after bringing into sharper focus the overall structure of the living organism and its relationship to

---

[18] Using this notion, the later restriction of *phýsis* to "physical" (belonging to physics) must then be overturned (see 3.1.3). The etymological similarity of the terms "physician" and "physicist" still highlights how common parlance transferred the focus of the term from the living body to inanimate physics.

the environment. In doing so, we can initially refer to the results of ecological and philosophical biology and their main protagonists, von Uexküll (1920), Plessner (1970), and Jonas (1966). At the same time, we can draw on biological system theories put forward by Bertalanffy (1968) and Maturana and Varela (1987), and on enactive approaches to life, as represented in particular by Varela et al. (1991), Thompson (2005, 2007), and Di Paolo (2005, 2009). Yet we can hardly restrict our analysis to the physiological-systemic aspect, since it is also important to incorporate the living organism as a subject.

3.2.1 Self-organization and autonomy

Living organisms can be first characterized as complex physical bodies or systems which through the continual functioning of their metabolism maintain their *form and structure* over time. This process of maintaining life is to be understood as *active self-organization* or *autopoiesis* (Varela 1979, 1997; Thompson 2007): an organism's form permits its matter not merely to flow through it, as a whirl does with the continual flow of a river. Rather, it submits matter to its own principle and purpose by incorporating and transforming it. In this process, the matter or substance attains new, *emergent* properties that it only acquires within the organic system. To give an example: iron contained in the hemoglobin of the blood behaves in a manner fundamentally different from iron occurring as a natural mineral: the former does not oxidize with irreversible effects, but is in a position to incorporate oxygen reversibly. This is a decisive precondition for the animal's energy balance. Equally, the metabolism functions as digestion to break down substances, resulting in a new synthesis and *transformation* of these substances. Food parts are thus converted into "vital matter," which is functionally integrated into the organism, or into materially bound energy, for that matter.

Thanks to its dynamic self-organization, a living being differentiates itself from its environment and achieves *autonomy* in differing degrees. In other words, its processes and behavior are no longer primarily determined from the outside. Rather, they are dependent on its fundamental dispositions as well as its current state (shortage or saturation, openness or withdrawal from the environment, performance or regenerative condition, etc.). Thus, external agents no longer have a direct or causal-mechanical influence on an organism (unless this is damaging or destructive). Their effect is rather that of *stimuli* which are *responded to* by reactions of the entire organism. Hence, Rosen has argued that organisms are different from machines because they are "closed to efficient causation" (Rosen 1991, 244). In an organism, but not in a machine, the catalysts needed for its operation are produced internally. In Varela's language, this corresponds to the difference between an autonomous system marked by "operational closure" and a heteronomous system defined by outside control (Varela 1979).

The basis of an organism's autonomy is the special *reciprocal relationship between the whole and its composite parts*. This is first manifest in the differentiation of part functions, that is, in the formation of subsystems and organs. An organism consists of the sum of its biological macro-molecules, cells, organs, and vascular and nervous systems. However, it is also more than these, albeit in a way different from a crystal, which also exhibits certain superordinate properties. The difference is twofold:

1. The organism as a whole is itself the *condition of its parts*, insofar as it first makes possible the existence of cells and organs. It produces and reproduces these as parts, of which it is composed at the same time. Self-preservation is continual self-reproduction down to the individual parts which outside of the organism soon decay into their components (within 1 year, 98% of the molecules in a human body are replaced; see Margulis & Sagan 1995, 23).

2. However, the complete living entity does not impact externally on its parts, as is the case with the crystal, or with a clock in which the meshing cogs have no autonomy. Rather, the presence of the living system is only *indirectly* manifested in its organs. These remain independent in their functioning, and precisely through this they contribute to the organism's life as a whole.

In his "Critique of Judgement," Immanuel Kant regarded it as a criterion of a "natural purpose" (*Naturzweck*) or of a living being that "its parts (as regards their being and their form) are only possible through their reference to the whole." Conversely, however, "its parts mutually depend upon each other both as to their form and their combination, and so produce a whole by their own causality" (Kant 1914, § 65, pp. 276–277). This type of "organized and self-organizing being" (p. 278) largely corresponds to the modern concept of *autopoiesis* put forward by Varela: a living system is constituted by a semipermeable membrane that delimits it from the environment, while at the same time allowing for the metabolic exchange by which the system constantly regenerates itself. At the most basic level, the system is a single living cell whose metabolic network continuously (re-)produces its membrane, thus creating the boundary which sets it apart from its chemical surroundings. Hence, an autopoietic system can be defined as a system, which "continuously produces the components that specify it, while at the same time realizing it (the system) as a concrete unity in space and time, which makes the network of production of components possible" (Varela 1997, 75; see also Weber & Varela 2002). Such a system, by virtue of its operational closure and autonomy over and against the environment, is equivalent to an organismic individual or a *self.*

3.2.2 Dependency and exchange between organism and environment

A self-organizing living system is fundamentally goal-directed, namely towards the goal of its self-maintenance. However, the autonomy or the self-organization of life is not possible in a state of autarchy. An organism's form gains sovereign control over its substance only at the price of dependence on the environment—that is, on account of its *neediness*. In order to maintain homeostasis, the changing matter must be repeatedly found and incorporated. By its needs, life is necessarily connected to its environment and depends upon exchange (Jonas 2001, 102–107). In this way, a living organism's *metabolism* is its primary connection with its environment. The cell membrane is not only a limit, but also a zone of interconnection. Moreover, metabolism is the foundation for the emergence of *valences and preferences* in living systems.

Its dependence on the environment renders life in principle *precarious*. The system has, so to speak, only a limited range of viability, of exchanges and interactions with the environment that will contribute to its self-preservation. There are also circumstances and influences that threaten its very existence. An advanced autopoietic system therefore needs to be able to deal with precarious situations, that is, *to anticipate* life-threatening circumstances and *to act* in order to avoid or revert them. This has been conceptualized as *adaptivity* by Di Paolo (2005, 2009), meaning the capacity of living systems to monitor and regulate themselves with respect to their conditions of viability, and to improve their situation when needed. Adaptivity implies the capacity of *sense-making*, which on the most basic level means the distinction of favorable and adverse circumstances in the environment, resulting in suitable, self-preserving actions. Sense-making turns the merely physical surroundings into an environment of significances and valences. Bacteria, for example, are able to move, by means of flagella, up the gradient of a glucose solution towards greater concentration (positive valence), or conversely, to move away from chemorepellents or negative valences ("chemotaxis," Eisenbach et al. 2004).

This adaptivity only emerges at the level of animal life. Plants are in an *open* and continual relationship with their environment, within which they are fully integrated. Their environment directly connects with the plant at its outer periphery, so that it gains everything required for its metabolism from the exchange at this boundary. In contrast to plants, animals display a *closed form of organization* (Plessner 1975). Here, the metabolism's outer surfaces of exchange are tilted inwards so that special internal organs and zones emerge, while the boundary layers with the environment are, in comparison, substantially reduced. Animal life therefore steps out of the immediate relationship with the environment and centralizes itself. Thus, it enters into a continuous process of *mediation* through space and time.

The clear demarcation from the environment requires, in turn, a *sensorimotor interface*, which re-establishes the exchange with the surroundings, albeit on an extended level: relevant objects for the animal are almost always located at some spatial distance that is overcome by perception and movement. This form of relation to the environment is linked to a specific organizational form of the body that reveals separate organs for sensory, motor, and connecting nervous functions. In this way, the closed form enables an animal's independent movement, but also necessitates it, namely to suspend the distance from the surroundings through the active search for objects. The loss of the plant's direct connection to the environment corresponds to a gain of *space to maneuver*, of increasing possibilities and degrees of freedom that are characteristic of animal life (Jonas 2001).

The centralization and internalization that is characteristic of the evolution of animal life is particularly represented by the development of the central nervous system (CNS). However, it is important to note that this developed from and within an already existing cycle of an organism–environment interaction: *directed behavior came before the brain* (Van Dijk et al. 2008). When the complexity and diversity of the interactions increased, an organ of integration became necessary that inserted itself between the sensory and motor functions placed at the periphery of the organism. The enormous acceleration of neural signal transmission as compared to the slow spreading of biochemical processes in fluids now enables the instant linkage of distant processes within the organism. Different sensory inputs have to be registered in the central organ and ranked in terms of priority of relevance; moreover, "multiple motoric output options need to be orchestrated in order to achieve fluent movements" (Van Dijk et al. 2008, 304). Thus, both in evolutionary terms and regarding the individual organism, the CNS *mediates, selects, and facilitates* organism–environment interactions. It is a secondary organ that intervenes in pre-existent functional cycles, modifies, and refines them, but does not create them as such.

By means of its sensorimotor interface, and mediated by the CNS, the living being is functionally coupled to a specific environment, or to its *ecological niche*. Anticipating current enactive concepts of cognition, Jakob von Uexküll (1920/1973) already described an animal's relationship to the environment as a feedback loop of "receptive" and "effective" processes (Figure 3.4). Each animal is connected to an object both by a sensory organ acting as a receptor (*Merkorgan*), and by a motor organ or effector (*Wirkorgan*). In this way, the animal discovers a corresponding "receptor sign" (*Merkmal*) and an "effector sign" (*Wirkmal*) in the

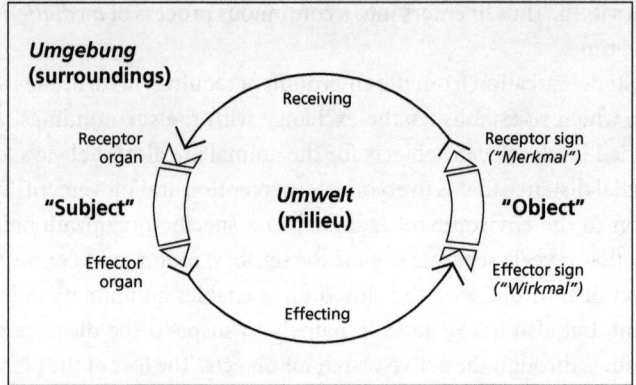

Figure 3.4 Functional cycle of a living organism, constituting its specific meaningful *Umwelt* (milieu). This *Umwelt* is different from the *Umgebung* or surroundings as seen from the perspective of an observer.
Adapted from Jakob Johann von Uexküll, *Theoretische Biologie*, DOI: 10.1007/978-3-662-36634-9. Copyright © 1928, Springer-Verlag Berlin Heidelberg. With permission of Springer.

object, or in other words, the animal attributes the *meaning* of a perceivable and treatable thing to this object. In consequence, an elementary coupling of "subject" and "object" has come about: receptor and effector possibilities of the living organism and the corresponding features of its *Umwelt* (milieu or perceived environment) are linked to form a *functional cycle*.

> Every animal is a subject which, thanks to its uniquely characteristic structure, selects specific stimuli from the general effects of the external world and responds to these in a specific way. These responses, in turn, are comprised of specific effects on the external world that also influence the stimuli. Thus, a closed cycle emerges that can be called the *functional cycle* of the animal. (von Uexküll 1973, 150; own translation)

The receptor and effector signs are the result of a living organism's needs, capacities, and potential achievements, which lend the objects their specific meaning or *affordance:* something *to grasp*, something *to climb*, something *to eat*, and so on. Indeed, something only becomes a receptor sign or stimulus by offering a certain possibility of reaction or treatment: "an animal can distinguish as many objects in its environment as the number of achievements it can perform" (von Uexküll and Kriszat 1956, 68). Thus, stimuli are not physical events that exist independently of an organism. On the contrary, each special sensorimotor organization—as the combination of a living organism's sentience and responsiveness—selects and makes accessible an associated section or aspect of the world. This is how the organism *makes sense* of its environment.

Consequently, the specific environment of a living organism is *first constituted* by the functional cycle, and thus receives a complementary character: "Where there is a foot, there is also a way. Where there is a mouth, there is also food. Where there is a weapon, there is also an enemy" (von Uexküll 1973, 153). The enactive approach to cognition takes a similar stance: a cognitive being's world is not a pre-given external realm, represented internally by the brain, but a relational domain created by that being's agency and coupling with the environment (Varela et al. 1991, Thompson 2007, 59). In other words: living systems *enact their world* as inseparable from their own structure and actions.

Therefore, the living organism does not enter into a relationship with the world, as if it could also exist before and independently of it. Rather, in a sense, *it is itself this relationship*, insofar as, according to its structure, it selects its specific *Umwelt* and is in constant exchange with it. The living being and its world *co-originate* and remain coupled to each other. Hence, a living organism is not in the world in the way a thing exists in physical space; instead, it is "towards the world," to use Merleau-Ponty's term. As the living organism selects and recognizes the appropriate, supportive, or detrimental elements within its environment, these are integrated into a *comprehensive system* consisting of the living organism and the environment. This living system is both subjective and objective in character.

3.2.3 Subjectivity

To what extent is the organizational form of animal life connected with the emergence of subjectivity? There is no conclusive answer to this question: namely, how far forms of subjectivity—which we recognize on grounds of our own experience, and our observation of mammals with their closely related organization and behavior—reach down in the series of living organisms. An elementary form of inner life or "inwardness" appears to be given per se with animals' autopoietic organization:

> The introduction of the term "self," unavoidable in any description of the most elementary instance of life, indicates emergence, with life as such, of internal identity. (Jonas 1968, 242)

Through semipermeable boundaries, metabolism, sensing, and movement, living beings actively produce and preserve an *inner/outer* or *self/non-self distinction*. "Life is thus a self-affirming process that brings forth or enacts its own identity" (Thompson 2007, 153), namely by maintaining an inner order and homeostasis against the entropic processes of the physical surroundings:

> The challenge of "selfhood" qualifies everything beyond the boundaries of the organism as foreign and somehow opposite: as "world," within which, by which, and against which

it is committed to maintain itself. Without this universal counterpart of "other," there would be no "self." (Jonas, 1968, 242–243)

However, this primary stage is not yet connected to consciousness. We can give several plausible conditions for the peculiar self-reference of a living organism, which manifests itself in basal conscious subjectivity:

1. The separation of the closed form from the surrounding environment, that is, the living being's *autonomy*: the interiority of life already prefigures self-awareness.
2. The formation of separate receptor and effector organs and the corresponding *sensory and motor capacities* which assign *meanings* to the environment and its objects: things attain (e.g., as tactile, visual, tangible etc.) specific relevance *for* the animal.
3. The formation of a *CNS*, which regulates and integrates the inner milieu, links receptor and effector organs, selects appropriate reactions from available behavioral repertoires, and thus functionally integrates the organism's overall state in relation to the environment.[19]

A living organism organized in this way not only lives, it also *experiences*, being conscious (to whatever degree), sensing, and reacting from out of its own center. The differentiation of perception and motion is, however, only achieved if there is an interruption between them, that is, an *inhibition* of the excitation produced by an external stimulus, which does not directly flow off in the form of motor reaction. For this purpose, the central organ or the brain must unite the functions of *inhibition and release*: "Reception is equivalent to inhibited, and effecting to released excitation" (Plessner 1975, 245). There would be no purpose at all in perceiving something, if this delay were missing. An object must *first* be perceived in order to *then* become the object of action, the latter not being effected out of blind necessity, but in accordance with the living organism's felt need or aim. *It is this inhibition and delay which opens up the space and time span for consciousness to emerge.*

However, we must not forget that the sensorimotor cycle is no end in itself, but only functions in the service of maintaining homeostasis. The coordinating

---

[19] According to Jonas, "the dissociation of moving and sensing, with neural mediation between them," reached in the metazoan stages of life, is the decisive step towards centralized self-control, and with it sentience and agency. "The nervous system, as a system of intercommunication distributed throughout the body, may then be said to constitute the 'higher level' we have indicated, and in this role provides a first answer to the question of who or what is the source of the control: it would be *the organism as a whole*, functionally integrated by its nervous system" (Jonas 1968, 246–247; emphasis added).

function of the CNS for an organism's interactions with the environment is therefore only possible in tight connection with its second function: *to register, integrate, and regulate the inner milieu*. Only then can sensorimotor activity be tailored to the organism's current metabolic needs. This regulation is (as we will see in 4.1.2) the source of a basic bodily self: an elementary vital feeling of well-being or discomfort, energy or fatigue, of basic drives such as hunger or thirst, and of mood states indicating the overall state of the organism with regard to its vital needs. More specifically, the *emotions* form an arc of tension that spans the time lapse between need and satisfaction (Jonas 2001, 103–104). The instinctive dynamics of lack, drive, longing, anticipation, fulfillment, and satisfaction—in short: conation (see 3.1.1)—is the experienced motivation fueling the functional cycles by which an animal lives. Last not least, emotions and moods lend vital meanings to objects in the environment, making them appear desirable, comforting, threatening, and so forth. The *valences* already given on lower stages (3.2.2) are now turned into *values*.

We can see here that the dichotomy between cognition and affection still common in cognitive science and psychology is no longer tenable. The living system *enacts* its Umwelt, a world of significance, and this is the basic "mark of the cognitive." However, this kind of cognition at the same time implies affectivity, in the sense that the living being is a subject of *concern* and thus never indifferent to its existence and environment. The process of sense-making implies both cognizing relevant aspects of the environment and evaluating or "feeling" them with regard to the organism's needs and concerns (Weber & Varela 2002). Hence, experiencing drives and emotions is also the means by which a living being evaluates and selects its possible *actions*.

Subjectivity thus bridges the gap between the organism's needs and fulfillment, as well as the gap between perception and motion, whose link is both inhibited *and* mediated by the CNS. This dual hiatus between present and future and between perception and action is bridged by subjectivity, on the one hand, through its time-spanning extension and, on the other hand, through various forms of selecting, meaning-attributing, emotionally oriented, and finally acting relationships to the environment. Nevertheless, subjectivity is not something "beyond" the organism, but may only be comprehended as *embodied*: it is the subjective body with its dispositions and capacities that establishes the original relationships with the world.

The progressive nervous centralization of the animal organism emphasizes the self that is opposed to the environment with increasing distinctness and awareness. Thus, pain and fear are specific psychic manifestations of self-preservation and demarcation against an intervening world. Correspondingly, the motor radius also widens due to the organism's distance from the ground and growing freedom of bodily extremities. At the same time, the senses for the distance gain increasing significance over the sense of touch, taste, and smell, thus multiplying the possibilities for action. A sensory world emerges through the differentiation of receptor species and their intermodal connection within the central nervous organ to a unified sensorimotor receptor and effector space.

The more pronounced the inhibition is against immediate reaction, and the longer the delay between a particular need and its satisfaction, the greater become the degrees of freedom. The organism's self-reference and the efficiency of its actions are especially magnified if perceptual and motor capacities are additionally linked in feedback and feedforward loops, so that self-perception continually guides and modifies actions. This is realized, for instance, in what Holst and Mittelstaedt (1950) identified as the *efference copy principle* (also known as "corollary discharge"): motor action is already signaled forward to the sensory system at the planning stage. This enables animals to distinguish whether changes in the relation to the environment emerge through their own or through external movements.[20] The animal's subjectivity reaches a new level, by integrating this self-referential aspect in the CNS. What it experiences is not only the external responses to its action, but also the monitoring and regulation of its own effecting behavior. The animal thus reveals the first signs of self-awareness in relation to objects.

The actual opposition of subject and object is, admittedly, only attained at the stage of "eccentric positionality" (Plessner 1975). Here, human beings step out of merely being integrated within the functional cycle. Now perceptual and motor processes can also be simulated by "trial action" in *imagination, anticipation*, or *fantasy*. This also inhibits emotional impulses, using the hiatus between perceptual and motor ability in a new way, namely in an "as-if" mode—including the option to reflect on the situation and the individual's place within it. This distancing of the subject from its present situation changes the merely subjective *Umwelt* of the animal into the shared, intersubjective, and therefore objective *world* of human beings: gaining distance from oneself also means being able to place oneself in another's position. Man's eccentricity is thus equiprimordial to his sociality.

---

[20] The principle may be illustrated by the example of the gaze: when the eyes are about to move towards an object in the periphery of vision, the visual region of the brain is "forewarned" of the eye movement and compensates for it. Perception is thus shielded from self-induced effects on the sensory input in order to achieve perceptual stability.

3.2.4 Summary

The closed form of an animal's organism creates a demarcation from the surroundings. At the same time, by transgressing its limits, the organism gains an active relationship with its environment. Peripheral organs and outer surfaces serve the purpose of continual exchange and communication with the environment, whether this occurs through the metabolism or the functional cycle of perception and motion. This cycle is mediated through special receptor and effector organs that, in turn, are linked through a central nervous organ. Such organization of the animal's existence forms a complementary relationship with its specific *Umwelt*: it selects the suitable elements from the physical surroundings and imbues them with the meaning of relevant stimuli, or of objects of perception, interest, and action.

This form of organization constitutes the animal's organism as a *bodily self*. It is aware in differing degrees; it perceives and reacts from its center and differentiates between what it perceives and its own actions. Subjectivity, which emerges with the formation of animal organisms in the world, in no sense signifies a transmundane inner life. Rather, it is always embodied and related to the environment, being present and effective within it. Subjectivity is the integral aspect of an organism's biological processes exhibiting a specific self-organization and self-referentiality as well as a sense-making relationship with the surrounding world. In all of its acts of life, the living organism is simultaneously revealed as physical and psychic, as "external" and "internal." However, inner and outer dimensions are not statically opposed, but constantly undergo processes of "externalization" and "internalization." This occurs through the dynamics of the metabolism of ingestion and excretion, through the cycle of receptivity and activity, and, ultimately for primates, also through the interplay of impression and expression in inter-bodily, facial, and gestural communication (Fuchs 2017d).

A physical body organized along these lines is thus not only the serving carrier of the CNS or brain as an information-processing apparatus that produces consciousness from within itself. Rather, an animal's body is organized and centralized in such a way that *as a whole* it is capable of bringing forth conscious acts of life. Subjectivity is not merely a by-product, an additional, and, if necessary, also dispensable partial function of a special organ. Rather, we may say: *in the same measure as subjectivity is necessarily embodied, so too, a suitably organized, living body is necessarily subjective*. The living body is a self, insofar as on one hand it is centralized, delimited towards the external world, and represents an indivisible functioning entity. It shows an inherent "inwardness." On the other hand, the living body transgresses itself by means of suitable boundary layers and organs and enters into a relationship with the environment, which thus attains significance. This type of living being has subjectivity: as sensation, feeling, striving, and becoming aware. First and foremost, subjectivity is life and vitality. All forms of experience are forms of life; there is a "deep continuity between life and mind" (Thompson 2007, 15), or, in other words, between living life and experiencing life, *Leben* and *Erleben*.

3.3 The circular and integral causality of living beings

In concluding the description of the organism's basic structures, I still have to analyze the specific causality distinguishing life. In bio- and neurosciences, the notion of monolinear physical causality still prevails: a cause, A, is followed by an effect, B, a stimulus by a response, and so on. Yet the specific causality of living organisms as self-reproducing autopoietic systems requires an autonomous concept. In association with synergetic models (Haken 1993), I will describe this concept as *"circular causality."* Here, two forms of circular or reciprocal relationships can be distinguished as *vertical* and *horizontal*, the one referring to the relationship of parts and whole within the organism, the other to the relationship of organism and environment. It will become clear later on that these circular relationships over time always unfold in "spiral-shaped" dynamics. In what follows, I first offer a fuller explanation of these terms, before linking them with reference to the idea of an *integral causality* of life.

3.3.1 Vertical circular causality

The vertical organization of living systems can be described as a hierarchy of different levels (see Figure 3.5), namely

1. The top level of the organism as a whole.
2. The intermediate level of partial systems and organs.
3. The basal level of cells.
4. The elementary or micro-level of material parts (macro-molecules, atoms).

Between these levels, in each case, there is a reciprocal relationship of the whole to the parts, as previously illustrated (see 3.2.1). Using Thompson's notion, this can also be described as *dynamic co-emergence*: the organism as a whole and its components (organs, cells, etc.) bring forth each other in a continual reproduction process (Thompson 2007, 60). At the same time, the whole entity assigns specific functions to the parts that present certain *restrictions* of their independent activity. For instance, to facilitate coordinated movement such as walking, various muscles must act together in a highly specific way (e.g., as agonists

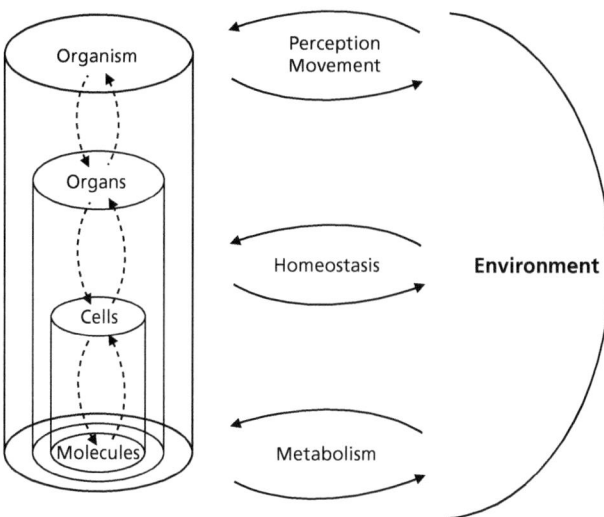

Figure 3.5 Vertical and horizontal circular causality.

and antagonists). For this, the unspecific excitability of muscular tissue must be constrained and integrated into a higher-order pattern. Other functions emerge, since the milieu and structure of the entire organism makes specific elementary processes *more probable* or *favors* them, especially in the case of basically chaotic processes, which characterize the immune system or also the neuronal system. The structuring influence, which a living system exerts on its parts, can be described as *formative* or also "*top-down*" causality (synonymous with "downward" or "global-to-local causality").

This kind of causality is often regarded as problematic or obscure, mainly for two reasons:

1. Since the whole consists of the parts, cause and effect may not be attributed to two different agents acting upon each other externally.
2. The causal effect of higher system levels either seems to presuppose unknown physical forces that contradict laws of energy conservation, or it falls prey to Occam's razor: it is simply dispensable for the actual causal effects (for this critique, see Craver & Bechtel 2007).

However, it is by no means necessary to restrict the notion of causality to effective causes (*causa efficiens*) according to billiard balls acting on each other. Macro-structures may well develop forming or organizing effects vis-à-vis the micro-elements in which they are realized, in accordance with Aristotle's *causa formalis* (Juarrero 1999, 125–128). This does not mean that new forces emerge which would contradict physical laws, for example of energy conservation.

Macro-structures, rather, are in a position, thanks to their form and configuration, to *select* specific properties and behaviors of their components and *block* others (Campbell 1974, Moreno and Umerez, 2000). Thus, these components acquire *emergent* properties, as for instance, the aforementioned capacity of organic iron to incorporate oxygen with a reversible effect. In inorganic nature, this would be an extremely improbable state. For this to occur, no physical "miracle" is required, but only a higher organizational structure (in this case hemoglobin) that selects and entrains or "enslaves" its own compositional elements (Haken 1997), that is, integrates them into specific behavioral patterns. Generally, the molecular processes within a living cell are designed so as to produce chemical reactions and molecules that defy the odds of natural occurrence by many orders of magnitude (Deacon 2006). Mediating catalytic reactions bias and constrain the formation and behavior of molecules in such a way that their composition occurs a billion times more likely than in non-organic environments.[21]

Analogously, mental processes, as embodied and integral acts of a living organism, can be effective in its physical behavior. Of course, the mental aspect does not affect physiological processes as an external force, but rather exerts a formative influence *in them* and *through them*. If I, for instance, speak a sentence, the muscles of my tongue and larynx display organized patterns of movement. Their proximate or efficient cause is the release of acetylcholine at the motor endplates of these muscles. Nevertheless, it is also correct to say that my tongue and larynx move in these ways *because I am speaking these words* and I am intentionally directed towards their content. This "because," however, no longer signifies an efficient, but a higher-order *selecting and forming cause*: the muscles are always ready for excitation, they could contract in manifold other ways, but they are drawn into a highly selective, superordinate dynamics. Thus, the organizing and encompassing cause of the muscle actions is my speaking (*top-down*) which in turn is realized by a complex, but constrained dynamics of physiological mechanisms (*bottom-up*).

However, the same applies to the neuronal activity in motor and other areas of my brain: they proceed in this determined way, because *I am speaking* these

---

[21] This may be illustrated by the formation of diamonds in contrast to the formation of hemoglobin: "Knowing the atomic structure of the carbon atom gives us a considerable ability to predict the probability that a diamond crystal will form. In contrast, knowing the atomic structure of haemoglobin provides almost no information about its probability of formation, its prevalence in certain environments, why it is found in context with certain other molecules, and why a normal distribution of related molecular forms is nowhere to be found" (Deacon 2006, 140).

words, consciously spanning the intentional arc of the sentence over time, and always anticipating ("protending," with Husserl's term) the next words to come. In other words, my embodied intentions and protentions are able to organize their physical implementation with the potential to even *achieve a future state that does not yet exist*.[22] *Qua* enactments of life, mental processes may thus be effective in the behavior of a living being without "acting on brain processes" externally. In order to avoid any connotation of such efficient cause, one could also speak of an *"implicational causality"*: *By way of* thinking or speaking, I— as a living being—also realize certain organized processes in which ordered neuronal and muscular activities are implied—they cannot help, as it were, similar to water molecules being drawn into a whirl that nevertheless consists of them.[23] The whirl as form or order *implies* their specific movements without acting upon them. Thus, the complete cause of my speaking is neither my tongue nor my brain, but I am this cause myself. In each conscious action— walking, speaking, writing, thinking—the human being as a whole acts as the forming, selecting, and organizing cause.

The living entity therefore asserts itself not through direct impact, but only in a constraining or *mediated* way through its parts. The absence of any compulsory central control corresponds to the autonomous nature and simultaneously high-caliber networking of sub-units. Their decentralized, yet interconnected activity precisely fulfills its functions and contributes to the life of the organism as a whole. The parts and partial processes thus have an impact on the preservation and functioning of the entire system, which can be termed *enabling* or "*bottom-up*" causality (synonymous with "upward" or "local-to-global causality").

The connection of both causal relationships may then be conceived as a *vertical reciprocal or circular causality* in the relationship between the whole and parts or between the hierarchically ordered levels of the system (Figure 3.5). This structure characterizes, for example, the relationships between the genes

---

[22] As I argued before (2.3.2), this capacity may not be attributed to a "predictive brain," since brain processes are never "ahead of themselves," but always only strictly present. This is not to say that protentions have no neural basis; there are various areas in the cortex, basal ganglia, and cerebellum which are involved in sensory or motor anticipation (Wolpert et al. 1998, Nijhawan 2008). Moreover, this anticipation seems to be associated with synchronized oscillatory activity in the cortical areas required for processing the respective event or the motor execution (Singer 2009). However, all these processes always remain within the momentary present. The overarching temporal synthesis that is at stake in drawing the intentional arc of speaking may only be attributed to my speaking as *conscious* speaking. For an account of its neurophysiological implementation with explicit reference to Husserl's concept of internal time consciousness, see Varela (1999).

[23] See Deacon (2006) for a more detailed dynamical systems account of emergence.

and the organism: the genetic structure of the cell core selects and controls the required composition of specialized cell organs and functions ("bottom-up"). Conversely, the entire configuration and function of the organism is involved in defining which genes of the individual cell attain any relevance at all for its development and regulation; this is the field of epigenetics ("top-down").

Vertical circular causality also characterizes the brain functions. Thus, noxious stimuli from the periphery lead, via central processing in the brain, to a reconfiguration of the nervous system, corresponding to the sensation of pain. Yet the overall disposition of attention and emotion in the respective situation decides whether the stimulus is "admitted" as a perception of pain or repressed by descending inhibitory tracts, for instance, in the case of a battle where the soldier's attention is distracted by his emotional arousal and an injury remains silent. Another example is that an emotional state can, on the one hand, be changed pharmacologically, that is, via local chemical influences upon the transmitter metabolism in the brain ("bottom-up"), and on the other hand, also by psychotherapy, that is, via a changed perception of the personal situation within a helpful dyadic encounter ("top-down"). Thus, fear can be influenced by sedatives or by a calming conversation. These relationships will be discussed in greater detail later on. For the time being, the proposal is the idea that the brain functions as a *transformer* for vertical circular causality. In other words, it transforms encompassing or high-level states (e.g., intentional directedness) and low-level (e.g., neurochemical) micro-states of the organism and, in each case, renders them effective on the other hierarchical level.

3.3.2 Horizontal circular causality

This second form of living causality emerges, on the one hand, from the multifaceted feedback effects *within* the organism, which do not occur hierarchically between different levels, but on a single level (e.g., reciprocal relationships between cells or organs, the cascade of blood coagulation, etc.). However, the feedback relationships and functional cycles of *organism and environment* also function horizontally, as previously outlined (see Figure 3.5). The circular relationship initially consists at a basal level in the *metabolism*, which is to be seen as part of the general regulation of *homeostasis* within the organism, under changing environmental conditions. The horizontal metabolism is linked with vertical, formative processes, which assimilate the absorbed substance, transforming it into the substance of the living organism.

The relationship of *perception*, *movement*, and *environment* also functions in a circular way: a living organism's reaction to external stimuli is responded to by the environment, which in turn has an impact on the organism, and so on, until the relationship of individual and environment attains a new balance. Here, the brain again functions as a *transformer* between the sensory stimuli and motor actions linked in feedback loops that extend into the environment (Varela et al. 1991). Here, too, vertical control processes intervene, as the living organism receives stimuli not only passively, but by itself selects the suitable excerpts from the environment, for example, by *attention*. Thus, central nervous, efferent influences on the peripheral sense organs co-define what is seen, heard, or touched (Liberman et al. 1990, Highstein 1991, Mikkelsen 1992).

Through these functional cycles, external physical events and their sensory and neuronal transducers become *carriers of meaningful relationships* between the living organism and its environment. As we have seen, the manifold transmission processes, such as those involved in visual perception, only consist of causally linked chains of individual physical events, if they are viewed in isolation. Light waves have no color and sound waves are not music or words. However, in contact with the living organism they become carriers of perceptual processes in which we see the color of a tree or hear a violin playing. Micro-processes, which occur between the environment and the living organism, are therefore *transformed* on a higher level into processes of attributing meaning, and ultimately into integral perceptions and movements.

3.3.3 Integral causality and its basis in capacities

The connection of vertical (internal) and horizontal (external) causality now leads to a notion of *integral causality*. Through this, a living organism realizes certain achievements in conjunction with a complementary environment that contribute to the continuation of its life: perceiving, desiring, or grasping something, walking towards a goal, speaking or writing, and so on. Such achievements represent acts of life which do not only relate to partial processes of the organism (as, for instance, absorbing oxygen through hemoglobin, the secretion of stomach acid, the patellar tendon reflex, etc.). Rather, they engage the organism as a whole. This means that in their realization, the living being is revealed in its dual aspect as a physical and a lived body—as a feeling, perceiving, desiring, and acting being.

This causality is based on a dispositional constitution of the living organism, which I wish to characterize by referring to Aristotle's notion of *capacity* (*dýnamis*).[24] By this, I denote a living organism's innate or acquired dispositions for the active realization (*enérgeia*) of its achievements in suitable environmental

---

[24] *Dýnamis* (potentiality, capacity, power) and *enérgeia* ("being-at-work," actuality, realization) are the complementary notions coined and developed by Aristotle mainly in book IX of his *Metaphysics*. Though referring to the general principles of movement and change in nature, they are frequently used by Aristotle to describe the dynamic potentialities and enactments of living beings. We will see that the CNS with its neuroplasticity is a crucial condition for the acquisition and realization of capacities in higher animals.

contexts. The concept of capacity thus describes a form of *integral potentiality*, which the living organism possesses as such, and which cannot be dissected into partial processes. This is true even if in each case there are special organic correlates for these capacities, primarily the central nervous structures, and relevant receptor and effector organs.

Acts of life such as perceiving, feeling, or acting are based on the condition of complex and integral macro-structures that are organisms. Below a certain scale, they vanish from view and can no longer be identified in physical micro-processes.[25] The same applies for the corresponding behavioral dispositions or capacities. However, in the life and cognitive sciences, anthropomorphic concepts, such as that of "capacity," tend to be avoided by generally adopting cybernetic terms such as "program," "regulation," or "control." A "program" is defined as "coded or pre-arranged information which controls a process […] so that it leads to a prescribed end" (Mayr 1979, 213). According to this view, genetic programs stored within the cell core, and perceptual, evaluator, or motor programs stored in the brain control the corresponding organismic functions. Of course, this eliminates everything that is experienced and intentionally directed—feeling, desiring, wanting, wishing, perceiving, and acting. On the contrary, the term "capacity" implies that these aspects cannot be severed from acts of life, lest we identify the behavior of a living being with that of pre-programmed machines such as torpedoes.

Capacities (such as the ability to write) are formed (1) through the development of the corresponding organic structures (e.g., hands, tendons, muscles, peripheral and central nervous systems, including suitable neuronal activation patterns), and (2) by embedding them in superordinate relations of organism and environment that imply specific attributions of meaning to objects (e.g., "pen," "paper," "words"). Thus, capacities bundle sub-systems and organs together in vertical causality to form cooperative units that are available to accomplish different functions. They *actualize* themselves as soon as a suitable situation arises: then, the *vertically joined* sub-units and partial processes cooperate and simultaneously connect with complementary counterparts of the environment in *horizontal feedback*. As a result, the living organism can realize the unified act of life or achievement—for example, grasping a pen and writing a letter. In realizing its capacity to grasp and write, a person acts in an *integral*, that is, vertical and horizontal combination of interlinked causality.

---

[25] 'A living body, seen at too close quarters, and divorced from any background against which it can stand out, is no longer a living body, but a mass of matter as outlandish as a lunar landscape, as can be appreciated by inspecting a segment of skin through a magnifying glass' (Merleau-Ponty 1962, 271).

This realization highlights even better what may be entailed by "capacity." The ability to write a letter is obviously not a capacity of the brain (although it is naturally a substantial requirement). Rather, it is the capacity of an embodied subject, whose environment makes available pens, paper, words, and text. Capacities therefore act as keys to the right locks in the surrounding world and imbue this world with meanings and affordances: "something to grasp," "something to write with," and so forth. They can only be described in terms of a *relationship* of the organism to its environment. Organically anchored perceptual and motor capacities form "*open loops*," as it were, which connect with suitable counterparts in the environment, so that at the moment of matching, a perception or action is achieved. In perceiving a tree, the dispositional knowledge of a tree is realized in contact with the real object, or in Aristotelian terms, *dýnamis* turns into *enérgeia*. Thus, on the basis of existing capacities a new *situational coherence* of organism and environment is created. "Our body, a system of motor and perceptual powers, [...] is a grouping of lived-through meanings which moves towards its equilibrium" (Merleau-Ponty 1962, 136).

3.3.4 The formation of capacities through body memory

The link of vertical and horizontal causality is modified and expands for human beings in the course of a biography, that is, through *learning and memorizing processes*. In the wake of horizontal interactions with the environment, recurring perceptual and behavior patterns are extracted and sediment as sensory, motor, emotional, or other *schemata* within an organism's memory substrate, above all in the brain. This formation of experience primarily affects the structures of implicit or "*body memory*" (Casey 1984, Schacter 1987, 1999, Fuchs 2000b, 2012a)—a term that indicates a system of embodied skills, habits, and dispositions which are acquired through practice and repetition. This is in contradistinction to autobiographical memory that enables us to recall specific experiences of the past. Body memory reproduces earlier experiences not as memories, but rather contains these as *experience or skill* in the form of perceptual and behavioral capacities, without needing awareness of their origin. It first includes motor habitual learning (procedural memory, for instance, how to ride a bike, how to dance a waltz), yet also perceptive, cognitive, and emotional capacities, for instance, recognizing objects (perceptual memory), finding one's bearings in a familiar dwelling or a town (spatial memory), or showing acquired emotional reactions to certain stimuli (emotional memory). Implicit memory functions are mainly based on subcortical brain systems, such as the basal ganglia, cerebellum, and limbic system (including the amygdala for classical and the *nucleus accumbens* for operant conditioning, respectively) (Graybiel 1998, Ennen 2003, Fuchs 2012a, Panksepp 2012).

Implicit learning thus causes repeated interactions with the environment to settle into dispositions, skills, and knowledge. A living organism expands its implicit knowledge and capacities not by filling an information store, but by the *change of its organic dispositional structure*, that is, by a process of growth and development. Now what is the locus of this embodied knowledge or body memory? According to the computational view of mind and brain, the process of learning writes bits of information into memory banks where they are stored and can be recalled at will. However, this representational and internalist view of memory does not fit with the dynamic interaction with the environment that takes place when bodily skills or habits are re-enacted. To be sure, as mentioned previously, this memory is also based on specific patterns of neural activation derived from earlier experience, mainly in subcortical regions of the brain. However, this does not imply any representational memory: instead of inner maps or models of external reality, the brain provides the *open loops* of potential interactions. As pointed out before, these loops are only closed to form full functional cycles by suitable counterparts in the environment that the body currently connects with, leaving no role for separate representations.

The term representation suggests that the brain activities could, at least in principle, be separated from the cycle, as if they were reconstructing or modeling inside what is outside. *But in a current sensorimotor coupling with the environment, there is no separate "inside" which could map, reconstruct, or represent the "outside."* In such an ongoing circular process, no segment can "represent" or "stand for" another (Fuchs 2011). Instead, the achievement in question is realized by the brain–body–environment system as a whole.

Thus, if "memory" means not some type of static inner depository, but *the capacity of a living being to actualize its dispositions acquired in earlier learning processes*, then this capacity is bound to the ongoing dynamic coupling between body and environment. An illustrative example is the attempt to find the keys for typing a certain word on an empty keyboard (i.e., where the letters have been removed from the keys) only by looking at it. Even for an experienced typewriter, this will be impossible—one usually has no representational knowledge of the position of the letters. However, at the very moment of having one's fingers set on the keys, they project their capacity onto the keyboard, and one can write the word immediately, without thinking. Here the knowledge is clearly an embodied know-how *without knowing that*, and the memory may well be said to reside in the "hands-on-the-keyboard," or to put it more precisely, the memory is an emergent dispositional property of *the whole system of organism and keyboard connected to each other*. Thus, since the locus of this memory is not the brain, "body memory" may not be regarded merely as a metaphorical term. Rather, it precisely describes the body in connection with the environment as the carrier of capacities, and thus, of habit or skill memory.

This memory now turns the circular relationship of organism and environment over time into a spiral-shaped development (Figure 3.6): each interaction changes—even if only minimally—the structure and disposition of the organism that, in turn, perceives or reacts to its environment in a modified way. The organism and its specific environment are thus linked in a process of continual *co-evolution*. In other words, the entire system consisting of organism and environment is reconfigured with each interaction, so that the respective presence of a living organism cannot be fully described without resorting to the history of its experiences.

The incorporation of experiences makes it possible for the living organism or human being to adapt to the experienced environment, and also to anticipate possible interactions. Our entire organism can, to a certain extent, be seen as a kind of implicit presupposition about the world. The phylogenetic learning history is already written into its genetic structure. The constitution of its organs and sensory, motor, and nervous systems thus contains an anticipation of the surrounding world in which it is to live and preserve life. This constitution defines its basic capacities and its corresponding ecological niche. The human brain now transfers this phylogenetic principle to ontogenesis, namely by its ability to incorporate within its principally plastic structures the organism's individual learning history. A person's capacities and dispositions therefore evolve in a way complementary to the environments, in which the person grows up, and

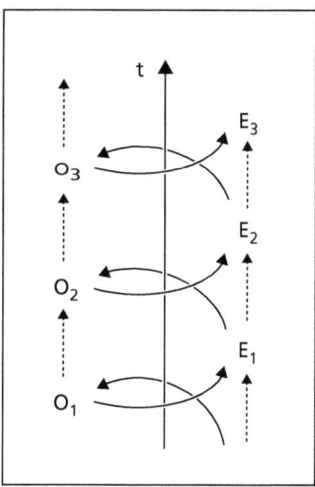

Figure 3.6 Co-evolution of organism (O) and environment (E) over time (t).

they presume future, similar situations. These interconnections will continue to preoccupy us in Chapter 4.

3.3.5 Summary

The brain's embeddedness in the organism and its environment assumes an understanding of life, which is not based on linear, but circular causality. Living organisms, as autopoietic systems, are both differentiated from their surroundings and interconnected with them. Their boundaries produce a fundamental *discontinuity* of inside and outside: physical causality "breaks" at these boundaries and cannot be prolonged in a linear fashion within the organism. Rather, due to their inner structure, living organisms themselves produce the section of the environment that becomes meaningful and effective for them. Linear causality is replaced by the specific link of stimulus and response, or of perception and reaction. Therefore, living organisms are not determined by the surrounding world of physical influences. Rather, they *respond* to perceived stimuli from their center by reconfiguring their entire system.

This relationship towards the environment may be described as *horizontal circular causality*, the relationship of hierarchical levels within the organism as *vertical circular causality*. Here, a first attempt to describe the brain suggests that it functions as an organ of mediation or transformation both within the functional cycle of organism and environment, as well as within the organism's functional cycle of the whole and parts.

Both functional cycles are interlinked in the *capacities* of living beings. Capacities refer to the ability of a living being as a whole to complete specific achievements, thus enabling a unique type of integral causality of life. A capacity functions like a key to the right locks within the environment, since this capacity has formed—phylo- and ontogenetically—in and through this environment. The brain serves as a central organ for this formation, to the extent that the organism's repeated experiences are "incorporated" into the principally plastic neuronal system. If the suitable opportunity emerges, the living organism can realize its capacities. In this case, inner-organismic processes (vertically) as well as organism and environment (horizontally) join together to form a cooperating unity.

As general dispositions, capacities do not function in a determinant fashion. Rather, they are enabling and oriented towards future situations. Thus, the corresponding achievements are not attained in a rigid and mechanical fashion, but they are always flexibly attuned to the requirements of the concrete situation in order to bring about the desired achievement. If I write a sentence on a piece of paper and then repeat it in larger letters on a board, both texts display the same personal features, even though entirely different groups of muscles were used—in one case, the hand, and in the other, the arm. Similarly, perception is in a position to recognize similar forms, such as a melody, which is played in entirely different pitches or on different instruments, even if this transposition was previously unheard. These laws discovered by gestalt psychology (Ehrenfels 1978) refer to the uniqueness of life's causality, which is precisely not grounded in complete determination. On the contrary, it allows for the flexible integration of organism and environment into an ever-renewing form and unity.

Even if partial processes of capacity-based functions can be described physically on an elementary level, their reduction to physical or neuronal event causality is prevented. This is because neither the capacities, nor the related memory and brain structures can be explained without the genetic reference to *earlier learning contexts* and without the functional reference to a *possible achievement*. A capacity may be regarded as an available "open loop," whose ends indicate complementary aspects in the environment. It is only in connection with these aspects that a capacity is realized. Therefore it is fruitless to want to localize motor schemata or other partial mechanisms within the brain without comprehending these from the outset as connected through the overarching function. Even put together, the fragments form no circle, but conversely the functional cycle *makes use* of the partial mechanisms and complementary aspects within the environment. The cycle is already inherent in a capacity as a potential or open loop, closing to form a real circle in an actual achievement. Capacities are always *dispositions of the living being as a whole*, and their realization in the relevant situation is the *activity of the living being as a whole*.

In the light of this exposition of the basic structures of life and its specific causality, we now proceed to a closer investigation of the brain's role within these structures.

Chapter 4

The brain as an organ of the living being

Overview

Chapter 4 at first takes a perspective on the brain as the central organ of regulation and integration, which is connected to the organism through various vegetative, endocrine, and autonomous regulatory feedback loops. The constant resonance between brain and organism furthermore is the basis of a background feeling of the body, a *"feeling of being alive"* that may be regarded as the foundation of all conscious experience. Further, emotions are states of the entire organism through which the living creature is specifically directed towards affective qualities of its environment (4.1). The relations of brain, organism, and environment are then portrayed by means of the *functional cycle of perception and movement*. The linear model of stimulus–response is replaced by the unity of organism and environment as a superordinate system in which capacities of the living being incorporated in the brain are joined together with suitable objects. As a crucial consequence of this extended or ecological model, consciousness is regarded as the *integral of the ever new closed functional loop between organism and environment* (4.2.1, 4.2.2).

In a next step, the biographical development of capacities is traced back to neuroplasticity. In this context, the main focus lies on an analysis of *implicit memory*, in which particular components of perception and movement are integrated into overarching patterns. These analyses serve as a foundation for the following investigation into the higher cognitive functions of the brain, which is mainly oriented towards gestalt perception. The central focus lies on the principle of *transformation*, namely of particular stimuli into neuronal patterns that resonate with appropriate objects in the environment (4.2.3, 4.2.4). This centerpiece of the chapter is followed by critical considerations of the notions of "information" and "representation" in the cognitive neurosciences, which are then contrasted with the alternative notion of "*resonance.*" In conclusion, these results are interpreted in terms of a Hegelian notion, namely that of "*mediated immediacy*" (4.2.5, 4.2.6).

Whereas the neurosciences, in the course of their progress, have found more and more detailed correlations between conscious experience and neuronal processes, in so doing they have usually described a brain without an organism. This corresponds to a reductionist understanding of the mind as a disembodied representational system within the skull. Accordingly, the classical cognitivist paradigm is based on the following assumptions:

1. Cognition is conceived as internal information processing and computation over neural representations of the external world.
2. The subject of cognition is not embodied or engaged in the world, but regarded as a detached observer.
3. The cognitive system is thought to be organized in the form of decomposable partial systems or modules which work with inputs independent from context variables (on a criticism of modularity, see 2.2.2).

In contrast, approaches of *embodied and enactive cognition* (Varela et al. 1991, Di Paolo 2005, 2009, Gallagher 2005, Thompson 2007, Stewart et al. 2010, Andersen et al. 2012, and others) no longer regard cognition and consciousness as an internal representation of an objective external world, but as components in circular interactions of organism and environment. On this view, the mind or, for that matter, the cognitive-affective system is seen as:

1. *Embodied* in the living organism and its specific constitution, which makes it necessary to regard the bodily realization of cognitive capacities as constitutive for their achievement;
2. Situationally *embedded*, that is, cognitive systems exploit the specific circumstances of their environmental context in order to increase their capacities;
3. *Extended* beyond the boundaries of the body in the form of ongoing feedback loops, thus being inherently connected with the respective natural, cultural, or social environment;
4. *Enacted*, that is, arising only through the active interaction of an autonomous living system with its environment.

In sum, this paradigm replaces the cognitivist "world-mirroring" by the concept of "world-making."[1] The purpose of the cognitive system is not to construct mental representations of external states, but to provide possibilities for embodied action within the world.

---

[1] "Consequently, cognition is no longer seen as problem solving on the basis of representations; instead, cognition in its most encompassing sense consists in the enactment or bringing forth of a world by a viable history of structural coupling" (Varela et al. 1991, 205).

This general approach, which accordingly has even received the name of "4e cognition," needs to be applied, however, not only to rather abstract autopoietic and cognitive systems, but also to the brain as embedded in the organism of a human being, as I intend to do in the following. Even more importantly, using a dynamical systems approach we should not lose sight of the fact that we are still dealing with an objectifying view of the life process, which may nevertheless not be comprehended without subjectivity. Taking the dual aspect of lived and living body into constant consideration is thus crucial in order not to fall prey to a "systemic reductionism." Dynamical systems or ecological approaches therefore need to be always complemented by phenomenological concepts, which start out from the subjective or lived body as the medium of our relation to the world.

On this basis, in what follows, I will first examine the brain's embeddedness in the human organism and then go on to the relations of brain, organism, and environment.

4.1 The brain in the context of the organism

4.1.1 The inner milieu

The primary environment of the brain or its organic milieu is the body. By means of the network of sensory, motor, and autonomic nerve fibers, but also by means of biochemical signals transmitted via the circulatory system, it is inextricably linked to the entire organism. Only via the body do all signals from the inner and outer environment reach the brain. It is not itself exposed to the world nor has an effect on it: hidden in the cavity of the skull, swimming in liquid, even insensitive to direct stimuli or pain, it entirely recedes into the background in order to better fulfill its mediating, regulating, and controlling functions.

The sensorimotor system usually stressed by embodied and enactive approaches is, however, only of secondary relevance in this context. The primary function of the brain is the regulation of the inner milieu, of vital needs and driving functions of the organism, which are prerequisites for its interactions with the environment. The brainstem and the diencephalon, especially the hypothalamus, are central organs for controlling neuroendocrine, visceral, and immunological processes. They regulate breathing, circulation, water and nutrition balance, body temperature, waking and sleeping, sexuality, and a number of further autonomic body processes. Central and peripheral functions are connected in complex horizontal–vertical feedback loops that, for instance, modulate the concentration of hormones, glucose, oxygen, or carbon dioxide in the blood. Already at this level, no clear dividing line can be drawn between the brain and the extracerebral body.

However, this close connection of brain and body has usually been neglected by cognitive neuroscience. Consciousness and mental functions were considered exclusively as products of the neocortex, to be grasped by functionalist, information-based, or connectionist models.[2] Only in the last two decades have concepts of a primary subcortical origin of consciousness been developed by affective neuroscience, above all represented by Damasio (1995, 1999a, 2010) and Panksepp (1998a). On this view, consciousness does not primarily consist of higher-order intentional functions such as perception, thought, reflection, and the like, but rather in a basic *affective sense of self and aliveness*. It emerges from the autonomous regulatory processes of the organism and provides the continuous background as well as the driving force for all higher cognitive functions.

Accordingly, the basal vegetative cycles mentioned earlier are elaborated at the level of the processes correlated with emotion and motivation, mainly located in the brainstem, diencephalon, and in the remaining limbic system. Organismic needs such as nourishment, hydration, rest, sleep, and reproduction are thus felt as a lack or a drive and lead to activities supported by basic affects such as seeking, fleeing, attacking, etc. For this, a constant feedback about the present state of the organism is required. Affective neuroscience regards such felt desires as homeostatic requirements of the entire body (Panksepp 1998a, de Cantazaro 1999, Craig 2002, Damasio 2011). It shows that, in order to adequately understand the brain as a regulatory organ, a "neuropsychosomatics" of homeostasis, vital feeling, interoception, drive, and affectivity is required, which models brain, body, and environment in their systemic unity.

On this presupposition, the usual separation of "higher" cognitions from affects and vital functions finds no basis in fact. All three domains are, much rather, closely linked by subcortical centers in the limbic system (basal ganglia, ventral tegmental region, and amygdala, among others). Even embodied and enactive approaches to

---

[2] "Most human psychological research, including cognitive and social sciences, typically focuses on the highest levels, commonly with little recognition of the lower levels" (Panksepp 2012, 10). The prevalence of functionalist and computational theories of the mind has favored the location of consciousness at the top of the brain, while neglecting its vital and affective basis. Typically enough, the embedding of brain functions in complex biochemical, humoral, and endocrine processes taking place in liquor and blood was regarded as cognitively irrelevant. Connectionist researchers built artificial neural networks that ignored the fact that signals in real brains constantly change from chemical to electrical and back to chemical transmission, and that they are constantly modulated by dozens of different neurotransmitters. All this is obviously not to be found in a computer.

cognition have so far largely ignored the energetic dynamics which fuel the sensorimotor cycles of organism–environment interaction. Perception, attention, action planning, and motor execution crucially depend on *conation*, that is, vital functions such as arousal, vigilance, drives, and basal motivational and affective states, which are mainly tied to centers in the brainstem and diencephalon, and which energize all higher functions of consciousness.

Thus, it already becomes clear that the unity of brain and organism at the vegetative level also includes the higher brain functions. Conscious activities such as perceiving, thinking, and acting are by no means based solely on neuronal processes in the neocortex, but equally on the continual vital and affective regulatory processes that involve the entire organism and its present condition. As such, the traditional cerebrocentrism of the cognitive neurosciences is based on a latent Cartesianism, which is not compatible with a systemic-biological perspective on the organism. Neither the brain nor consciousness can be separated from the living body as a whole. I will now take a closer look at the connection between brain and organism as the basis of embodied subjectivity.

4.1.2 The feeling of being alive

In section 3.1, the embodied foundation of subjectivity was described from a phenomenological perspective: life always precedes its becoming conscious. It manifests itself primarily in a basic sense of the "deep body" as the source and origin of conscious experience, not as its object. As we will see in the following, we find similar insight in neurobiology: consciousness arises on the basis of the interaction between body and brain, in such a manner that the body not only secondarily becomes the object of consciousness, but is *constitutive for its origin*. This idea guides current neurobiological concepts of a primary, vital or interoceptive consciousness, such as those of Damasio (2000, 2005, 2011), Panksepp (1998a, 1998b), or Solms (2013).

According to Damasio's theory, a "protoself," a primary sense of self, arises from a complex of neural activation patterns in the upper brainstem "which map, moment by moment, the state of the physical structure of the organism in its many dimensions" (Damasio 1999, 154).[3] This encompasses the proprioceptive, visceral, vasomotor, endocrine, and other afferences from the internal body (e.g., heart rate, blood pressure, blood oxygen, glucose, pH level,

---

[3] As a reservation to the following description, it should be noted that Damasio's entire terminology and theory of the self, by using notions such as "mapping," "images," or "representations," fully remains within the representationalist paradigm. Nevertheless, I render his conception in this terminology, my critique is given later in the text.

temperature, intestinal movements, vestibular sensations, muscle tension, etc.), transmitted mainly via the spinal cord, the cranial nerves, and the area postrema.[4]

In this way, the inner milieu is continuously registered as *interoception* (Craig 2002, 2003). Conversely, the inner milieu and its homeostasis are constantly regulated by the brain via descending innervations (parasympathetic and sympathetic nervous system, regulating the functions of the inner organs) as well as via hormone secretions from the hypothalamus and the pituitary, including the vegetative reactions to the composition of the blood. Taken together, these processes create what may be called an "interoceptive loop."

Looked at more closely, the multifarious interactions of brain and organism are processed and integrated in somatosensory structures of the upper brainstem, above all in the *nucleus tractus solitarii* and *nucleus parabrachialis* (Damasio 2010, 64, 66), with close relations to the region of the *periaqueductal gray* in the midbrain, as emphasized by Panksepp (1998a, 1998b). These interactive regulatory processes convey what can best be described as "the feeling of life itself, the sense of being" (Damasio 1995, 150), with the hue of comfort or discomfort, pleasure of displeasure, relaxation or tension, or other basic moods. The feeling of being alive corresponds to a *basic bodily self-affection* or a *minimal form of subjectivity* (Fuchs 2012b). On this view, homeostatic regulatory processes between body and brain can be seen to lie at the root of consciousness:

> The deep roots for the self, including the elaborate self which encompasses identity and personhood, are to be found in the ensemble of brain devices which continuously and nonconsciously maintain the body state within the narrow range and relative stability required for survival. (Damasio 1999, 31)

The primordial feelings of existence, sentience, or being alive "reflect the current state of the body along varied dimensions, for example, along the scale that ranges from pleasure to pain, and they originate at the level of the brain stem rather than the cerebral cortex" (Damasio 2010, 21). Granted, our attention is mostly directed towards directed emotions, perceptions, imaginations, or thoughts, so that one might be led to consider these functions the actual activity of consciousness. But all higher intentional feats remain embedded in the basal bodily sense of self: "The background body sense is continuous, although one may hardly notice it, since it represents not a specific part of anything in the body, but rather an overall state of most everything in it" (Damasio 1995,

---

[4] This is an area in the upper brainstem in which the otherwise given brain–blood barrier is suspended so that the humoral milieu of the organism (concentrations of hormones, blood composition, pH level, temperature, etc.) enters into the processing.

152). The feeling of being alive thus constitutes the affective backdrop of every conscious state, again confirming the inherent linkage of processes of life and processes of experience, or *Leben* and *Erleben* (Fuchs 2012b).

A similar model of the emergence of primary process consciousness has been developed by Panksepp (1998a, 1998b), putting more emphasis on *conation*, that is, on basic instincts and corresponding motivations. Interactions between the brainstem and the peripheral body play an essential role here as well. However, the resulting interoceptive self-experience is mainly bound to the already mentioned periaqueductal gray and implies a number of basic motivational affects which are crucial for vital and emotional survival: *Seeking, Rage, Fear, Panic, Lust, Care*, and *Play*, as they are termed by Panksepp. The *Seeking System*, for example, generates the arousal and energy which awakens our interest in the surrounding world and fuels our attention. This seeking is undetermined and objectless at first, and only differentiates itself under the influence of experiences (searching for nourishment, sexual partners, change of environment, and the like). Other motivational systems bring forth elementary impulses and reactions such as "fight" or "flight." In this view, we share with all mammals a set of primary affective experiences that guide our living at an instinctive level.

Apart from certain differences, it is a central claim of both conceptions that primary consciousness is not a product of the neocortex, but ultimately originates from the vital regulatory processes taking place between brainstem and organism, in other words, that it emerges as an *embodied subjectivity* from the very beginning. The self is not a result of cognitive sophistication or reflection; rather, it arises with the affective and motivational instincts that serve the organism's vital needs.[5] This view fundamentally contradicts the prevailing concept of the "cortical mind"; it therefore seems so counterintuitive that it seems worthwhile to support it by looking at an extreme anomaly, namely at *hydranencephaly*. This unfortunate condition affects children which as a result of a major stroke *in utero* are born without a cerebrum (cortex, thalamus, basal ganglia), leaving only the brainstem intact. Though scans show almost the whole skull cavity being filled with cerebrospinal fluid, these children are not only awake and conscious, but also clearly show expressions of basal feelings (see Figure 4.1), a primitive, subcortically mediated kind of sentience and corresponding behavior. For example, they

---

[5] See also the instructive overview on interoception by Cameron (2001), with the conclusion: "The body and subjective awareness of the body, including visceral awareness, instantiates the 'self' and provides the intermediary by which the nervous system interacts with the external world" (p. 708).

Figure 4.1 Expression of joy in a girl with hydranencephalus when her baby brother was placed in her arms.

Reproduced from Mark Solms, The Conscious Id, *Neuropsychoanalysis*, 15 (1), Figure 4, p. 11, http://dx.doi.org/10.1080/15294145.2013.10773711, Copyright © 2013 Routledge. Reprinted by permission of the publisher (Taylor & Francis Ltd, http://www.tandfonline.com).

> crawl toward a spot on the floor where sunlight is falling and where the child will bask in the sun and obviously draw benefit from the warmth. […] They tend to be fearful of strangers and appear happiest near their habitual mother/caregiver. Likes and dislikes are apparent, none so striking as in examples of music […] they can respond to different instrumental sounds and different human voices […] In brief, they are most joyful when they are touched and tickled, when preferred music pieces are played, and when certain toys are shown in front of their eyes. (Damasio 2010, 81)

As we can see, children with this severe malformation show nearly all of the basic affects emphasized by Panksepp, related to felt rather than intentionally perceived environmental situations. A basal affective consciousness thus arises already through the integration of bodily states on the level of the brainstem, as a "feeling of live itself"; for this, the cortex is not necessary. In contrast, lesions of the upper brainstem rapidly and totally destroy consciousness (Solms 2013). Thus, the cortex, though contributing to intentional consciousness (see below), depends on the integrity of subcortical structures, not the other way round.

4.1.3 Higher levels of consciousness

The elementary protoself originating at the level of the brainstem now becomes, according to Damasio, the basis for two further levels of conscious self-awareness, namely the core self and the autobiographical self:

> The first step is the generation of primordial feelings, the elementary feelings of existence that spring spontaneously from the protoself. Next is the core self. The core self is about action—specifically, about a relationship between the organism and the object. The core self unfolds in a sequence of images that describe an object engaging the

protoself and modifying that protoself, including its primordial feelings. Finally, there is the autobiographical self. This self is defined in terms of biographical knowledge pertaining to the past as well as the anticipated future. (Damasio 2010, 22–23)[6]

Leaving the autobiographical self aside, I will only explain the second level in more detail, which also tells us more about the role of the cortex. The primary interaction of brain and body is continuously further processed and integrated in higher brain centers (thalamus, cingulate cortex, superior colliculi, anterior insula, and somatosensory cortex; Craig 2002, 2003). At the same time, it is connected with sensorimotor experiences directed towards external objects. Through these processes of higher-level integration and interaction with the environment, the *core self* is formed, more or less corresponding to the phenomenological notion of *pre-reflective self-awareness*, and including a basic sense of agency. It arises from the interaction of subcortical and cortical brain functions which simultaneously are in resonance both with the entire organism and with the environment.

Experience on the subcortical level remains instinctual, impulse driven, and undirected or objectless: one may think of the newborn baby searching for something to satisfy his hunger, without knowing yet what a breast is. The role of the cortex now consists in *establishing the intentional direction of basic affective consciousness to external objects*. This happens both in the dimension of cognition and emotion:

1. The cortex mediates the sensorimotor interactions with the environment, through which intransitive, bodily self-consciousness is turned into transitive, object-directed consciousness. Over time, these interactions are sedimented in an *embodied memory* of the objects and their affordances.[7] This enables remembered objects of desire to be activated in the form of *image schemas* (Lakoff & Johnson 1999), that is to say, to be vaguely imagined once the instinctual motivation sets in again: the hungry baby now searches for the breast which had quenched his hunger before.

2. Under the influence of specific experiences of objects, the basic affectivity, which is marked by undifferentiated drives, pleasure, and displeasure, changes as well and differentiates into more specifically directed *emotions*. By these, the object relations mediated by the cortex receive affective

---

[6] It should be noted that protoself and core self are not exclusively human, that is, they may be present in non-human animals and do not depend on language, reflection, and higher forms of intersubjectivity.

[7] On implicit or body memory, see 3.3.4 and further descriptions later in this chapter (4.2.3).

significance, that is to say, the objects are valued in different ways: *I feel like this about that* (Solms 2013).[8]

Cortical activity thus has a twofold effect: on the one hand, it is crucial for the constitution of the objects that we have learnt to perceive and to handle from birth onwards with increasing skill. On the other hand, under the influence of these interactions with the environment, undirected affectivity is gradually turned into specific intentional emotions. By these, the organism is motivationally directed towards future states to be achieved, and towards objects to be reached or avoided. Above all, interest and attention are the expression of the conative energy that directs consciousness towards the world. This energy ultimately originates from the basal endogenous affectivity of the "deep body." *Thus, the self-affection and conation of the body are the foundation and source of our intentional directedness towards the world:* if there is something "it is like" to have intentional conscious experience (2.1.2.1), and if indeed all experience contains a sense of "mineness" (Zahavi 1999), then this felt dimension is derived from the organism's elementary *self-affection* which precedes every sensorimotor contact with objects mediated by the cortex.[9]

Taken together, the concepts of Damasio, Panksepp, and Solms may be illustrated by the following schema (Figure 4.2): processes of autonomous regulation, which serve the homeostasis of the inner milieu, are in continuous interaction with brainstem nuclei (nucleus tractus solitarii and parabrachialis). This interaction is the basis of the most basic bodily self-affection or *feeling of being alive*. The resulting *protoself* is then further enriched by basic affects and drive motivations (desire, anger, lust, etc.), which may be attributed, above all, to the periaqueductal gray. Higher-level integrative processes then include the remaining limbic system and the cortex. They direct the endogenous affective energies towards the environment and its objects, thus enabling the emergence of the *core self* as a pre-reflective "being-(and acting)-towards-the-world" (Merleau-Ponty). These intentional relations of consciousness to its objects are

---

[8] To note, we are not yet dealing here with a reflective or "ego"-experience, which only arises on the third level of the autobiographical self.

[9] This concept comes obviously close to Michel Henry's (1963, 1975) emphasis on a pre-intentional auto-affection or "pathos" of life on which all intentional consciousness is based. For Henry, life refers to a primordial "self-appearing" that is radically distinct from the ek-static, world-directed nature of intentionality. Somewhat pointedly, one might say that Husserl's primary focus is on intentional acts of consciousness, Merleau-Ponty focuses on the "operative intentionality" of the sensorimotor body mediating our relation to world (the body as "surpassed," in Sartre's terms), while Henry brings into view the pre-intentional life of the body itself.

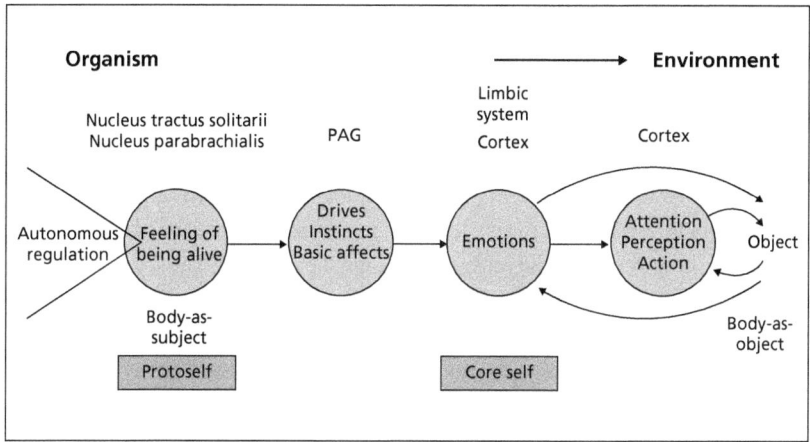

Figure 4.2 Basal and higher levels of consciousness (PAG = periaqueductal gray).

established, above all, by *attention, perception, and movement*. At the same time, through the organism's perceptions and action tendencies, the *emotions* direct themselves towards the environment and their affective values. In turn, it is the emotions that lend attention, perception, and action their motivational dynamics, purpose, and interest.

The basal processes of consciousness thus emerge deep inside the organism itself, to then direct themselves, on higher levels of integration, increasingly towards the environment. It is important to note, that this latter development is no longer predetermined genetically, but proceeds under the influence of early environmental and socialization experiences (see Chapter 5).

This "nested hierarchy" of consciousness thus described is also the biological foundation for the ambivalent status of the body as treated in section 3.1.2. On the one hand, the basal feeling of being alive corresponds to the internal, deep body or "*body-as-subject*" (Figure 4.2), that is to say, the endogenous source of experience that cannot itself become an object. On the other hand, the body re-appears on the level of directed, sensorimotor relations to the environment, namely as an *object* of proprioceptive, tactile or visual perception, that is to say, as "*body-as-object*" (a special object though, as it remains always present).[10] The internal body conveys the background state of being-conscious-of-something or being-directed-towards-something—the body as medium—whereas the

---

[10] In a similar way (and following William James's famous distinction), we may speak of the protoself and core self as the *self-as-subject*, whereas the reflective, autobiographical, or narrative self developing in the second year of life could be termed "*self-as-object*."

external body is the body that we become conscious of. This conception corresponds to the phenomenological notion of embodied subjectivity: the lived body (*Leib*) as a "natural subject" (Merleau-Ponty) is at the roots of all reflecting acts of consciousness, not least including its own objectivization as physical body (*Körper*).

Despite this convergence of phenomenology and affective neuroscience, some important critical remarks have to be added here: neither Damasio nor Panksepp actually escape the representationalist paradigm. Thus, the brainstem too is said to "map," "mirror," or "represent" the states of the body as if we were dealing with an imaging relation. Damasio assumes that the "mental images of the body produced in body-mapping structures" constitute the protoself (Damasio 2010, 21) and even presumes to claim:

> The human brain is a born cartographer, and the cartography began with the mapping of the body inside which the brain sits. (Damasio 2010, 64)

Of course, this is meant metaphorically, but the question arises: what could then be a non-metaphorical cartography? The brain can hardly establish the "as-if" relation between map and landscape—as we have already seen, this is at most possible for a neuroscientist (see 2.1.2.2). Moreover, in none of the structures of the brainstem can we find a cartographic, spatially differentiated "image" of the body (e.g., of the heart, the viscera, the muscles, etc.)—the bodily signals arriving here are rather integrated in global formats. But apart from that, Damasio himself describes the relation of body, brain, and environment in a way that *basically excludes a representational opposition*:

> [Neural] maps are constructed when we interact with objects, such as a person, a machine, a place, from the outside of the brain toward its interior. I cannot emphasise the word *interaction* enough. (Damasio 2010, 63–64)

But how could the inside of the brain represent the external world when the brain is *constantly interacting with it*? A representational relation presupposes that the representing and the represented may be separated from each other—just like a picture of Marilyn Monroe still represents her though she has long since died. *In an ongoing interactive cycle, however*, as I have already argued before (3.3.4), *no segment can "stand for" another or represent it*—and in such an interactive process also consists the relationship between the brain and the peripheral body. The structures responsible for the protoself, according to Damasio,

> are literally and inextricably *attached* to the body. Specifically, they are attached to the parts of the body that bombard the brain with their signals, at all times, only to be bombarded back by the brain and, by so doing, creating a resonant loop. This resonant loop is perpetual, broken only by brain disease or death. (Damasio 2010, 21)

This resonance corresponds to a circular interaction of brain and body. Both influence and modulate each other constantly so that the brain–body system as a whole maintains a homeostatic equilibrium. Damasio even speaks of a

> looped circuit where the body communicates to the central nervous system and the latter responds to the body's messages. *The signals are not separable from the organism states where they originate.* The ensemble constitutes a dynamic, bonded unit […] this unit enacts *a functional fusion of body states and perceptual states*, such that the dividing line between the two can no longer be drawn […] the signals conveyed would not be *about* the state of the flesh but literally *extensions of the flesh*. (2010, 273; my emphasis)

Within such a looped circuit, however, *there is neither place nor time for a separate representation*. The intertwined processes are in constant flux and do not allow for any "mapping." Instead of an imaging or mapping relation, we should therefore rather speak of a *resonance* between brain and body.[11] This is not just a quarrel about terms; we are dealing with the decisive question whether basal consciousness is ultimately still created "in the brain," or whether the *body* actually is the "rock on which the protoself is built," as Damasio also writes (2010, 21). If that is the case, then consciousness can no longer be localized anywhere in the brain. Rather, it is rather a manifestation or *the "integral" of the overarching process of life* which encompasses the whole organism.

> In algebra, the integral enables the calculation of an area that is bounded by a function over a certain basis. I use it as a metaphor to signify the integration which consciousness achieves over an extended basis, without being separable from that basis as a "representation." Similarly, even if the brain is the *conditio sine qua non* for this integration, it does not become the separable "seat" of consciousness. I will return to the notion of "integral" in section 4.2.2.

The same argument is also valid for a frequent objection against the concept of the embodied mind (e.g., Block 2005): we should distinguish, the objection goes, between physical processes that are *causally effective* for mental states (enabling them as external causes), and internal brain processes that are actually *constitutive* for mental states. But if the interconnected loops of brain–body interaction are ongoing and inseparable, if they even show a "functional fusion," as Damasio argues, and if cause and effect are thus interdependent and commutable—then the neat separation of causal and constitutive processes no longer works. Where should we draw the dividing line? At the level of the diencephalon? At the brainstem? At the upper spinal cord? Where does consciousness begin or end? All this seems rather arbitrary.

---

[11] The notion of resonance will be further elaborated in section 4.2.5.

One might object that neurological conditions such as paraplegia or locked-in syndrome do not affect conscious functions. But even in these cases, most of the autonomous, interoceptive, and neuroendocrine interactions are still preserved. Hence, we are not allowed to functionally disconnect the head from the body when looking for the basis of vitality. Why exclude the heart and its most sensitive contribution to interoception (Cameron 2001, Pollatos et al. 2005) from being co-constitutive for the feeling of being alive? Why disregard the baroreceptors in the peripheral vessels which crucially contribute to the autonomic regulation of blood pressure and thus to the global background feeling of the body? Why neglect the continuous flow of breathing, the gastrointestinal motility, and the huge enteric nervous system (often called the "gut brain")? The integration covers all major domains of the organism. Or let us finally think of sexual arousal: is it really located in the brain? Certainly not—peripheral stimulation of the genitals (erection, lubrication, etc.) and sexual pleasure mutually enhance each other, thus creating the integral phenomenon of sexual arousal. On the other hand, erectile dysfunction results precisely from the circular interaction of deficient erection and lacking pleasure (Laan et al. 1995, Chivers et al. 2010).

In sum, basic self-affection is a condition and process of the whole body, a "manifestation of the flesh." Of course, one would not go as far as to contend that the background feeling of being alive integrates, for instance, even the single movements of the white blood cells in the spleen. But as we have seen, at least the extension of the nervous system and its countless receptors and effectors over the whole body, including the modulating function of neuroendocrine and humoral processes on signal transmission in various areas of the peripheral and central nervous system—all this can count as a plausible basis for the constitutive embodiment of subjectivity, thus being extended far beyond the brain.

4.1.4 Embodied affectivity

The above description has already shown that not only basal affects, but also intentional emotions (joy over ..., anger at ..., shame about ...) should be seen as *embodied reactions* which are directed towards appraised situations. This view has been highlighted in particular by Colombetti:

> [E]motion should be conceptualised as a faculty of the whole embodied and situated organism. Evaluations arise in this organism in virtue of its embodied and situated character, and the whole organism carries meaning as such—not by way of some separate abstract cognitive-evaluative faculty. (Colombetti 2010, 146)

Moods and feelings, seen from an enactive perspective, are states of the entire organism, which involve virtually all subsystems of the body: the central and autonomic nervous system, the endocrine and immune system, visceroception (heart rate, circulation, respiration, intestines; see Wiens 2005), and proprioception (muscular tension, posture, facial and gestural expressions). Every experience of feeling is inextricably linked to changes of this "bodyscape," and to the reciprocal interactions of brain and body. This link was already stressed by William James: the body and its emotional reactions form a "sort of sounding board" which every emotion, however slight, "may make reverberate" (James 1884, 191).

> If we fancy some strong emotion, and then try to abstract from our consciousness of it all the feelings of its bodily symptoms, we find we have nothing left behind, no 'mind-stuff' out of which the emotion can be constituted, and that a cold and neutral state of intellectual perception is all that remains. [...] A purely disembodied human emotion is a nonentity. (James 1890, vol. 2, 452)

Damasio has developed a comprehensive neurobiological theory of the interrelations of body and feeling, which is to be briefly sketched here (Damasio 1995, 127–135; 1999, 67–72). According to it, emotions arise as physiological states in complex feedback loops between various body systems and brain centers, serving to ensure the sustenance of the organism and its homeostasis. *Primary*, innate emotions such as fear or anger are triggered in response to relevant situations by subcortical-limbic structures such as the amygdala or periaqueductal gray. They cause—mediated by the autonomic nervous system and neuroendocrine signals—a reaction of the entire organism (change of heart rate, blood pressure, respiration, gastrointestinal motility, perspiration, muscle tension etc.). Various basic emotions are thus associated with different physiological profiles. These peripheral reactions, in turn, are registered primarily in the somatosensory regions of the right brain hemisphere (insula, parietal lobe) and ultimately lead to the conscious experience of these coordinated reactions as *feelings*. The body thus is the actual "theatre for the emotions" (Damasio 1995, 155).[12]

Secondary emotions such as shame, sadness, or envy are in humans triggered by perceptions, thoughts, and imaginations, of which the correlates in the prefrontal cortex are connected with prior emotional experiences and corresponding appraisals. By means of the amygdala and the cingulate gyrus they, in turn, activate a cascade of bodily reactions, which is again registered by

---

[12] Damasio's terminological distinction between subpersonal "emotions" and conscious "feelings" seems problematic, however; for a critique, see Panksepp (2003). Moreover, Damasio even refers to experienced feelings as "varieties of maps (images)" or "spontaneously *felt* images" (Damasio 2010, 76), which seems a rather adventurous category mistake, for what could a state of joy or anger be an "image" of?

somatosensory brain regions. This feedback loop is closely linked to Damasio's theory of "*somatic markers*" (Damasio 1995): not only the perception, but also the imagination of possible situations causes bodily, especially visceral, reactions (the well-known "gut feeling"), which then resonate in the somatosensory regions of the brain. This resonance influences—mostly unconsciously—our decision-making by supporting appraisals of possible scenarios. The accompanying bodily reactions function as an *emotional body memory* that provides rational deliberations in decision-making with a gradient of value.

> Damasio then goes on to describe patients with prefrontal brain damage that exhibit aimlessness and irrational behavior, even though their cognitive ability to rationally assess situations is by no means impaired. Due to their damage they are no longer capable of "letting their body resonate" and thus assess the imagined options emotionally. Even simple decisions such as making an appointment can lead to endless deliberations, since the lack of value distinctions, so to speak, levels the decision making landscape (Damasio 1995, 205–208).

Thus, emotions involve a sensation of the entire visceral and musculoskeletal condition of the body, while this changes under the influence of certain perceptions. Here the organism as a whole functions as a sounding board, and its emotional resonance causes corresponding "reverberations" in the brain. As such, emotions can also be seen as an example of vertical circular causality: in reaction to a perceived situation, emotional bodily reactions are triggered (top-down), which then feed back into the brain and influence felt experience (bottom-up). Central and peripheral reactions mutually modulate and enhance one another. Emotions thus are integral expressions of life, in which the entire organism is directed towards specific environmental situations, both in an evaluating and motivating way.[13]

Drawing on more recent concepts of embodied affectivity (Colombetti 2013, Fuchs & Koch 2014), we can also describe emotions in terms of a *circular* or *feedback relation* between a living being and its present situation with its particular affective qualities, values, and affordances—as a cycle of embodied affectivity (Figure 4.3). I will briefly present such an embodied and extended model of emotions:

---

[13] A frequent objection to this embodied concept of emotions is that patients with paraplegia or other forms of disconnection of brain and peripheral body are still able to feel emotions. However, while the intensity of emotions may be diminished by a lack of proprioceptive and kinesthetic sensations, in all these cases the basic bodily resonance via the autonomous nervous system, visceroception and neuroendocrine interactions is still preserved, leaving the deep bodily basis of emotions untouched.

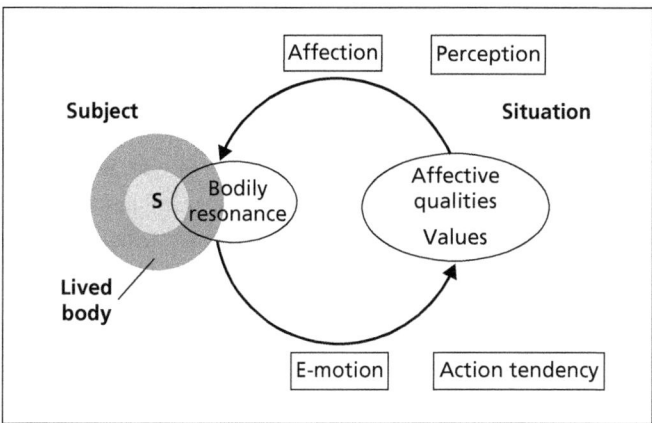

Figure 4.3 Cycle of embodied affectivity.
Adapted with permission from Thomas Fuchs, 'The phenomenology of affectivity', in KWM Fulford, Martin Davies, Richard Gipps, George Graham, John Sadler, Giovanni Stanghellini, and Tim Thornton, ed., *The Oxford Handbook of Philosophy and Psychiatry*, Figure 38.1, p. 623, DOI: 10.1093/oxfordhb/9780199579563.013.0038 © 2013, Oxford University Press.

1. Emotions emerge as specific forms of a subject's bodily directedness towards the affective qualities and values of a given situation. They encompass subject and situation and therefore may not be localized in the interior of persons (be it their psyche or their brain). Rather, the affected subject is engaged with an environment that itself has affect-like qualities (e.g., attractive, comfortable, repulsive, frightening, uncanny, etc.). For example, in shame, an embarrassing situation and the dismissive gazes of others are experienced as a painful bodily affection (blushing, "burning") which is the way the subject *feels* the sudden devaluation in others' eyes. The emotion of shame is thus extended over the feeling person and her body as well as the situation as a whole.
2. Emotions further imply two components of bodily resonance:
    ◆ *A centripetal or affective component*, that is, being affected, "moved," or "touched" by an event through various forms of bodily sensations (e.g., the already mentioned blushing and burning of shame). This resonance in turn influences the affective perception of the situation—for example, feeling one's heart beat and a shortness of breath increases the frightening impression of a dangerous situation.
    ◆ *A centrifugal or "emotive" component*, that is, a bodily action readiness, implying specific tendencies of movement (e.g., hiding, avoiding the other's gaze, "sinking into the floor" from shame; flight reaction in anxiety).

In emotions, we are *moved to move* (Sheets-Johnstone 1999) (*emovere* = to move out).

3. On this basis, emotions may be regarded as *circular interactions* or *feedback cycles* between affection, perception, and movement (see Figure 4.3). Being affected by the value features or affective qualities of a situation triggers a specific bodily resonance ("affection") which in turn influences the emotional perception of the situation *and* implies a corresponding action readiness ("e-motion"). Embodied affectivity consists in the whole interactive cycle which is crucially mediated by the resonance of the feeling body.

4. In this way, bodily resonance, as James already emphasized, serves as the *medium of our affective engagement in a situation*. It taints and permeates the perception of the situation without necessarily stepping into the foreground. In Polanyi's terms, bodily resonance is the *proximal*, and the perceived situation is the *distal* component of affective intentionality, with the proximal component receding from awareness in favor of the distal (Polanyi 1967). This may be compared to the sense of touch which is at the same time a self-feeling of the body (proximal) and a feeling of the touched surface (distal); or to the subliminal experience of thirst (proximal) which may first become conspicuous as the perceptual salience of a creek flowing nearby (distal).

As we can see, the resonance or "affectability" of the body is a crucial component of emotions. This has been confirmed by research indicating a positive relationship between interoceptive awareness and the ability to perceive one's own emotions (Pollatos et al. 2005, Dunn et al. 2010), and conversely, by impaired processing of interoceptive signals in mental disorders characterized by emotional dysregulation, such as major depression, depersonalization, alexithymia, or borderline personality disorder (Paulus & Stein 2010, Herbert et al. 2011, Terhaar et al. 2012, Schultz et al. 2015, Müller et al. 2015).[14]

If this bodily resonance is modified in specific ways, it will also change the person's affective perception accordingly. Thus, a *lack of resonance* will impede perceiving affective affordances in the environment. In Parkinson's disease as

---

[14] Cognitive or appraisal theories of emotion (Lazarus 1966, Cacioppo et al. 2000) usually consider bodily resonance as merely unspecific arousal which at best contributes to the intensity of emotions. Research on embodied affectivity as shortly presented in what follows contradicts this assumption. Moreover, as Lewis (2005) has pointed out, the separation of systems for appraisal and for arousal is not tenable on the neurophysiological level. For a more extensive critique, see Colombetti (2010).

well as in severe depression, the rigidity and freezing of facial and other bodily expressions results in a reduced intensity of emotions, to the point of being no longer moved and affected by situations or other persons at all (Mermillaud et al. 2011, Fuchs 2013a).

> The modulation of emotions through inhibited bodily resonance has also been confirmed by a number of studies in embodiment research (see Niedenthal 2007, Fuchs & Koch 2014, for an overview). Thus, Strack et al. (1988) demonstrated that the inhibition of smiling—by asking participants to hold a pen between their lips—caused them to judge cartoons to be less funny than when smiling was activated (by holding the pen between their *teeth*). To give another example: the injection of botulinum toxin in the frowning muscles impairs the understanding of negative semantic content which is normally facilitated by a slight frown (Havas et al. 2010). This connection has already been used to successfully reduce symptom severity in patients with depressive guilt-feelings by administering botulinum toxin for several weeks (Wollmer et al. 2012).

Conversely, *increasing* a certain bodily sensation or expression favors the correlated emotions as well as emotional attitudes towards objects or persons: for example, Williams and Bargh (2008) showed that holding a hot cup of coffee elicits a "warmer" (more generous, caring) impression of a target person than holding a cup of iced coffee. Bodily felt warmth thus directly affected the interpersonal impression of warmth. This linkage also works the other way round: Zhong and Leonardelli (2006) found that people estimated the room temperature as being colder than before after they had experienced social exclusion from a group. Interpersonal coldness was thus felt as physical coldness.

As we can see, the different components of the affection–resonance–emotion cycle mutually influence each other. Only through their ongoing circular interaction do they create the fully fledged phenomenon of emotional experience which therefore cannot be located "in the brain."

4.1.5 Summary

Our position so far can be summarized as follows: the brain is not an isolated organ that produces its own world within the skull and, on this basis, sends signals into the body. Much rather, it is an organ of *regulation and perception* for the entire organism. The body is the actual "player on the field": Its homeostasis and relation to the environment is crucial and its inner states can best give indications as to appropriate reactions and behavior. Center and periphery are therefore closely connected and influence one another in constant circular feedback loops.

At the same time, it becomes clear that consciousness is not a product of the isolated brain, even less of the cortex, but has the organism as a whole as its basis. The one-sided focus of the neurosciences on cognitive functions could, for a long time, give the impression that the organism was of importance to the brain only as a "carrier." However, functionally the brain does not end at the brainstem, but extends via the spinal cord, the sensory nervous system, and neuroendocrine functional circuits into the entire body. An examination of the basal affective functions of consciousness shows that they develop from vital regulatory processes that occur between brain and the periphery of the body and keep the inner environment of the organism constant. *The continual "resonance" of brain and organism is the precondition of conscious experience.* Basal consciousness consists in feeling alive, in mood and attunement—it forms an integral of the corresponding state of the organism itself, as embodied subjectivity.

In the evolution of consciousness, integrated affective states and corresponding action tendencies occurred long before cognitive or reflective capacities. They enabled an awareness of organismic imbalance and prepared the organism for balancing it out by *means of appropriate interaction with the environment*. As such, it is to be assumed that basal feelings initially developed from global reactions of the body to deficiencies, but also to external stimuli such as heat, cold, touch, light, and so on, as the example of anencephalic infants has shown. The phylogenetically original form of world experience consisted in the immediate, affective self-awareness of the body in its respective states within the environment.

The specifically directed, intentional emotions only developed in connection with situations perceived with increasingly differentiated appraisals, especially social relations. This was enabled, above all, by the development of the cortex, which oriented the global primary consciousness more and more to specific environmental objects. But also after the emergence of higher emotional and cognitive functions, basal affective experience remained the indispensable foundation of foresight, planning, and goal-directed intentionality. More than other kinds of experience, feeling alive, moods, and affects show us that we are beings incarnate—creatures of flesh and blood.

4.2 **The unity of brain, organism, and environment**

The previous section examined the vertical functional circuit of brain and organism. Now we turn to the horizontal relations of the living being to the environment, insofar as they are mediated by the brain, especially the functional circuit of perception and movement.

4.2.1 Linear versus circular organism–environment-relations

The division of brain and body, as is still common in cognitive neuroscience, corresponds to a further distinction: organism and environment are seen as two separate systems, the boundary between which is drawn at the skin. Accordingly, "inside" and "outside" are fundamentally separate from one another. From this follows the common conception of their relation (see Figure 4.4):

1. The senses receive stimuli from the external world, which are encoded in the form of action potentials, transmitted via the nerves, and forwarded to the brain.
2. The brain internally processes these signals according to cognitive algorithms and thus constructs neuronal *representations* or *internal models* of the world outside the brain.
3. This central processing of stimuli finally results in the organism's motor reaction and effect on the environment.

In principle, this results in the following linear sequence: sensory input → internal cognitive processing → motor output. As such, the physical principle of cause and effect is transferred to the living being. Conscious perception results as a side effect in the brain, but it only plays an epiphenomenal role: if organism and environment are two separate systems, the relation of consciousness to the environment can only be that of an internal representation. As such, consciousness remains a passive inner observer of information processed in the chain of physical processes. This traditional, linear input–cognition–output picture of the mind has been aptly described by Hurley (1998) as the "sandwich model" of the mind.

But is there even a fixed boundary between the living creature and its environment? Previously, in section 3.2.2, I have portrayed the coupling of organism

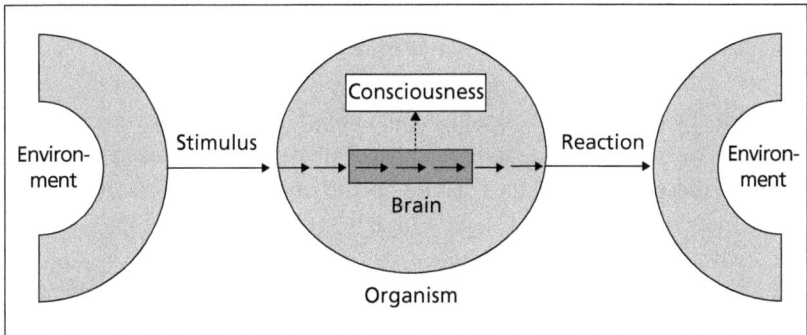

Figure 4.4 Linear causality in the stimulus–cognition–reaction scheme.

and environment as *two components of an overarching system*. We can now return to these results. Let us take the example of an instrumental action such as writing a letter. In order to do so, I pick up a pen that was previously outside my perception, but had already been preconceived by my imagination. It is also suitably shaped for being held by my fingers and has an expected weight. In other words, my lived body already anticipated the pen through its habits and protentions.[15]

Now my hand moves the pen across the paper, but the locus of my sensation is in its tip: I feel it scraping on the paper, that is, the pen has become embodied and I perceive the paper's surface "through it." Paper, pen, hand, and my entire organism form a functional unit—so where are we to localize writing? In the pen? In my hand? In the brain? Or maybe in my consciousness, in which the written words are formed? But I am actually writing, and the words "flow into my fingers." My writing cannot be dualistically split into a mental and a bodily writing. *In a realized function, no distinction between inner and outer can be drawn*—just as it would be nonsensical to ask whether the air breathed still is part of the external world or already of the organism.

From this it follows that we cannot conceive of an activity such as writing as an interaction of two separate systems, that is, of organism and environment, *because in the achievement we can no longer distinguish them from one another*. Already the capacity to write and the corresponding organic dispositions only exist as complements to suitable structures in the environment (pens, paper, words …). Before I write, my body has already anticipated the pen in the sensorimotor functional cycle, that is, it has formed a general scheme of its appearance and use (a *Vorgestalt* or "pre-gestalt," according to Conrad 1947, or an *image schema*, according to Lakoff and Johnson 1999), into which the actual pen only needs to fit in. With Heidegger's term, it has to already be "ready-to-hand," so that I can use it. Likewise, the realized function of writing itself is only possible in the functional cycle of perception and movement, which ties organism, pen, and paper together into a dynamic unit.

This results in the following altered conception (Figure 4.5): in the actual achievement, the relation of organism and object does not entail that one system influences another, but rather that the whole system of organism and environment is reconfigured on the basis of a pre-existing complementary relation. The activity of the individual is not caused by a stimulus—the sight of a pen— but, conversely, *itself is what brings forth the stimulus*. For only based on the

---

[15] Thus, electromyographic experiments have shown that the grip force required to lift objects is pre-adjusted according to their anticipated weight (Johansson & Westling 1988).

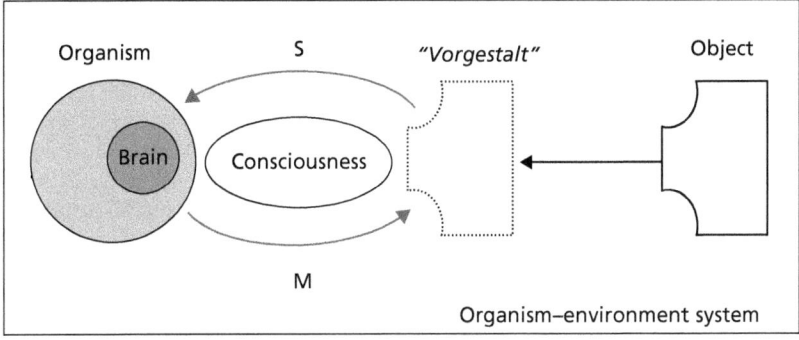

Figure 4.5 "Open loop": anticipation of the object in the functional cycle within the organism–environment system (S = sensory system, M = motor system). Organism and environment form an overarching system, in which the organism always already has outlined suitable objects as preconceptions (*Vorgestalten*) of his sensory and motor systems (suggested by suitable "niches" in the object to which perception and movement can dock). Thus, organism, sensory and motor system form an *"open loop,"* into which real objects can be *inserted*. Conscious perception is not located within the organism, but is based on the current individual connection of organism and object, in which the open loop is closed (see also the later text).

capacity formed in the organism to perform this activity, the stimulus becomes its trigger; only then is it perceived *as* a stimulus. The explanation of a behavior thus does not lie in the specification of the relevant trigger—this only occasions and does not cause it—but in comprehending the *shared history of organism and environment*, which has created the preconditions for the present functional cycle.

So a series of preconditions forms the "open loop," into which the stimulus, the pen seen, is only inserted in order to close it (Figure 4.5). The stimulus does not, as in the linear model, exist as an independent physical event which causes perception; rather, the organism takes it as an occasion to actualize its perceiving. Like a missing puzzle piece it steps into an open position in the already existing functional cycle that has been formed in the overarching system of organism and environment. As such, perception is not a linear, but a circular process, always already connected with pre-existing know-how and possible action.

The dichotomy of organism and environment had already been questioned by American pragmatism, in particular by John Dewey. In his classical paper "The reflex arc concept in psychology" (1896), Dewey criticized the separation of stimulus and reaction in two units of action. It is not the stimulus as such that works; instead, the active organism grasps and interprets the stimulus as an occasion for possible action:

> Upon analysis we find that we begin not with a sensory stimulus, but with a sensorimotor coordination [...] and that in a certain sense it is the movement, which is primary, and the sensation which is secondary, the movement of the body, head and eye muscles determining the quality of what is experienced. In other words, the real beginning is with the act of seeing: it is looking, and not a sensation of light ... [In audition] the sound is not a mere stimulus, or mere sensation; it again is an act [...]. It is just as true to say that the sensation of sound arises from a motor response as that the running away is a response to the sound. (Dewey 1896, 359)

Perception and action are thus connected in circular loops; that is to say, cognition may not be described apart from action, but arises in the course of the sensorimotor coupling of organism and environment. This conception corresponds to the dynamic or enactive sensorimotor theory of perception put forward mainly by O'Regan and Noë (2001), Hurley and Noë (2003), and Thompson (2005). On this model, perception is not an internal state in the brain, but a skilful activity of the organism, which is shaped by (1) *sensory variance contingent on movement* and (2) the *implicit, practical knowledge of an object*.

1. With each eye movement the sensory stimulation of the retina is changed in a certain way, similarly so if the body moves back and forth etc. Without this feedback between movement and perception, we could not recognize anything: even when fixating an object, the eyeballs are constantly, even if imperceptibly, moving.[16] In the case of touch, sensorimotor interaction is even more obvious: the feeling hand determines what is felt and, conversely, the felt object directs the movement. The familiar patterns of dependence between sensory stimulation and bodily action—the moving eyes, the grasping hand—enable the skilful exploration of the environment in which perception consists (O'Regan & Noë 2001).

2. We do not perceive objects neutrally, but always in a context of possible action and significance. As already mentioned (see 1.3.2), we perceive a house as a spatial whole, implicitly anticipating its back side, although we always see only one aspect of it. We can go towards or around objects, they are "ready-to-hand" and afford certain possibilities: the stairs to be climbed, the apple to be eaten, the pen to be used for writing, etc. This corresponds to the connection of sensory and motor systems at the neuronal level: if one, for

---

[16] If these movements (microsaccades) are counteracted, thus completely stabilizing the retina image, our perception of stationary objects even fades completely, due to neural adaptation (Martinez-Conde et al. 2006). Hence, there is no visual perception without movement.

instance, sees a tool, the same neurons are activated in the premotor cortex that would also be activated in *using* the tool (Grafton et al. 1997, Gallese & Umiltà 2002). Perception always calls up patterns of interaction that had been established in prior experience with the object. In other words: *to recognize an object means to know how to use it.*

As I have already indicated in Chapter 1, enactive approaches to cognition generally regard perception as a process of active *sense-making*: by interacting and coping with the environment—moving their head and eyes, touching a surface, walking towards a goal, grasping a fruit, etc.—living beings make sense of their surroundings (Varela et al. 1991, Thompson 2005, 2007). In these interactions, the brain functions as the *organ of mediation*: through its networks it provides open loops that are closed by appropriate elements in the environment and become actualized functional cycles (Figure 4.6).

For the emergence of these open loops, the already mentioned neuronal structures of implicit memory (see 3.3.4) are of crucial importance. Due to its plasticity, the brain is capable of transforming repeatedly occurring links between organism and object into sensorimotor couplings that are the basis of the corresponding functions. Thus the brain becomes a *matrix for all possible preconceptions* (*Vorgestalten*) or, in other words, an "*organ of possibilities*," of capacities or potentials.

An impressive example of this action-dependent plasticity is provided by work in music psychology by Bangert and Altenmüller (2003): if one practices a certain series of notes on a piano, over time, a coupling between tonal and motor sequence is established. Subsequently it is then sufficient to merely hear the melody to activate in the brain the corresponding movement pattern of the fingers. Conversely, if one plays the learnt sequence on a silent piano, the corresponding tones are imagined or "heard." This sensorimotor coupling can already

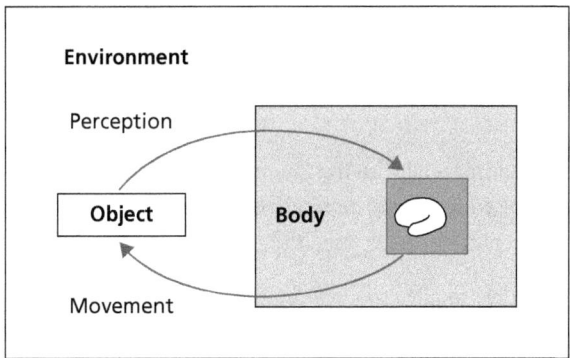

Figure 4.6 Sensorimotor functional cycle.

be demonstrated after 20 minutes of practice by means of electroencephalography (EEG) showing synchronous activity in the temporal and parietal lobes (as the correlates of acoustic and motor processing). The melody is now heard in an embodied way: it involves imagining what it would be like to play it. Or in other words, it evokes precisely the motor sequences in the brain which could produce it. That is to say, the melody heard has been imbued with the *additional significance* of a motor sequence and vice versa.

Even more in the case of experienced pianists, the cortical regions for listening to music have constantly been synchronized with those for playing it, so they virtually move their fingers when they hear a melody. Conversely, they anticipate the sound before their fingers produce it on the piano (Münte et al. 2002). Further neural correlates of motor patterns are found in the cerebellum and the basal ganglia, which enable elaborate motor sequences in dealing with an instrument ("sequence memory"; Ennen 2003).

So the organism of the pianist is prepared to close the open loops, that is, to link listening, playing, and imagined inner melody in enaction. In so doing, the player, the instrument, and the music become part of one dynamical process, the components of which are not separable. We can no longer say "This is the pianist here and that is the piano there," but would rather have to talk of "the-pianist-with-his-piano-in-the-soundscape."

> To give another example of such adaptation: As we have already seen in section 1.2.2, a tool becomes an extension of the hand in both a physical and a perceptual sense. Neural correlates of this extension of the body schema have also been found in studies by Iriki et al. (1996) and Hihara et al. (2006) who trained macaque monkeys over 3 weeks to retrieve distant objects by using a rake. As a result, somatosensory neurons in the monkeys' parietal cortex became bimodally active and responded also to visual stimuli projected from the visual brain centers. These neurons thus extended their receptive fields to include the entire length of the rake and to cover the expanded accessible space. In this way, the rake was literally assimilated as an extension of the monkey's intermodal body schema, or as part of his felt body.

These findings chime well with the intentionality of the lived body, as analyzed by Merleau-Ponty, which is not dependent on conscious planning: "as a system of motor or perceptual powers, our body is not an object for an 'I think,' it is a grouping of lived-through meanings which moves towards its equilibrium" through connecting itself with every situation (Merleau-Ponty 1962, 177). The brain, by forming the substrate of memory, certainly provides a key component of this whole, however, without localizability of function as such. The measurable activation of certain brain regions during enaction of a function entails only—to use an image—that the electric circuit has been closed, leading to increased activity in the region. But what is needed for the "current" is the entire circuit, that is, an active organism situated in a complementary environment.

A consequence of this conception affects the notion of *representation*. As already mentioned, it is based on a principled division between organism and environment and, as such, on a theory of perception, which does not connect us to the world, but merely with internal images or constructs. But brain states as such do not have representational contents, they cannot "describe" the world; for they merely *participate* in the functional cycles from which those contents result. Seen in isolation, the brain state is merely a *fragment* of the entire functional circuit, which assigns significance to certain elements of the environment, and respectively contains the open loops, into which these fit. Without this circuit every neuronal state would remain a meaningless process of excitation. Thus, the basis for representation in the sense of mirroring the external world in the brain crumbles. The notion of representation, already criticized under the aspect of subjectivity in 2.1.2, also proves inadequate from a functional-biological perspective.[17]

The concept of representation gets its seeming plausibility not least from our ability to imagine objects, to remember or anticipate them. Different from perception as a *presentation*, the former means re-presenting something which is *not present* (for which there are certainly also neuronal correlates). But this specific human ability to represent one's own experiences as such, detached from the environment, is secondary to the lived relations to the environment. Perception does not serve to "represent information," which is then viewed by the subject as if on a movie screen, in order to draw conclusions for its actions. It does not provide images or models but *opens up action possibilities* for a moving, embodied, and situated creature. Moreover, in the skilful dealing with objects there is no need for a representation of the object or goal as such, for the lived body itself possesses suitable dispositions in order to establish functional coherence with the environment: "The world is its own best model" (Brooks 1991).

An experienced tennis player, for example, need not consciously apprehend the ball, the position of the racket, or the direction of play vis-à-vis the other player. Much rather, her arm spontaneously finds the right position and punch

---

[17] Edelman, too, critically remarks "that much of cognitive psychology is ill-founded. There are no functional states that can be uniquely equated with defined or coded states in individual brains and no processes that can be equated with the execution of algorithms. [...] Intentionality and will, in this view, both depend *on local contexts in the environment, the body, and the brain, but they can selectively arise only through such interactions*, and not as precisely defined computations" (Edelman 2004, 111; emphasis added).

in completing the prefigured *Vorgestalt* of a "good blow," which results from the conjunction of the approaching ball, the field, the net, and the running opponent. The movement fulfills exactly what is needed to realize this highly complex gestalt, but this gestalt is not what one can represent as a goal in consciousness and translate into calculated movement (Dreyfus 2002). Insofar as the lived body is linked to the environment, it possesses implicit knowledge that is not explicable and, as such, also cannot be represented. No internal reduplication of the environment would enable our bodies to react to its changes and demands with such immediate and dynamic coherence.

The brain is often seen as the "conductor" of the organism. But, in fact, its role does not consist in determining behavioral routines, but in providing variable sensorimotor schemata for the interaction of organism and environment and constantly adapting them. It modulates the respective movement dependent on the constant feedback from the organism in the field, on muscle mass, muscle tension, gravity, resistance, and so on. Thus, the brain functions as a flexible control unit and not as a store of fixed and complete motor programs. We find a parallel in genetics, where observing the interaction of nucleus, periphery, and environment of a cell has increasingly taken the place of an alleged determinism of the genes. Matters are similar in the case of the relation between brain, body, and environment, which continually regulate one another. This process does not proceed linearly and hierarchically from a control center to the periphery, but as a circular dynamism between organism and its complementary environment. Thus, the brain is not the conductor of the body; rather, it is like a musician in a group of jazz musicians jointly improvising on the basis of certain chords.

4.2.2 Consciousness as integral

In the linear model (Figure 4.4), conscious experience is located at a certain place within the causal chain of stimulus, afference, central nervous system, efference, and muscle activation, namely in the brain. Of course, the fewer gaps remain in this detected chain of physical processes, down to the complexities of neuronal networks and synapses, the more consciousness is marginalized and seen as an epiphenomenon. Since it is not a link in the chain itself, it seems to remain causally inefficacious, due to the axiom of the causal closure of the physical. This does not change if conscious experience is identified with certain brain states. In any case, it remains an internal space separated from the real world.

In the ecological model (Figure 4.5 and Figure 4.6), however, consciousness cannot be localized in any place within the organism. It is neither the product of certain neuronal processes nor is it identical with them. Instead, the continuous resonance between brain and body, as we have seen, is constitutive for the embodied background consciousness tied to all experience (see 4.1.2). What is further needed is resonance between the organism and the current situation. Conscious experience in its full sense *only arises in the overarching system of organism and environment*, on the basis of the dynamic interaction of various components, of which the central and peripheral organs are parts equally as much as the suitable "counterparts" in the environment. Perception and action do not come about without their complementary objects. Thoughts, feelings, and wishes receive their meaning only in virtue of their relation to possible objects and other persons. As such, the foundation of the psychic is not the brain itself, but much rather an overarching *process of life*, in which the brain is, of course, centrally involved—namely as the mediating organ for the dynamical relations of the organism to its natural and social environment. This process of life is not restricted to the physical organism, let alone the brain. It constantly transcends the boundaries of the body and involves the complementary potentials of the environment.

The thought experiment of a disembodied "Brain in a Vat," which could, if suitably nourished and stimulated by a sophisticated computer program, produce an illusory experience that cannot be distinguished from our everyday reality, can be refuted for the following reasons: in order to create the feeling of bodily existence as well as basic mood states, the device would first need to exactly simulate the homeostatic self-regulation of the organism, coupled with the vat-brain, and mutually regulating each other. Such a coupled system would already resemble a body much more than a vat. Moreover, in order to then produce the illusion of bodily being in the world, the device would have to construct all continual interactions and sensorimotor feedback loops between brain, body, and environment—which would only be possible through a *mobile* body device organized in form of sensors and actors. In the end, the experiment would require an apparatus that is nothing else than *a quasi-living body interacting with the environment.*[18]

---

[18] See Gallagher and Zahavi (2008, 131) and Cosmelli and Thompson (2010) for an extensive refutation of the brain-in-a-vat argument, with the conclusion: "any vat capable of performing the necessary functions will have to be a surrogate body that both regulates and is regulated by the nervous system. In other words, the vat will have to exhibit a level of complexity at least as high as that of a living body" (2010, 378). Here, defenders of brain-centrism might take refuge in Searle's argument: the brain-in-a-vat thought experiment is imaginable, he says, simply because "we are in fact brains in vats," nothing else (Searle 2015, 77). Certainly, the brain is housed in the body, but it does not matter where it receives its input from—consciousness is in the brain. Against this unquestioned conviction, I have already shown that basic self-affection is not only a *causal derivative* of internal body processes signaled to the brain, but *constitutively based* on the ongoing regulatory processes between brain and body (4.1.3). The same applies to the sensorimotor loops as well.

Only as an organ of a living being can the brain mediate subjective experience. Therefore already the presupposition of the thought experiment is misguided: *there is not something within ourselves that perceives, feels, or thinks*—neither a Cartesian ghost nor a disembodied brain. Consciousness is not an inner state or a "tunnel," but an enaction of life of an animate, living being.

To that extent, the brain as such does indeed not contain more consciousness than, for example, the hands or feet; *only as a whole is the living creature conscious, does it perceive or act*. The brain is indispensable for the emergence of consciousness, since, in it, all circular processes come together, are linked, modified, facilitated, or selected—just like railway tracks are joined and the traffic is coordinated in a central train station. If the station or parts of it are destroyed, the railway service of course breaks down, whereas individual tracks in the periphery can be shut down without affecting the main service. But, to expand on the metaphor, *the traffic is neither produced by the central station nor is it localized there*. Rather it uses the system of tracks with its many switches, junctions, and its central coordination unit in the main station in order to enable smooth transport processes. So even if in the central station we find a higher density of tracks and trains (this would be the equivalent of measurable neural activations in the brain while performing a certain action)—the train traffic remains tied to the entire system of tracks. Analogously, *conscious activity forms, in each case, an "integral" of the current relations between brain, organism, and environment*.

So the brain is the central, but still only a necessary condition for the emergence of consciousness. A *sufficient* condition is only provided by the existence of a living organism with a peripheral and a central nervous system that is constantly interacting with the whole body and with its environment. A certain brain state is a necessary condition for being in an experiential state, but to which kind of experience this brain state corresponds is not determined by its own microstructure, but only by its particular relations to the body and environment. *Thus, there cannot be circumscribed and in itself sufficient "neural correlates of consciousness"* (NCCs). Only as integrated into the actual "bodyscape" and embedded in the ecological relations with the environment can brain states become "carriers" of conscious experience. The brain is not a standardized Turing machine with fixed modules or NCCs, but an extremely plastic, embodied, and embedded organ, or rather: *a dynamical process*. Looking for meaning or consciousness in the brain as such is therefore not merely a category mistake, but also lacks any biological basis.[19]

---

[19] At this point, *dream consciousness* might be raised as a possible objection. Here the organism's relation to the environment is indeed highly reduced through sensory decoupling

These claims may hardly seem plausible to some—too deeply entrenched has the reification of conscious experience become in our everyday understanding of science and ourselves. Should it then be the case that the brain is not the locus of consciousness? Here everything depends on whether we can overcome our dualist intuitions of an inner sphere of consciousness, which we must then ascribe to the brain, or whether we can instead see conscious experience and action as an enaction of life, as "*activity-of-a-living-organism-related-to-its-environment*." As soon as individual elements are taken out of this long expression, only fragments remain, from which the process of life can no longer be pieced together again. Mental processes are living processes; whether affective or cognitive, conscious or unconscious, they "loop through the physical, social and cultural environment in which the body is embedded" (Silverstein 2006, 209).

In order to clarify, let us take a look back: in the first part it was shown how the Cartesian dualism of mind and body lives on in brain research, that is, as a combination of subjective idealism, on the one hand, and physicalist materialism, on the other. On this is based the neuroconstructivist epistemology with its central notion of the internal representation of the world in the brain. Such a position objectifies experience and ultimately leads to neuro-solipsism. Its basic problem lies in the exclusion of life and the resulting short circuit of consciousness and brain. In contrast, I have proposed the alternative conception of a "biological aspect-duality," according to which both psychological and physiological processes have the *living creature* as their carrier. But under this

---

and motor inhibition. Is dreaming then restricted to the brain after all? For the following reasons, this view can be rejected:

1. Even during dreaming, the multiple interactive cycles between brain and body remain effective, underlying the persisting bodily awareness, the basic mood states, as well as the intensive emotions that are felt while dreaming.
2. Thus, the dreamer's entire body is wholly involved in the dream state, discernible also by the eye movements in rapid eye movement sleep, changes of breathing and heart rate, tension or movements of limbs, unintended verbal utterances, and the like.
3. The contents of dreaming are derived from the dreamer's lifeworld, in particular his intersubjective relationships, and gain their meaning only through their intentional relations to it.
4. Only the dreamer himself can report his dreams afterwards, whereas there are only indications for a dream state in the brain.

Dreaming is therefore not an activity of the brain, but an enactment of an embodied person: "the body, as perceptual focusing in general, as relation to dramatic situations is the subject of dreams, rather than the 'imaging consciousness'" (Merleau-Ponty 2010, 148). On a recent functionally embodied concept of dreaming, see also Windt (2015).

precondition neither aspect remains fixed: they rather intertwine with one another insofar as they both have a fundamentally relational structure:

- The subjective phenomena leave the inner sphere and become experiences of an *embodied subject belonging to the world*. As the body-as-subject, the organism is the medium of "being-towards-the-world." At the same time it is the "nature that we are," in contrast to a nature seen from the objectivizing perspective of an external observer.
- The physiological processes, in turn, are not describable in a physicalist linear causality, but only as circular relations of organism and environment, which together form a self-sustaining and self-reproducing *system*. The living body is not confined to the skin—it is always already beyond its boundaries, a body-in-relation.

That I can write a letter, that is, translate non-spatial thoughts by means of body movements into spatial-material signs, is based, on the one hand, on the fact that my thoughts themselves already are enactments of life. They are not purely "mental events," but thoughts of an embodied and living subject. On the other hand, my acts of writing are based on the fact that the physiological processes at its root are not purely of a physical nature (i.e., described by physics). Rather they are dynamical interrelations of the organism with the environment, or more precisely: reconfigurations of the entire system of organism and environment.

This perspective does not do away with the difference between phenomenological and physiological description. Both forms of description are complementary, that is, we cannot reduce them to one another. The *personalistic attitude* anchored in the lifeworld and the *objectifying attitude*, on which they are based respectively, stand in a relation of mutual obfuscation, just like the aspects that we grasp and describe by means of these attitudes. If I am talking about light waves emanating from the surface of a tree, about stimuli on the retina, or neuronal processing in the brain, my perception itself remains hidden. Conversely, perception knows little of waves or neurons. But the conception defended here establishes a structural similarity between both systems of description, in that it

1. Conceives of the phenomenal level as embodied, that is, as the intrinsic relation of a bodily subject to the world;
2. Describes the physiological level as systemic or ecological, that is, understands the relation of brain, organism, and environment in such a manner that it can, in the first place, become the foundation of the phenomenal relation of the subject to the world. This is particularly enabled by embodied, embedded and enactive approaches to cognition.

Thus, the *phenomenology of bodily being-towards-the-world* and the *ecology of the organism-in-its-environment* correspond to one another, however, without being identifiable with one another. This is also captured in the polarity of lived body and physical body, or *Leib* and *Körper*.

Counter to the project of "naturalizing the mind," the dual-aspect conception conceives of both the mental and the physiological as essentially *living*. As such, it establishes a joint point of reference for psychological and neurophysiological description: insofar as both refer to living creatures in the environment, they may not have the same *intension* (meaning), but they do have the same *extension* (reference). In order to do so, the living creature respectively the person has to be seen under the complementary aspects of *Körper* and *Leib*, of physiological processes and integral enactments of life. Then it at least in principle becomes possible, beyond the merely external correlation of neuronal and conscious processes, to also understand the *incorporation of experience*. It is the decisive precondition for the development and differentiation of the conscious functions and will be our topic of concern in the next section.

4.2.3 Neuroplasticity and the incorporation of experience

The brain is embedded in the organism and linked to the environment by its various, mainly sensorimotor, interactions. This only becomes comprehensible under the aspect of development: the human brain is not only the most complex, but also the most adaptable organ that we know of. Due to its high degree of plasticity, it can incorporate the learning history of the organism since its first intrauterine stages of life; it epigenetically develops into an organ that is complementarily structured with regard to the individual's environment. Thus, all our experiences, perceptions, and interactions with the environment continually modify our neural structures throughout our lives.

This *incorporation* of experience in memory structures is based on the functional activity of the brain constantly changing its own microstructures. Other than in the case of a computer, here function ("software") and structure ("hardware") cannot be distinguished. This entails a *spiral-shaped relation* of brain and environment, which mutually change one another (see 3.3.4, Figure 3.6). The environment shapes the neuronal structures, but these, in turn, influence future stimulus processing and, as such, the perception of the environment. This *reciprocity or process and structure*, too, has already been seen by Ludwig Feuerbach:

> Only through thinking is the brain formed as an organ of thought and adapted to thinking; it is modified and determined through the habit of thinking this or that, one way or another […] But only through the fully shaped organ of thought thinking itself becomes

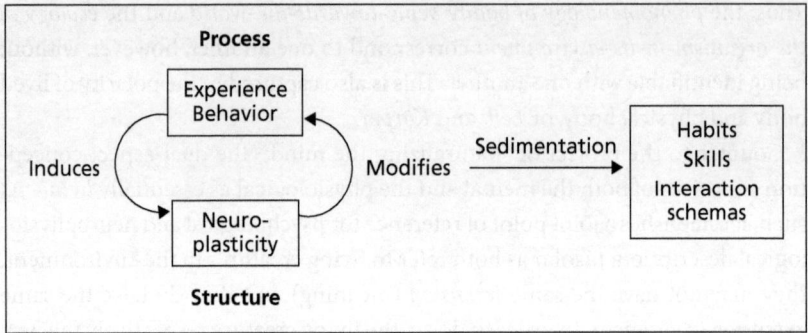

Figure 4.7 Reciprocity of process and structure: learning as transformation of experience or behavior into organic dispositions.

> erudite, skilled, secure [...] What was effect, becomes cause, and vice versa. (Feuerbach 1985b, 201–202, own translation)

Figure 4.7 illustrates this once again: every experience and behavior is sedimented in the plastic, neurally anchored memory of the developing organism. From this sedimented memory, in turn, results a continually changing experiencing and acting. Over time, experiences become organic dispositions, habits, and schemata of interaction. Human beings "teach their organs" and thus shape their own organic being as well as their abilities.[20]

Let us take a closer look at this and in so doing, first of all, bring to mind the neuronal basis of plasticity.

At first, in its early embryonic stages, the brain is determined genetically in the development of its rough neuronal structures. Its further maturation has been described by Edelman in his theory of "neural Darwinism" as an intra-individual evolutionary process (Edelman 1987, Edelman & Tononi 2000, 79–84): a significant part of neurons, of which there is, at first, a surplus, dies in the late embryonic stages and in the first months after birth due to lack of use ("apoptosis"). Only those neurons activated in constant interaction with the environment are selected and survive. Similarly, in early childhood, twice as many synapses are formed than are ultimately needed for the further epigenetic formation of the brain, a selection process that is also termed "pruning."[21] The persisting neuronal structures are thus the result of a *selection from an excess of*

---

[20] This was already Goethe's insight: "The animals are taught by their organs, said the Ancients; to this I add: humans equally, but they have the advantage to teach their organs in turn" (Goethe, letter to Humboldt, 18 March 1832).

[21] See Markowitsch and Welzer (2009, 115). For a detailed account of this early selection process in the visual cortex, see also Sur and Rubenstein (2005).

possibilities which the brain's growth provides. Here we find again the principle of *formative or downward causality* which we have considered in section 3.3.1 within the organism: the child's embodied experiences, as situated in the environment, constitute the superordinate process, whose recurrent patterns select the neuronal links on the micro-level, just as hemoglobin radically reduces the behavioral options of iron molecules.

This epigenetic, experience-based process of selection shapes the remaining anatomical neural network up to the end of the third year of life (Markowitsch & Welzer 2009, 87). Its microstructure, however, remains alterable throughout the entire lifespan in the form of synaptic sensitivity and network patterning, regulated by changes of gene expression, signal transmission, and receptor density. Dendrites, too, can, to some extent, still develop or recede (Serres 2001, Lee et al. 2006), and even the formation of new neurons in the adult hippocampus is possible (Björklund & Lindvall 2000). Just like muscles grow through exercise, but atrophy without activity, so too do neural networks grow or degenerate, depending on the execution of the superordinate function ("use it or lose it"). So in the adult brain, too, there is a constant growth and elimination of neural networks and patterns (experience-dependent-plasticity).

>The basic rule of these synaptic adaption processes was first formulated by Hebb in 1949. If at first the pre-synaptic neuron fires and then the post-synaptic one, the connection is strengthened—the term here is "long-term potentiation." In the case of asynchronous activation, however, the connection is weakened. So the joint activation of neurons that correspond with current stimulus patterns in the environment provides a crucial guiding force of brain development. Interestingly, experiences tied to intensive desires and emotions have a particularly structuring force, since they influence higher cholinergic and dopaminergic modulation systems in the mes- and diencephalon (Kilgard & Merzenich 1998, Bao et al. 2001); in other words, emotions function as a particular order parameter which amplifies learning processes.

The significance of insights into neuronal plasticity can hardly be overestimated. They reveal that *it is the interaction with the environment that creates the necessary conditions of experiencing this environment*. This applies in particular to the maturation of the cortex: "The higher brain, namely neocortex, is born largely tabula rasa, and all functions, including vision [...] are programmed into equipotential brain tissues" (Panksepp 2012, 8). This conclusion is based, among others, on findings demonstrating that the visual cortex of mammals is largely developmentally shaped, not genetically dictated.

Thus, Mringanka Sur and his group could induce in newborn ferrets a far-reaching cortical reorganization (Melcher et al. 2000, Sur & Rubenstein 2005). They severed the ferrets' optic nerves, so that the stump grew together with the part of the thalamus that usually transmits impulses from the auditory nerve to the auditory cortex. Now visual stimuli reached a brain region that usually processes acoustic signals. But surprisingly the brain adapted to these new stimuli: in the course of several weeks, the *auditory cortex became a visual cortex*; it even developed orientation-selective cells that are characteristic of the visual cortex, so that the ferrets were finally capable of seeing with the respective eye (even if not quite so well).[22] As it turns out, it ultimately depends on the sensorimotor interaction and its specific patterns of neural excitation, which tasks a cortex region ultimately takes on. This may be expressed by the following principle: it is not the brain that creates a function, but conversely, *the function creates the conditions of its own realization, or its appropriate cerebral organ*.[23]

Similar cortical reorganizations can be observed after injuries or accidents. The brain's capacity for the restitution even of complex functions is impressive: language and orientation functions, even after large brain lesions, can be taken over by other structures, for instance, by the other hemisphere. These feats of restitution, once again, bear out the primacy of function over structure. But also in normal learning processes, cortex areas often change in macroscopically measurable ways (Elbert & Rockstroh 2004). If a person learns to play the violin, the motor representation of his left hand increases in size and becomes more differentiated, since it needs to be moved in a more complex manner (Elbert et al. 1995). Taxi drivers, who need to carry out particularly complex orientation tasks, show a significant growth of the hippocampus (Maguire et al. 2000).[24]

---

[22] Strictly speaking, the sensory stimulation *as depending on the organism's own movement* has to be considered here, that is, the sensorimotor functional cycle: the co-variation of sensory input and motor output enters into the formation of neural patterns. What turned the ferret's auditory cortex into a visual cortex was not only its connection to the retina, but its integration into the superordinate sensorimotor dynamic that is characteristic for visual function (see O'Regan & Noe 2001). Learning to see presupposes *mobility*, as the experiment of Held and Hain with newborn kittens has shown (see 1.2.1).

[23] As Sur and Rubenstein summarize the findings, "brain pathways and cortical regions that are established during early development depend on their inputs for physiological and behavioural instruction" (Sur & Rubinstein 2005, 809). Neuroanatomist Brodmann, in his "Comparative doctrine of localization in the cerebral cortex" (1909), already expressed the Aristotelian idea that "*the function creates its cerebral organ.*"

[24] It should not be forgotten that neuroplasticity may have negative effects as well, for example through maladaptation resulting in phantom pain or pain memory. Constant or intensive states of pain, as in chronic back pain, lead to an enlargement of the corresponding somatosensory cortical area, so that even normally non-painful stimuli may reach above-threshold levels and trigger considerable pain (Flor et al. 1997).

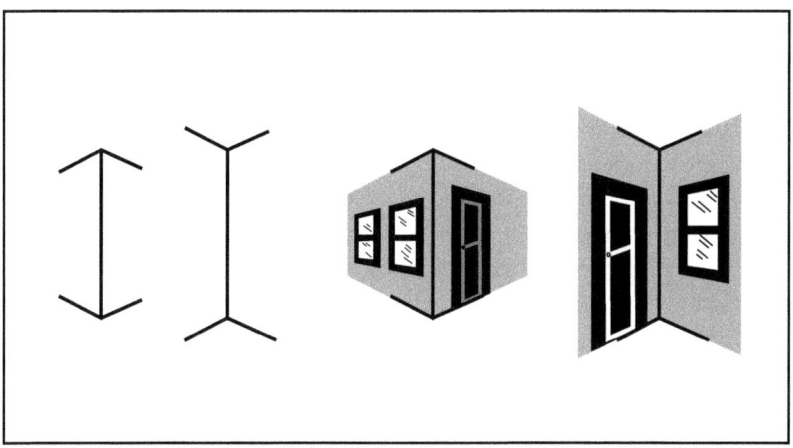

Figure 4.8 The Müller-Lyer illusion. (Left): the distance between the arrowheads appears to be shorter in the left figure, although both are in fact equal. (Right): basis of the illusion in the adaptation to cuboid objects.

This opens up a field of far-reaching influences of culture on brain development which may even concern the fundamental organization of perception. One would, for example, assume that visual illusions are due to innate visual processing in the brain. However, it turned out that the well-known Müller-Lyer illusion (Figure 4.8) is not equally effective in all cultures. Following Gregory's (1966) widely accepted explanation, the illusion is based on the adaptation of the sensorimotor system to the spatiality of cuboid objects (Figure 4.8, right) whose edges either protrude (houses, cupboards) or recede (indoor spaces). The resulting spatial proportions are adjusted for by the brain during early development. Such cuboid structures, however, are characteristic of *urban cultures* and rarely found in natural environments. As it turned out, in members of African round hut cultures the Müller-Lyer illusion in fact does not occur, or at least much less frequently (Segall 1963, Deregowsky 1973). This shows impressively how the cultural environment determines the development of the brain, even including the organization of spatial vision.

The basic principle of plasticity is the adaption of the brain to the interaction with the environment in the sense of *optimal coherence*. In this manner, the sensorimotor and neuronal structures become *media* that enable a relation to the world as immediate and free of interference as possible. If one, for instance, wears a new pair of glasses, signal processing in the visual cortex adjusts within a few days so that an undistorted perception becomes possible again and the glasses are no longer noticed. This adjustment goes so far that, if carrying prism goggles, which turn the image of the environment upside down, after a while visual perception, despite wearing the glasses, inverts itself again (Kohler 1951). In so doing, it follows the actual body position represented in the vestibular organ. Here the brain re-establishes *intermodal coherence* between the senses (in this case the sense of balance proves to be "stronger").[25]

Thus, neural plasticity ensures that the functional cycles of organism and environment are as permeable or *transparent* as possible. It enables the "mediated immediacy" (see 1.3.2) of the relation between lived body and environment, in which the brain as the mediating organ itself remains invisible. Thus, it enables us to be immediately directed towards the world, its objects, and meaningful contents. If, for example, blind persons learn Braille, their fingertips feel the raised dots, while they try to grasp the shape and the meaning of the letters, until they can immediately read the dots *as writing*. In the process, the region in the sensory cortex related to the fingertips of their index fingers becomes enlarged (Pascual-Leone & Torres 1993). At the same time, new connections to the language center, specifically Wernicke's area are established. Thus the sensations of touch are imbued with new meanings—similar to the case of the pianist already mentioned (see 4.2.1)—and become transparent for words and language.

In Chapter 3, the acquired capacities of the organism were discussed in relation to *implicit memory*. Now we can also understand this from a neurobiological point of view: in the establishment of neuronal networks, repeatedly occurring sensorimotor interactions are incorporated and become an *implicit form of know-how or knowledge*.

1. Let us, at first, take a look at *motor habitualization*: someone who plays the piano or learns to write with a keyboard, at first, explicitly assigns to each key a sound or a letter, in order to then gradually habituate the fingers to these associations, that is, to *forget* them again. The implicit ability is now "in the hands" and one can no longer say how one does what one does. The body has incorporated the instrument into its body schema[26] and this immediate coupling operates much more efficiently than any conscious imagination or planning, so that we can focus on higher-order goals while enacting it. The intention in typing is directed towards the words, and the appropriate fingers move of their own accord. In a similar manner, a pianist can, beyond the movement of their fingers, focus on the music and "listen to himself

---

[25] For more recent research on the brain's adaption to prism goggles with right-left inversion, see Barton et al. (2009).

[26] "It is literally true that the subject who learns to type incorporates the key-bank space into his bodily space" (Merleau-Ponty 1962, 167).

while playing." Correspondingly, imaging studies show that experienced musicians and athletes, compared to beginners, have significantly *reduced* brain activation in exerting their skills, especially in the basal ganglia and other subcortical regions (Milton et al. 2007). Obviously, the involved motor networks are organized more efficiently due to extended practice, or in other words: optimal coherence needs less energy.

2. A parallel process in *perception* is, for instance, exemplified by reading. When children learn to read, they gradually connect the letters, initially known only individually, to word gestalts. When their brains have established the relevant connections, the children recognize the words in an instant, up to the point when they immediately grasp the meaning of a sentence in fluent reading. Here, too, forgetting the individual elements enables the intentionality of perception: via the letters, which initially were individual particulars, the child now focuses on the gestalt and the meaning of the words. Of course, our Braille example has also made clear that reading is also possible with sensing fingers. So perception, movement, and recognition are not, in principle, separable from one another. All perception is based on the active recognition of gestalts that we have learnt and that have become part of our implicit knowledge.

Implicit know-how only becomes accessible in the process of action—for this reason one speaks of "procedural memory" (Schacter & Tulving 1994). Only by means of my fingers and the keyboard can I realize the learnt piano piece, and not by means of an "image" of movement, as though my memory were contained in my mind or brain in isolation. The ability to play the piano only exists with reference to the entire closed functional circuit, within which it is actualized. As I have already pointed out in section 3.3.4, memory, as an acquired ability and disposition of a living being, is not to be localized exclusively in the brain. This holds equally true for skilled movement as for skilled perception: nimble hands know how to tie a knot in a manner one could not explain in words. An experienced doctor's ear recognizes cardiac defects, whereas a student that has read everything about them, only perceives dull thumping. *Memory and intelligence are always already extended over the body and the environment.* This is also the precondition of all external symbol and memory systems that have been developed by humans, from wax tablets to libraries and the Internet.

4.2.4 Transformation and transparency: the brain as a resonance organ

Let us now more principally describe the discussed phenomena of implicit memory and habitualization, as enabled by brain plasticity. This will lead us to a fundamental concept of the brain as a *resonance organ*, which mediates the sensorimotor interaction with the environment as the basis of our conscious being-in-the-world.

The basic structure of these phenomena obviously consists in the fact that, in them, individual elements of perception or movement are integrated into wholes, that is, *they are transformed into gestalts*. The intention is, via A_1, A_2, A_3, ... A_n, directed towards the whole B (as via the letters to the word). Individual elements are "overlooked," whereas it is only through them that B is realized. Between A_1, A_2, A_3, ... and A_n there is an *implicit coupling* which, as such, remains unconscious in a manner that the individual elements function as the medium, through which we perceive the whole. So the transformation results in *phenomenal transparency*: the individual elements are merged in the perceived holistic gestalt and therefore recede into the background. In other words, perception has the structure of *mediated immediacy*: individual elements become "transparent" for the gestalt, or put differently, they take on the *meaning* of the gestalt for the perceiving subject.

> This structural analysis follows Polanyi's already-mentioned theory of "*tacit knowledge*" (Polanyi 1967). It is, however, not to be understood in such a manner that whole gestalts need to be first composed from individual elements. Much rather, our perception *primarily* grasps wholes, impressions, and situations, from which we can only isolate individual elements in secondary analysis. So even the explicit synthesis of individual letters in learning to read and write always already presupposes the whole of a word or meaning as acquired in interactive contexts.

The couplings and respectively the resulting gestalts can be of different kinds, for instance, sensory, motor, sensorimotor, sensory-affective. I give a few examples here:

- The connection of stimulus and *vegetative response* in classical conditioning may be conceived as a *coupling of meaning*: if in Pavlov's famous experiment the sound of a bell was repeatedly connected with food intake, this coupling conferred a novel meaning on the stimulus for the dog, namely that of "meat." In principle, similar couplings of stimulus and *affect* are established in conditioned fear or disgust, as well as in operant conditioning.
- In recognizing a face or in understanding a facial expression, we are, via the individual anatomical features, directed towards the characteristic appearance, the expression or physiognomy, without being able to explicitly explain their details in words. The body of others thus becomes *transparent* for their appearance, it becomes *their lived body*.

- In speaking, we are via the individual movements of the larynx, the tongue, the mouth, and the individually articulated syllables directed towards the content of what is said. So here motor enactions become transparent for intentional contents. If this fails, one starts to stammer, that is, the intended whole falls apart into fragments. In learning a foreign language, the pronunciation of syllables alone can be effortful and require focused attention until the tongue speaks "of its own accord." We have also already looked at other examples of sensorimotor coupling, such as playing the piano or reading Braille.

With regard to our bodies, implicit coupling entails that what is "proximal" becomes the medium for what is "distal" (Leder 1990, 113). If we want to drive a nail into the wall, we are not directed towards the movement of the muscles or the hand, but towards the goal of the action. The body, as the medium of all functions, itself becomes transparent; it is hidden in its implicit functioning. Of course, I can focus my attention on my body (for instance, if I hit my thumb with the hammer). But the more proximal the mediating organs are, the more elusive they are to reflective attention. I can still feel the sensing surface of or the muscles of my hands, the seeing eye can be painful, vision can become blurred. But I cannot bring to consciousness the nerves in my sensing hand—much less the brain, the organ which, lying at the heart of all mediations, remains hidden beyond consciousness.

Based on these considerations, we are now in a position to formulate the following basic theses about the higher (primarily cortical) functions of the brain:

1. *Principle of "open loops"*: the central principle of higher brain functions consists in the fact that they form "*open loops*" by means of neuronal couplings, which, in their actualization, lead to a coherence of organism and environment, manifested in conscious experience.

2. *Pattern formation*: open loops originate from a sedimentation of recurring configurations of stimuli or elements, with which the organism interacts, namely in the form of complex *patterns of neuronal excitability*.

3. *Resonance*: if these established patterns now correspond to a current constellation of the environment, they are activated, resulting in an overarching systemic state or *resonance* of brain, organism, and environment, which are manifested in the perceptual and motor gestalts of our conscious experience. Put differently, *the open loops are completed by the interactive coupling of organism and environment*.

4. *Transformation*: here the central function of the brain consists in *transforming constellations of individual elements into wholes* and thus enabling integral perceptions and movements.

5. *Transparency*: the neuronal couplings underlie the implicit couplings by means of which we perceive the configurations of individual elements as

gestalts, that is, what is "proximal" as "distal." Thus, the transformation at the neuronal level enables phenomenal transparency. As a result, the body becomes *a transparent medium of the embodied subject.*

We can illustrate and further elaborate these theses using the perceptual example shown in Figure 4.9.

After a few moments, we can make out a Dalmatian in this pattern of irregular black blots, that is, we no longer see the blots individually or scattered but in their configuration to one another *as* a Dalmatian. How are we to explain this in more detail? Several components are needed for this example of conscious perception:

- Motivated by an *interest* to recognize the picture, we have to actively search for its meaning, applying our *attention* and trying out various possibilities.
- While our eyes are scanning the picture in search for a meaningful gestalt, various image schemas or *Vorgestalten* are implicitly actualized that could match the black-and-white configuration. On the neural level, the "noise" of blots is constantly matched with available neural activation patterns in the visual association cortex, until finally a suitable pattern—the Dalmatian—is extracted and activated.

Figure 4.9 Gestalt formation and pattern resonance.

- The prepared pattern derives from earlier, similar perceptions and the corresponding categorization of Dalmatians. Thus, a neural "open loop," an image schema of a Dalmatian was formed, into which the current object can now be fitted (see Figure 4.5). In dynamical systems theory, such a pattern may be described as an "attractor," that is, a preferred oscillation state at the lowest energy level, at which a chaotic system such as the central nervous system evens out (Haken 1993, Kelso 1995, Cosmelli et al. 2007).
- In the moment of perception, a *resonance* between the neuronal pattern or attractor and the stimulus constellation of the picture arises and, as such, a new *coherence* of the sensory system with the environment.
- By means of pattern resonance or "transient oscillation" of the total system of organism, brain, and picture, the individual stimuli from the picture are *transformed* into an integral gestalt, corresponding to a circular causality of parts and whole.
- The conscious perception of the Dalmatian and, as such, grasping the *meaning* of the picture corresponds to this overarching formation of coherence. Ontologically speaking, however, this perceptual experience may not just be equated with the coherence, since we are always considering here the two *complementary*, not identical sides of the dual aspect.

Let us take a closer look at these matters, which are crucial for a comprehension of the brain, first at *attention*. As we have seen in section 4.2.1, perception is not a passive reception of stimuli. Organisms are motivated to actively "look for" elements of the environment that are significant for their purposes. According to their current motivational state, they anticipate possible objects or events of interest, whose categories selectively direct their attention. Emotions such as desire, longing, interest, curiosity, anxiety, suspicion, and so on, guide this anticipation and selection process. Thus, without brain areas responsible for motivated attention and interest such as the amygdala and anterior cingulate cortex, the processing of afferent stimuli in the visual cortex or occipital lobe alone cannot result in conscious perception (Mack & Rock 1998, Faw 2003). Efferent and afferent processes have to come together.

Generally, attention is a function of an organism's affectively charged purposes and motivations aimed at maintaining homeostatic balance (see 4.1.2: "The Seeking System generates the arousal and energy which awakens our interest in the surrounding world"). As we saw, cortical activity directs this basic affective energy originating in the brainstem and midbrain towards specific objects in the environment (see 4.1.3, Figure 4.2). However, this is not only connected with selective attention and motivated search, but also with an activation and projection of *endogenously produced image schemas* into the perceptual field. When looking for an object in our environment, we already form a rudimentary image that we are prepared to see (of course, it is similar in listening or touching). This anticipatory image projected into the environment enables the subject to detect and see the object more readily when it is presented (Posner and Rothbart 1998, Ellis & Newton 2008, 117–121).

> A personal observation can illustrate this: when searching for porcini mushrooms in the forest, one needs a specific "search image" (*Suchbild*) or image schema of the mushrooms which are only difficult to recognize in the undergrowth for the untrained eye. After a longer intensive search, one can notice how the image schema still constantly arises in imagination even when one has long left the forest. The anticipation may also lead to illusions, for example, when mistaking a shiny leaf for a mushroom, or when expecting to meet an acquaintance and mistaking another person in the crowd for him.

It may be assumed that a continuous endogenous production of Vorgestalten takes place in the brain, corresponding to an unconscious activity of "fantasy systems" under the influence of drive-oriented needs and wishes (Aurell 1989, Brown 2001). The perceptual field is thus searched for desired images of objects, sometimes creating "wishful illusions." Dreaming may be regarded as the uninhibited production of affectively charged images, released from sensorimotor and environmental feedback. On the contrary, if these schemas enter into resonance with presented afferent stimulus patterns, the conscious perception of objects such as the Dalmatian occurs.

Let us turn now to *pattern formation*. As has become clear, one of the main functions of the cortex is "to detect" relevant patterns of stimuli in the environment, which is achieved by the activation of prepared neuronal assemblies specifically fitting to these constellations (Singer 2001, 2009). This detection of image schemas presupposes prior neural pattern formation: if similar constellations are repeatedly presented to the brain, it is capable of extracting regularly correlated features or prototypes from these experiences. This happens, in accordance with Hebb's rule (see 4.2.3), in the form of strengthened coupling between the neurons that react to the correlated features.[27] In this manner, potential patterns of activation are prefigured, which lie at the basis of the perceptual pre-gestalts or image schemas. At the phenomenal level, the *implicit coupling* of individual elements to gestalt patterns, which we have

---

[27] This explanation is supported by research into artificial, computer-based neural networks which adapt to repeatedly presented patterns of stimuli by modifying their synaptic weightings. In other words, they are able to extract prototypes from configurations of stimuli; see Edelman and Tononi (2000, 113–115).

already described, corresponds to the neuronal coupling. Gestalt formation is supported by linguistic or conceptual categorization (e.g., "Dalmatian"). Importantly, the gestalts as well as the more general categories are always primary as against the elements or tokens.[28]

The next and crucial step is the emergence of *resonance*. If the organism is presented with objects or configurations of stimuli in the environment, the neuronal system sets itself in transient oscillation by activating its most suitable patterns, thus establishing a resonance to the environment.

> This basically proceeds according to the principle of parallel processing: take the example of welcoming a friend, simultaneously seeing, hearing, and touching him. This is mediated by the differential processing of the incoming signals in specialized regions of the brain, where they are further analyzed according to certain categories (e.g., color, contour, contrast, intensity, etc.) and then matched with stored neuronal patterns or "attractors." The results of this highly distributed processing in different subsystems have to be integrated at higher levels, thus enabling my perception of *my friend* instead of separated fragments or categories. The problem of how this integration or "binding" is achieved (which is nothing else than Aristotle's problem of the *sensus commmunis*) is still not solved. In any case, the most probable explanation consists in long-range, phase-locking neural synchrony between widely separated regions of the brain. The participating networks start oscillating in phase, thus entering into resonance to each other as well as to the configurations of environmental stimuli (Varela 1996, Thompson & Varela 2001, Uhlhaas et al. 2009).

In the case of the Dalmatian, this means that the individual stimuli (blots) registered by the visual system are matched with visual patterns stored at a higher level. This matching proceeds as an interaction of bottom-up (stimulus-driven) and top-down (concept-driven) processing in the visual system (Mechelli et al. 2004, Beck & Kastner 2009), that is, in a *vertical–circular* relation of analysis and synthesis, which, in the case of ambiguous patterns, takes more time than usual. The ultimate disambiguation of the Dalmatian pattern (or the stabilization of the corresponding attractor) is, according to current research, enabled by synchronized oscillation of the involved neuronal assemblies in the primary and associative visual cortices, particularly within the gamma (30–80 Hz) and

---

[28] In early childhood development, categorizations are acquired in a descending manner, that is, by increasing differentiation of basic categories (e.g., the child learns to distinguish between "mobile and immobile" objects, going on to "animate vs inanimate," then to "human vs animal," "dog vs cat," and finally to "Dalmatian vs sheepdog").

beta (15–30 Hz) frequency band (Rodriguez et al. 1999, Singer 1999, Uhlhaas & Singer 2006, Uhlhaas et al. 2009).[29] Moreover, recent results indicate that this kind of synchronization is highly susceptible to another top-down influence already mentioned earlier, namely to *conscious attention*:

> [F]ocusing attention on a particular stimulus or on a particular modality increases the synchrony of responses in the neuronal networks that process the attended stimulus. (Singer 2009, 193)

This points to a particular role of top-down or vertical causality in establishing intermodal binding and gestalt formation. Varela (1996) had already proposed that every cognitive act corresponds to the formation of a transient spatiotemporal pattern of synchronous neural activity. This seems indeed to be realized by internal resonance:

> For example, one resonant assembly could transiently bind together the different populations of neurons involved in analyzing the shape, color, and motion of a visual object, and this temporary assembly would constitute a neural substrate for the transient perception of a visual object. (Cosmelli et al. 2007, 737)

The internal resonance within the brain is thus embedded in an external resonance, namely between an external stimulus constellation and patterns of neural oscillations. *This superordinate resonance corresponds to the conscious perception of the Dalmatian*—without, of course, us being able to further explain this subjective phenomenon at the purely physical level.

The result is equivalent to a *transformation* of individual elements into holistic perceptual gestalts, that is, the separation of figure and ground, or "gestalt closure."

---

[29] In such tests, subjects are usually shown either upright or upside-down "Mooney figures," that is, black-and-white faces, which are easily perceived as such when presented upright but appear as meaningless figures when upside-down. Neural phase synchrony in the gamma range is observed when the subjects report seeing faces (Rodriguez et al. 1999).

In another study, Melloni et al. (2007) compared the electrophysiological responses related to the processing of visible and invisible (i.e., only subliminally presented) words. Both conditions caused an increase of local gamma oscillations in the EEG, but only *perceived* words induced a long-distance synchronization of these oscillations across widely separated regions of the brain. Thus, the critical process mediating conscious perception seems to be the global increase of phase synchrony in the gamma frequency range.

These matching or "detection" processes may also be accounted for by recent models of "predictive coding" (e.g., Hohwy 2013, p. 13–40). However, as pointed out earlier (see 2.3.2), there is no actual "inference" or "prediction" involved, but rather a stochastic (Bayesian) process of selecting the adequate pattern of neural activation, with ongoing matching against the incoming data. So instead of postulating "hypotheses," "predictions," or even "prediction errors" of the brain, a better notion would be the match or mismatch of neural forward models or *open loops* with the current environment. For an enactive account of "predictive coding" involving the whole brain–body–environment system, see Gallagher and Bower (2017).

It is achieved once there is a sufficient match or resonance of central patterns of neural activity and peripheral patterns of stimulation. The circular interaction of top-down and bottom-up brain functions reflects the insights from gestalt psychology of the last century, which dealt with general structures of coherent perception brought to the sensory material by the organism. Gestalt formation proceeds according to holistic categories such as contour, proximity, similarity, closure, and similar motion of the elements. Above all, it applies the perceptual schemes available from earlier experience. As such, perception always means a "remembering present" or a *re-creation*. It is essentially based on a process of self-organization, that is, the vertical–circular causality of the emotional and sensory system connected with the environment, not on a passive internal imaging or representation.

Let us summarize: first, the organism must actively "look for" the object if it is to consciously perceive it. Only motivated attention produces the efferent processes, in particular the activation and imagery of possible objects, which enable the alignment of endogenous and exogenous sources of perception. Even in the case of the sudden appearance of new stimuli, the attentional system is geared towards their detection and recognition. This recognition is based on stored image schemas or pre-gestalts of object categories.

When presented with a new stimulus configuration in the environment, a circular interaction occurs between afferent and efferent processes: the patterns of stimuli are analyzed through bottom-up and top-down processing in the brain and integrated to higher-level wholes, that is, to synchronized patterns of neuronal activation that form the basis of perceived gestalts. Afferent data, when resonating with emotionally motivated anticipatory imagery, are thus transformed into higher-level transient oscillation, corresponding to conscious perception. The degree of resonance corresponds to the clarity and vivacity of the perceived object and makes it clear that now there is a perception and no longer mere endogenous imagery (Ellis & Newton 2010, 129).

We have now explored the various components and overarching integration which underlie conscious perception. Let us complement this analysis by a brief look at *motor action*.

Similar to perception, action presupposes an endogenous source which I have termed *conation* (3.1.1, 4.1.2): action wells up from within the organism, driven by basic motivational affects. Moreover, action is preceded by an implicit bodily protention of what it will "feel like" to perform an action, or by what Jeannerod (1995, 1997) has called "motor imagery." This is connected not only to a preparation of motor action in the premotor cortex, cerebellum, and basal ganglia, but also to electromyographic activity in the muscles concerned, and even to preparatory increase of heart and respiration rate (Jeannerod & Decety 1995). Motor imagery is also the basis for mentally rehearsing and training a certain motor skill, as in sports. During action imagination, the efferent motor signals are still inhibited at the level of the supplementary motor area (cf. Jeannerod 1997). In order to realize the imagined action, the inhibition has to be suspended and the movement released, which is connected to a subjective sense of *agency*.[30] The resulting action then corresponds to a resonant loop between the participating neural assemblies and the peripheral movements including the continuous sensorimotor feedback from the effector organs.

The movement gestalts or "kinetic melodies" (Sheets-Johnstone 2011) are likewise based on learnt patterns, that is, couplings of movement components through repeated practice. On the neural level, this is realized by cortical links between neuromuscular units and subsequent storing of these motor sequences in the basal ganglia (Graybiel 1998, 2005). But it is also possible to couple motor regions with linguistic ones, as in the case of learning to write. As a result, it is then sufficient to intend to write the word "apple" in order to automatically connect the patterns of activation in the linguistic regions with the corresponding patterns in the premotor cortex, which activate the movement. Hence, starting motor execution also requires a resonance or synchronized oscillation of spatially distributed neuronal subsystems, which is reflected, among others, in the "readiness potential" in the supplementary motor cortex (Deecke et al. 1969, Libet 1985; see also Aoki et al. 2001 on neural synchronization in visuomotor tasks).

Edelman and Tononi (2000) have suggested the suitable concept of "functional clusters" consisting of integrated neural assemblies that are formed for brief periods (50–100 milliseconds) and recruit various areas for the specific task and context. Functional clusters are not cognitive or motor modules with fixed or prescriptive programs; rather, they serve as *temporary attractors*, which bias the inclination of the brain–body–environment system towards fluent performance, as it was described earlier by the example of the tennis player (see 4.2.1). A perturbation of the neuronal system landscape—for example, a mismatch between "predicted" and factual motor feedback, subjectively experienced as a deviation from the optimal spatiotemporal gestalt of body–environment coupling—results in a shift towards a new attractor that balances out the occurring instability, corresponding to a suitable corrective movement of the body (see Cosmelli et al. 2007).

Let us summarize once more: both perception and movement depend on an ongoing endogenous activity which provides (a) the motivating energies and

---

[30] See Gallagher and Zahavi (2008, 158–166) for an extensive phenomenological and empirical account of agency.

(b) the image and action schemas that prefigure the successful sensorimotor coupling of organism and environment. Moreover, perception and movement are equally based on processes of resonance and transformation, which, in vertical circularity, integrate elementary events or stimuli into higher-order patterns of resonance. Of course, both systems may not be conceived apart from each other. As we have seen, perceiving an object as a meaningful gestalt implies imagining what it would be like to deal with the object in the context of purposeful action. This involves brain areas related to motor activity, which are necessary for grasping the affordances of the object (a ball to grasp or to throw, a dog to play with, or to flee from, etc.). To recognize an object always involves knowing how to deal with it. Granted, most of these affordances are not explicitly imagined on any given occasion, but "all of this imagery, sedimented from numerous past experiences of performing such actions relative to similar balls, goes into our current understanding of what a ball is" (Ellis & Newton 2010, 29). Conversely, no movement may be conceived without continuous kinesthetic and other sensory feedback, or its felt "kinetic melody."

Looking back at the localizationist and the holistic theories of brain function (see 2.2.2), we can now say that each represents one aspect of what happens: local and integral activities mutually depend on one another in circular causality. On the one hand, processing is analytic, specialized, and localized; on the other, the high degree of interlinking between subcenters and the superordinate oscillatory synchrony enable the integration of partial processes. This entails, however, that all sensorimotor activity has to be conceived as an integral process which is not composed of separable modules or decoupled representations, but instead unites neural activity, body, and world.

There is agreement today that nowhere in the brain is there to be found a Cartesian "center" of such integration, no brain region to which all other brain regions ultimately "report" and in which the partial processes are conjoined into a "perceptual image," let alone into conscious perception. Much rather, the entire system constantly re-arranges itself according to the patterns with which it is in contact, until its self-oscillations resonate with these situational patterns and provide for fluid interaction of organism and environment. Successful synchronization, then, corresponds to the "experience of fulfillment," for instance, when recognizing the Dalmatian. Similarly, no precedent "motor image" or "central representation" is required for bodily movement; instead it proceeds in ongoing adjustment of neural patterns to bodily and environmental demands.

In the preceding section (4.2.3), the creation of an *optimal coherence* between organism and environment was stressed as the fundamental principle of neuroplasticity. It also applies to each current sensorimotor achievement. The resonance of patterns of brain activation with complementary situations in the environment has a tendency towards *closure* (in terms of dynamical systems theory, towards an attractor at the lowest energy level)—only thus do the mediating processes enable transparency. From this result everyday "phenomena of closure" such as the suppression of the blind spot in the perceptual field or the completion of gestalts, for instance, of words with missing letters.

More impressive are the clinical *neglect* phenomena, in the case of which patients with a lesion in the parietal lobe do not notice losing half of their visual perception and, for instance, only draw the right half of a daisy thinking that they have reproduced it completely. In *anosognosia*, such patients also lack awareness of paralysis of one hemisphere of the body and incorrigibly take the paralyzed limb to be an alien object that they try to throw out of their beds (Ramachandran 1995, Ramachandran & Blakeslee 1999, Appelros et al. 2007). So the perception of reality is not "halved" in these cases, but is itself maintained as a whole at the expense of utmost distortion. Also, delusional phenomena in schizophrenia can be regarded as deficient substitutes of normal reality constitution, which establish a coherent experience despite disorders of basal perceptive processes (Fuchs 2017b). That in all of these cases of closure phenomena the patients lack awareness of the distortion can be explained very well by the proposed theory of brain–environment coherence. If conscious experience is the integral of functional loops between brain, organism, and environment linked by means of resonance, then in the case of a failure of single loops, coherence is established on a reduced basis, but still integrally. Or, put differently, it is maintained at the price of a loss of which there can be no awareness precisely in order to establish coherence.

4.2.5 Information, representation, and resonance

The following section serves to conceptually and theoretically clarify a number of issues arising from what has been said so far. Let us once again consider the conventional neurocognitive theory of perception: distal events lead to proximal stimuli that, as *information* coded in the form of action potentials, are passed on to the brain via nerve fibers in order to be processed there. The results of this processing become accessible to consciousness as *representations* of the external world. Thus, in perception we only have access to representations and not to reality. Two central notions of this theory, however, lead to category mistakes, that is, "information" and "representation." In what follows, I want to outline this and replace them with the notions of "pattern" and "resonance."

4.2.5.1 Information

> There is a strong and, it seems, almost irresistible tendency in the human mind to interpret human functions in terms of the artefacts that take their place, and artefacts in terms of the replaced human functions […] The use of an intentionally ambiguous

and metaphorical terminology facilitates this transfer back and forth between the artefact and its maker. (Jonas 2001, 110)

The computer metaphor of the mind as being a program of symbols and information is still the leading paradigm of contemporary cognitive sciences and popular notions of the mind. Transferring the technical notion of information to neuronal processes is its central maneuver. But is such a transfer even permissible? If we take a closer look, this notion remains vacuous without *subjects* of information.[31] When looking inside a computer, or a brain, we don't see or even detect information—we only see physical stuff: voltage levels in a computer, biochemical processes in the brain. On the contrary, "information" means *informing somebody of something*, that is, it primarily denotes a type of human communication, in which the partners are jointly directed towards semantic contents. As the result of an informational process somebody is "informed" about something. This does not mean that his brain has "stored certain pieces of information," but that *he* has acquired the capacity to reproduce a certain kind of knowledge. Information can be passed on by means of spoken words, but also by other symbols or signals, for instance, if communication is mediated by newspapers, telephones, or television. But even so, it remains tied to a receiver who *understands* the message, that is, interprets the signs *as* signs.

Thus, encoding and transfer to a carrier medium themselves do not constitute information, which is ascribable to the carrier as such, but only *potential* information that is *realized* when understood by a person. A computer, too, does not contain information. There is no semantic content, no "data" along the causal chain of electronic processes which one could point to and say: here is information. Rather, the computer "computes" only from the perspective of a user—seen in itself, it merely transforms electronic patterns into others.[32] Consequently one can only speak of "symbol manipulation" or of an "information processing system" from the perspective of a human being that interprets the system as such. Even more problematic is to apply the notion of information

---

[31] As is well known, the term "information" as used in telecommunication engineering (Shannon–Weaver model) only signifies the statistical distribution of a sequence of electronic impulses, rendered in a binary code and measured in "bits." Any semantic, meaningful, or contextual aspects are explicitly excluded.

[32] "The problem with the concept of 'information processing,' is that information processing is typically in the mind of an observer. For example, we treat a computer as a processor and a bearer of information, but intrinsically, the computer is only an electronic circuit. […] The electrical state transitions of a computer are symbol manipulations only relative to the attachment of a symbolic interpretation by some designer, programmer, or user" (Searle 1998, 1941).

to natural objects such as genes or neuronal activity, which were not, of course, created and programmed for human purposes.[33]

In Chapter 3, we investigated how living creatures, due to their vital drives and the appropriate organs of perceiving and effecting, imbue their environment with *meaning*. In this sense, one can say that animals receive non-linguistic information about their environment. But to speak of "information flow" in their sensory or neuronal systems reinterprets a causal relation as a semantic one—a category mistake to which Searle has drawn our attention in his Chinese Room argument (2.1.2.2). Of course, causal processes can be correlated with intentional relations, they can even be their necessary vehicle. But the intentional relation to objects of perception cannot be reduced to physical processes, because they, taken by themselves, do not contain information.

No engineer or computer scientist has encoded information in the excitation of the optic nerve, and there is no homunculus in the brain with the purpose of decoding these excitations. There is no end of the causal chain where everything arrives and then conscious experience happens. Neither is the brain as such in a position to "decode information" in order to make "inferences" about external objects beyond the retina. The transmission of sensory stimulus events within the neuronal system remains a physical process, which may well have certain *patterns*, but which is never translated into "information" or "meaning." The search for a "neuronal code" is in vain, for in nature there are neither genetic nor neuronal "codes" nor "descriptions." Even if a neuroscientist could in detail recognize the causal correlation of events in the environment with patterns of neuronal activity and thus "decode" them, it would be only he who conceives of these patterns as codes.

So the sensory system does not process information. It reacts, as a transducer, to stimulus patterns and transforms them into neuronal excitation. The central integration of these excitations, that is, their resonance with prefigured

---

[33] Talking about "genetic information" or "information processing in the brain" has become so ubiquitous that it seems nearly futile to question it. However, at least in scientific theory we should be aware that we are only dealing here with metaphors. Giving up the (inter-)subjectivity of information would indeed render the notion completely arbitrary: why should not an oxygen molecule contain "information" that is transferred to other molecules, which, depending on their "information processing," associate with the former or else take their hook? If all order structures and order transitions in nature constitute "information" without a researcher being necessary to interpret them as such, then even a snowflake is constituted by information.

activation patterns and their linkage to substrates of memory, evaluation, and so on, is what then produces the brain's complex state of order, which, in turn, is correlated with conscious perception. But it is only this conscious state that contains meanings or "information" about what is perceived, for instance, whether something is green or red, light or heavy, slow or fast. Neurophysiological states are not "meaningful" let alone "experienced" states—only in the context of an organism's history and the current environmental situation can they become *carriers* of such states. These ecological preconditions of meaning as being tied to an embedded subject have, following up from Putnam (1979, 222), been poignantly expressed by McCulloch in the following syllogism:

(a) Meanings just ain't in the head (in accounting for meanings, we must advert to factors in the agent's environment).

(b) Meanings are in the mind (meanings and grasping meanings are conscious phenomena).

(c) The mind just ain't in the head (an adequate characterization of an agent's consciousness must advert to factors in the agent's environment). (McCulloch 2003, 11–12).

From this follows that a given brain state or process is not sufficient to determine the corresponding state of the mind, as the theory of supervenience claims (Davidson 1980, Kim 1993); for it can be related to different contexts, and only this relation completes the organism's process of sense-making.[34] Alva Noë (2009, 3) has given an illustrative comparison: imagine a group of scientists examining a dollar bill with all available technology, including analysis of its subatomic structure. However, nowhere in the examination will the value of the bill show up—it only emerges within the context of a larger system of symbolic relations, established by shared intersubjective intentionality. Similarly, brain states do not have intrinsic meaning—they only contribute to meaningful experience as components of a human being's enactment of life. This ecological dimension of consciousness as an activity of a living being in its environment is, however, missed by the second central notion of the cognitive sciences, that is, representation.

---

[34] Since there are different notions of the "extended mind," it remains to be added, that the so-called content externalism only regards intentional meanings as necessarily embedded in environmental contexts, but not at all conscious states as such. Different from enactivism and from the concept defended here, externalists usually locate mental states in the brain (see, for example, Clark 2003).

4.2.5.2 Representation

Certainly, many neuroscientists would ask why the patterns of neural excitation described previously should not be called a "representation" of the Dalmatian dog? Well, this notion has already been repeatedly criticized (2.1.2.2, 4.1.3, 4.2.1); I repeat the main results:

The representational relation—something stands for, or refers to some other thing—requires a subject who establishes this relation and for whom it exists. We can see certain configurations of colors in a frame on the wall *as* a picture or recognize hollows in the snow *as* tracks, that is, we can take something present as referring to something absent. The color configurations, as such, however, represent nothing and the hollows in the snow are hollows, nothing more. *We* have to establish the relation of depiction or reference *ourselves*. But this relation is a special mental feat added to certain perceptions, it cannot be transferred to perception as such. We see *trees*, not pictures of trees. Perception is not a representation, but a *presentation* of objects (see Searle 1983, 46).

Speaking of neural representations therefore runs into homunculus problems: Who in the brain should recognize the excitation patterns as patterns of a dog? One might argue that neural patterns are not only causally connected to earlier input (as the tracks in the snow are) but are also functionally connected to adequate behavior (e.g., recognizing and calling one's dog). They could then be called representations because they fulfill a function for the living system. However, the perception of the dog is only accomplished through the interaction of neural activations, eye movements, and environment forming a closed loop. There is no component within this ongoing cycle that represents another one, in the sense that it stands for it while it is absent. As I have already argued (3.3.4), the term representation suggests that the brain activities could be separated from the cycle, as if they were remodeling inside what is outside. But in the ongoing perception–action cycle as described earlier there is no inside and outside any more. Instead, perception is enacted by the brain–body–environment system as a whole.

The neurocognitivist notion of representation implies an inner mirroring and thus a doubling of reality, which cannot escape the homunculus fallacy. This leads to contorted formulations of the following kind: "Once someone experiences his mental states, he in fact perceives his brain states, indeed the brain actually perceives its own states" (Tetens 1994, 124). But no human being has ever seen their own brain states, and no brain has ever perceived anything. We neither see inner images nor do we, unbeknownst to us, somehow perceive our own neurophysiological states. But such fallacies are all too easily suggested by the notion of representation.

This aporia does not diminish if one, with Northoff (2004a, 116), conversely speaks of an "inability of the brain to directly detect and recognize its own brain states as brain states." This inability should be responsible for the fact that we have "no direct epistemic access to the First-Brain Perspective" (2004a, 117), but only to the phenomenal first-person perspective experienced by the brain instead. For this alleged and seemingly astonishing incapacity of the brain, Northoff even coins a specific term, namely "autoepistemic limitation," and goes on to explain "the epistemic illusion of the mind" (2004a, 304) by means of the brain's attempts to, despite its own limitation, make sense of the origin of its mental states: "our brain suffers from a knowledge gap because it remains unable to perceive itself directly as brain […] as a result, our brain cannot do otherwise but 'posit' the concept of mind" (Northoff 2004b, 484).

Thus, the brain inside the skull becomes its own homunculus, unfortunately stricken with self-blindness. However, this pitiable brain can be helped: for purely logical reasons alone it need not and cannot perceive anything, for perception is a conscious activity which can only be attributed to a living being. In order to recognize neuronal or, for that matter, any kind of states, a conscious being already needs to be presupposed—a brain alone, alas, is not sufficient for this. Brains or brain centers cannot *a limine* know anything; they need neither hide this inability from themselves nor from others.[35]

To this epistemological aporia corresponds a hardly less grave aporia at the neurobiological level: a center, in which this internal perception occurred, if it existed, is nowhere to be found in the brain. In the light of this, a cleverer form of representationalism has emerged in the meantime, which believes to be able to sidestep the homunculus fallacy. In order to do so, it uses a notion that has also played an important role in the phenomenological position defended here, namely "transparency"—of course, in a very different sense:

> According to this contemporary form of representationalism, the internal representations mediating visual perception are not the immediate objects of perception; rather,

---

[35] Already Schopenhauer got entangled in the same brain conundrum: "But in so far as the brain knows, it is not itself known, but it is the knower, the subject of all knowledge […] On the other hand, what knows, what has that representation, is the brain; yet this brain does not know itself but becomes conscious of itself only as intellect, in other word as knower, and thus only subjectively" (Schopenhauer 1966, vol. 2, 259). Well, the brain in fact knows nothing, because the only knower is the living human being. Granted, this human being usually does not know much more about its brain than about its spleen, but that is not a big deal. Most of our organic processes are not available for introspection, precisely because they are only *enabling or mediating* processes. That is why eyes do not see themselves, and brains do not know themselves.

the object of visual perception is the distal physical world. Internal representations are what enable the perceptual state to be directed towards the world—they represent the world to the perceiver in a certain manner. Thus it is often said that we do not see our representations; rather, we "see right through them," as it were, to the distal world. (Thompson 1995, 221)

This variant of representationalism has become known mainly through Metzinger's (2003) "self-model theory." I have already criticized his notion of representation detached from subjectivity (2.1.2.2); here, however, I want to concern myself with his notion of transparency. For Metzinger, a representation primarily is "a process by which some biosystems generate an internal depiction of some parts or reality" (2003, 15). Such internal representations may become accessible to the system as "mental" or "phenomenal representata," that is, as conscious experiences. The possibility of directly perceiving physiological processes is, of course, rejected by Metzinger:

[I]t is not the basic neural process as such that is mental or that becomes the content of consciousness, it is a specific subset of […] neurally realised "data structures," which are generated by this process […]. If you now look at the book in your hands, you are not aware of the highly complex *neural* process in your visual cortex, but of the content of a phenomenal mental model […], which is first of all generated by this process within you. (2003, 22)

[This process] … is transparent in the sense of you looking currently through it. (2003, 334)

Of course, the question remains whether anything is gained by saying that, instead of neuronal processes, I am said to be aware of my neuronal "data structures," for instance, if I am perceiving a book. In any case, the neuronal processes underlying perception elude introspection according to Metzinger. Therefore, similar to Northoff, he also uses the term "autoepistemic closure," meaning that "conscious experience severely limits the possibilities we have to gain knowledge of ourselves" (p. 175), since we have no introspective access to the neural representations in our brain. Transparency thus results from our "inability to recognize a self-generated representation *as* a representation" (p. 292). That the book in my hands seems to immediately present itself to me results from the inability of my consciousness to register the neural processes occurring at far too fast a pace behind my perceptual experiences, thus creating the "illusion of substantiality" (p. 23).

Transparency then leads to the "naive realism" of the everyday world, in which we believe to see real books or trees and to be in direct contact with the world, while we are really dealing with mere constructs or illusions:

[P]henomenal experience […] will always be characterised by naïve realism. This realism can now be interpreted as a kind of hallucination that proved to be adaptive for systems like ourselves. (p. 250)

The transparency of the representata or the "illusion of givenness" here holds primarily for perception and less for other modes of mental experience, which can more easily be recognized as constructs:

> It is much more difficult to recognise the book you are currently holding in your hands as an internally generated state than the thoughts and feelings arising while reading it. (Metzinger 1999, 64, own translation; see also Metzinger 2003, 250)

But a book that one holds in one's hands, feels, and sees is anything but an "internally generated state"—it is a real book that I can also show to others so that they can take it out of my hand and themselves re-read the passage to which I have drawn their attention. *It is one and the same real book that is transferred from my hand to theirs.* If it only were an internally generated state, it would indeed be a kind of illusion or hallucination; and if it existed for each of us only within our respective brain, we could never consensually intend the object as a shared reality. But this would be the end of any scientific attempt to even investigate the brain and its processes. So the book I am holding in my hands cannot be "internally generated." If at all, it could only be my *entire experience* including myself, my perception of the book *and all other human beings*—and that is indeed the consequence Metzinger finally draws in his book: even a subject of a brain-in-a-vat could have the same experience of another's loving gaze as a real person:

> SMT [i.e., the self-model theory] makes the claim that even for consciously experienced intersubjectivity of this type, it is true that an appropriately stimulated brain in a vat could activate the same phenomenal content. (2003, 602)

In consequence, the other would forever remain inaccessible to me, for what I see, if I see him, what I touch, if I give him my hand—this would all just be mental images generated by my brain. Hence, in the final analysis the self-model theory leads to a *neuro-solipsism*. We are even worse off than Plato's cavemen who had at least their common prison—Metzinger's cavemen are each enclosed in their own cave. As such, it is no surprise that in Metzinger's most detailed descriptions of his theory of subjectivity, scarcely any mention is made of the other, respectively of intersubjectivity. For the other is not given to me solely as a mental object, but as a being-in-itself beyond all its appearance-for-me. Perceiving him as himself must of necessity cancel out all alleged virtuality of perception.

Metzinger's notion of transparency fulfills the function of explaining the allegedly illusory or virtual character of perception: our impression of seeing the real book is founded on the transparency of the neuronal models, which obfuscate their computational origins. But this runs counter to the very meaning of "transparent": something can only allow "light to pass through so that objects behind can be seen" if these objects are real—otherwise it would be a mirror or a film. Metzinger's subject, however, "sees through" the representata, but in so doing it sees only appearances or, rather, it looks *into nothing,* for reality always lies outside its field of vision, in the data-producing processes of the brain.[36]

The *phenomenological notion of transparency* defended here is obviously of a very different kind. According to it, it is the transparent medium that makes reality itself accessible for us—as *mediated immediacy.* Through seeing individual blots, I see the Dalmatian. Through individual finger movements, I realize the piano piece. And if the body of another person becomes transparent *with regard to herself* in a manner that she appears in her body and expresses herself, then this is no illusion—on the contrary: if I only see her body in an objectifying perspective, I lose what is essential. Whereas the ophthalmologist can see the bodily structures of the eye, he can no longer see the gaze of the other.

As such, the phenomenological notion of transparency also entails a possible change in attitude respectively attention. I can direct myself towards distal objects, but also backwards to my proximal bodily states. I can *explicate* the implicit structures of my experience, for instance, by paying attention to the individual letters in a text instead of to the meaning of the words. But the explication can only cover the perceived environment or the periphery of my felt body, no longer its center: I cannot take a step back behind my perceptions themselves. The activations of the optic nerve or of the neurons in the visual cortex remain hidden from me. The neuronal processes are not just the latent structure of my experience, which is inaccessible to me for reasons yet to be uncovered. As conditions of enablement, they, from the start, *lie beyond it.* As such, the notion of transparency only makes sense in relation to a subject of perception or movement *that lives and realizes itself in the transparent medium*—and this medium is its own embodiment. Neuronal states, however, are indeed "transparent" for no one—they can only be accessed from the third person perspective and, as such, are irrevocably opaque. The notion of transparency cannot salvage the neurocognitive notion of representation.

4.2.5.3 Patterns and resonance

In any case, let us recall that the transformation of individual stimuli into integral patterns of neural activation (see 4.2.4) obviously leads to a "matching" or "correspondence" of brain states and environmental contexts—recognizable by the fact that prefigured patterns of neural activity "snap into place," for instance, in the case of closure phenomena in gestalt perception or the experience of

---

[36] Hence the apposite title *Being No One* (Metzinger 2003): the self-model too is only an internal representation of the entire neural system through which we see, thus believing to be ourselves, whereas we actually see—no one.

fulfillment in recognition. If one wants to find a term to denote this coupling of brain and environment, and which avoids the pitfalls associated with the notion of representation, the concept of *resonance of patterns* offers itself, which has already been repeatedly used. Neural networks do not represent static objects or situations in the external world, but rather they resonate with environmental stimuli in a coordinated manner, insofar as these are aligned with prefigured neural patterns.[37]

As we have seen, recurring configurations of external and internal stimuli are sedimented in complex patterns of neural activation readiness, which may then be actualized by appropriate situations in the environment. So the familiarity of a perceived object is equivalent to the degree with which such patterns match with isomorphic stimulus constellations. The coordinated transient oscillation of neural assemblies or "attractors" in turn affects other associated systems to resonate with it. This can ultimately lead to the activation of motor centers, for instance, to the impulse to grasp or flee.

Thus mediated by the body, brain and environment mutually *resonate* with one another; they are linked by dynamic isomorphic patterns of oscillations. The justification for the notion of resonance, on the one hand, lies in its reference to these synchronous oscillations, which also represent a proposed solution to the problem of intermodal "binding": Converging evidence suggests that the binding of sensory attributes as well as the overall integration of various dimensions of a cognitive act, including associative memory, emotional tone, and motor planning, are achieved through large-scale dynamical patterns of neural activity over multiple frequency bands (see 4.2.4). Coherent perceptual and motor gestalts presuppose highly synchronous or resonant reverberations of the networks involved, in continuous alignment with the environmental stimulus configuration, that is, both *internal and external resonance*.

On the other hand, the notion of resonance is suitable to lead us out of the aporias entailed by the neurocognitive concept of representation, which always ends up as a mirroring and doubling of reality, ultimately making it susceptible to the homunculus fallacy. The notion of representation, of "inner images" or "mirrors," is derived from the sphere of visual perception, which, out of all sense modalities, is strongest in establishing a *static opposition* between perceiver and what is perceived. It is based on a theory of perception that does not connect us to the world, but only with constructs generated from sensory data. Applied

---

[37] Gibson, in his "ecological theory of perception," also speaks of the sensory system resonating with global changes in the perceptual field: "In the case of the persisting thing, I suggest, the perceptual system simply extracts the invariants from the flowing array; it *resonates* to the invariant structure or is *attuned* to it" (Gibson 1979, 249).

to neural processes, "representata" become seemingly discrete, localizable, and thus reified entities.

By contrast, the notion of resonance takes root in acoustics and the mechanics of oscillation; it refers to bodies and systems that are attuned to one another by their own vibrations and, above all, are *currently connected with one another*. Resonance contains a dynamical as well as a rhythmical element and thus establishes a *temporally* overarching relation between the systems involved. Other than representandum and representation, or original and image, "resonandum" and "resonans" thus cannot be separated. Whereas representations can easily be isolated and seen as carriers of consciousness or even be identified with it, the notion of resonance does not permit such a division: only in *synchronization*, as connected systems, can brain, organism, and environment become carriers of consciousness. Insofar as the brain, according to the ecological conception, is inseparably, dynamically, and flexibly linked to the organism and the environment, the notion of resonance characterizes this relation much better than that of representation. As such, *the brain can be conceived of as an organ of resonance*, the rhythmical oscillations of which continually establish a coherence between organism and environment.[38] In this sense, synchronization and resonance may be regarded as the fundament of all perception and cognition, for the coherence of the experienced world and the unity of consciousness itself.

The principle of resonance may finally be formulated on the basis of a realistic knowledge theory. Recently, Dreyfus and Taylor (2015), in a fundamental critique of representationalism, have pointed to the alternative of the Aristotelian theory of perception as a "contact theory": perception is not mediated by mental ideas or representations, but it is a direct relation to reality itself. For according to Aristotle's *De Anima*, actualized knowledge and perception, in a sense, become one with its object: "knowledge and sensation [*aísthesis*] are in a manner identical with their respective objects" (*De Anima*, 431 b22). Of course, they thereby do not become identical with the *material* object; for "it is not the stone which is present in the soul but its form [*eídos*]" (431 b29).

Eídos means the form which shapes the object, but also the gestalt or pattern, the model or prototype. This gestalt can also be received by the mind; for the *nous* is for Aristotle the "form of forms" (431 b 31), the most general potentiality (*dýnamis*), that is to say, it can potentially receive all forms. In the *realization*

---

[38] This does not exclude the possibility that the patterns of resonant activity, once emerged, become the basis for further internal processing within the brain, thus forming "resonances of resonances" which would be required for the emergence of, for example, imagination or thinking. The profound self-referentiality of the brain, which manifests itself in the reentry mechanisms or recurrent feedback loops described by Edelman and Tononi (2000, 113–116), might also be the presupposition for higher or self-related forms of consciousness. However, at the present stage of knowledge, this remains highly speculative.

of the form (*enérgeia*), that is, in the actual recognition or perception, the mind becomes one with the object. In other words: in the perception of an object, the *nous* is *in-formed* by the same *eídos* which also shapes the object: "When I see this animal and know it is a sheep, mind and object are one because they come together in being formed by the same *eídos*" (Dreyfus & Taylor 2015, 18). The physical object is not alien or incommensurable to the mind, since it displays a form that is potentially mental and thus capable of *informing* the mind.

The principle of form now may be understood as *order pattern* in the sense developed earlier. Then the brain could be conceived as a matrix, which like the mind is able to "receive all forms," that is to say, to take them over in its own structure as neural patterns or potentials. In the actual perception "mind and object become one," corresponding to an encompassing resonant system state in which the same pattern or form is activated in the brain as it is displayed by the object. The principle of pattern resonance thus may be regarded as a reformulation of the Aristotelian theory of perception by means of an organic substrate of which Aristotle himself had no sufficient knowledge yet. He assumed the organ of unitary perception, which integrates all sensory modalities into a "sensus communis" (*koinē aisthēsis*), not to be the brain, but the heart.[39]

4.2.6 Conclusion: mediated immediacy

This chapter has put to use the conception of circular causality of living beings in order to interpret higher brain functions. The linear-causal model of stimulus, information processing, representation, and reaction proved to be unsuitable for modeling the overarching unity of the brain–organism–environment system. In its place, derived from von Uexküll's "functional cycle," we developed a model of "*open loops*," which the organism, as it were, casts into the environment according to its own structure and its prior experience.

Now these loops can combine with the fitting elements of the environment— its relevances and affordances—in such a manner that *coherence* emerges within the entire system: thus the patterns or gestalts prefigured by the organism are actualized. A major consequence of this conception is that conscious experience can no longer be ascribed to a single section or partial process of the functional cycle. Much rather, it forms the *integral of the entire brain–body–environment nexus*. If visual perception, for example, is based on a pattern resonance between object, eye, and brain, or in Aristotelian terms, on the resonance of a form that shapes both the brain and the object, then in the moment

---

[39] See *De iuventute et senectute* ("On youth and old age") 469 a, *De somno et vigilia* ("On sleep and waking") 455a, and *De partibus animalium* ("On the parts of animals") 670 a 23–25.

of perceiving a tree, this particular object, the rays it emits, and the body with its sensorimotor functions including the brain form an encompassing, coherent system. But if that is the case, then it is also true in a precise sense that *the tree which I perceive becomes part of my conscious experience, and that conversely my conscious experience literally extends to the tree itself.* Neither the tree nor a copy of it needs to get into the brain, for in seeing the tree I am in contact with the tree itself. As such, consciousness is no longer an inextended inner state or a representational sphere, but an extended relation, namely a *coexistence with the objects and with the world.*

The idea of an extended consciousness might still be not immediately plausible. Let us therefore consider three possible objections:

- Searle derides the idea by arguing that subjectivity would then "float around," and when we see a table, the table molecules would be "realizing consciousness," which he thinks "is not worth serious consideration" (Searle 2015, 51). However, it is of course not the table as a molecular structure, but as an object displaying a certain *form and appearance* which—via the pattern of light it emits—indeed contributes to realizing perceptual consciousness.

- From an idealistic point of view, one could object that I only see the tree as an "object-for-me," which would have to be distinguished from the tree itself. A similar distinction is also found in Husserl: my act of seeing (*noesis*) is directed towards the intentional content of my seeing (*noema*), but this content, the tree as seen by me, is not the physical tree in the world (Husserl 1950). However, this comes near to the representationalist duplication of reality. Of course, my seeing-the-tree is different from the tree as such; nevertheless what I see is the tree itself and not a "seen tree" or an immanent content of my consciousness (on this fallacy, see also Searle 2015, 24–29). In other words: in the case of perception, *noema* and real object coincide.

- A third (rather physicalist) objection against the extension of consciousness would be that I could hardly be connected to the rising sun, which due to the long distance covered by its light has actually risen 8 minutes ago, or to the stars, which might already have perished when their light reaches me. However, this temporally extended distance does not contradict the copresence of perceiver and perceived. Granted, I perceive the sun or the star across time, but still *as they show themselves to me*. The medium of this showing (the emitted light) is part of the connection, regardless how much time the transmission needs.

In the further course of our investigation, the formation of open loops, and thus the specific capacities of the human organism, was traced back to the plasticity of the brain matrix. It enables the incorporation of experience and the formation of highly differentiated capacities over the course of an individual's development. This is achieved mainly by the *implicit coupling* that forms between sensory, motor, and affective modalities, leading to procedural and other types of embodied memory. At its basis lie connections between neural dispositions for excitation in various brain regions, especially transcortical and cortical–subcortical connections. Thus not only integral perceptual and motor gestalts are formed, but these are also coupled with meaningful contents in such a manner that individual elements can become carriers of overarching intentionality, as was illustrated by the examples of speaking, writing, or playing the piano. [40]

Now this dispositional structure of implicit coupling is actualized in *current perception*, as we investigated in the case of visual gestalt closure. The coherent neural patterns established by prior experience here *resonate* with appropriate constellations in the environment, which leads to the overarching system states that underlie experienced perceptions and movements. In these processes, the system is always geared towards an *optimal coherence* between organism and environment, which manifests itself in the unity of conscious experience, even in the case of severe failures of partial functions.

Accordingly, the central function of the brain for the experiencing and acting living creature consists in transforming configurations of individual elements into *resonant patterns* that form the basis of integral acts of life. Thus the brain becomes the organ of mediation between, on the one hand, the *microscopic* world of material-physiological processes and, on the other, the *macroscopic* world of living creatures. As a general matrix for patterns, capable of receiving all forms and being shaped by them, and as a resonance organ for the current patterns in the environment, it opens up the perceiving and acting access to the world for the living creature. This access is not possible directly, but requires mediating processes. The forms and gestalts cannot simply be detached from matter and, as such, be immediately transferred. As much as their existence in the world is tied to a material substrate, as much does their mediation to perceiving creatures require carrying and resonant material media. The physical processes serving this purpose (e.g., light waves, acoustic waves, etc.) are transformed by the sensory organs and the brain into the organism's own activity, which does not turn them into internal models, however, but continuously resonates with the patterns of the environment. In other words: *the transfer of forms proceeds by pattern transformation and pattern resonance.*

---

[40] The coupling with affective and evaluative modalities or patterns shall be left aside here just for reasons of simplification.

The result is neither a mere construct nor a grasping of the "thing in itself," but the *mediated immediacy* of the relation between the subject and the world, as it has been described by Plessner. According to him, the direct connection between subject and object is necessarily only possible as an indirect, mediated one (Plessner 1975, 48). "Something real can, *as* real, be in no other way related to the subject than […] as appearance (*Er-scheinung*)," that is, as "mediated immediacy" (p. 329; own translation). Thus a form of connection is denoted, "in which the mediating link is necessary in order to establish the immediacy of the connection" (p. 324); in this consists the basic structure of the living creature in relation to the environment.

Of course, in contrast to animals, humans know of the indirectness of their relation to the world: due to their capacity for reflection, their relation is given to them also "as a mediated one" (p. 325). This must lead to a situation, "in which man begins to doubt the immediacy of his knowledge and the directness of his contact with reality, as it exists for him in absolute evidence […]. Of course, it is then argued, the subject believes to grasp reality and to have it itself. But this is only true for the subject. In fact, it merely moves among contents of consciousness, representations and sensations" (p. 329). Grounded in this feeling of uncertainty lies idealist respectively neuroconstructivist epistemology: we never arrive at reality; all that we have, are models and representata. "We never really advance a step beyond ourselves" (Hume).[41]

In contrast, the notion of mediated immediacy or transparency offers a possibility to found the actual being-in-the-world of the embodied subject. But there is still another reason for realism. For precisely the relativization of the mere impression that man can and must make due to his "eccentric position," on the other hand, enables the *objectivity* of human perception. For it is in this way that the human subject "keeps the distance which is required by reality, should it reveal itself, and has the leeway (*Spielraum*), in which reality can appear" (Plessner 1975, 331).

However, this objectifying feat of human perception is not only, as assumed by Plessner, due to an individual distance, but much rather due to an *implicit intersubjectivity* (see 1.3.2). The objects perceived by me are always also in principle perceivable by others. Through the implicit participant perspective or the "we"-perspective, my subjective perception receives its fundamental, even if refutable, objectivity. Neuroconstructivism ignores the objectivization, of which human perception is capable in the form of implicit intersubjective perception: the book I see is always the book potentially seen by others; and if I give

---

[41] David Hume, *A Treatise of Human Nature*, Book I, Part II, Section VI.

them the book, it is at the same time the book attended to, seen, and received by them. Thus the manifold processes of mediation and transformation, which underlie my perception, can also become the foundation of a *shared* reality. They do not merely create momentary glimpses or specters *for me*, but present the objects themselves, *for us*. So we can replace the notion of a "naive realism" by a *realism rooted in the shared lifeworld*.

The intersubjectivity of perception is not, of course, just pregiven: as we will see, it needs to be acquired and learnt in early childhood. Thus, human perception is not a purely natural process, but is a form of *socialized and cultivated perception* enabled by joint attention and shared intentionality. This holds equally true for human feeling, thought, and action. This cultural development of specific human capacities cannot leave the brain unaffected, quite the opposite: it can only perform its highest functions as an organ of a creature that lives together in a community with others. We will turn now to this relation between brain, organism, and the social world in more detail.

Chapter 5

The brain as an organ of the person

> **Overview**
>
> Chapter 5 examines the socially and culturally scaffolded development of the human brain, especially in early childhood. Beginning with the prenatal period and the early intersubjectivity in the dyadic relationship between mother and child, I first focus on interactive forms of implicit memory (5.1). As the neurological basis of this development, I then present and discuss the attachment system and the social resonance system ("mirror neurons") (5.2).
>
> In what follows, I turn to secondary intersubjectivity, which manifests itself towards the end of the first year of life, among others, in the development of joint attention. Understanding others as intentional agents lays the foundation for later perspective-taking and thus for the "eccentric position" of human beings. On this basis, language acquisition is examined as the anchoring of an embodied interpersonal practice, connected with the biological resonance system of mirror neurons. After reviewing the further development of perspective-taking, reflection, and thus the eccentric position, the chapter closes with some fundamental considerations concerning brain and culture (5.3, 5.4).

From the early hominids to the emergence of *Homo sapiens*, the ratio of brain size to body size has tripled, with a tremendous acceleration within the last million years (Ruff et al. 1997). The brain size of the great ape species closest to humans in evolutionary terms, such as chimpanzees and bonobos, is only 25–35% of that of humans, although body size is comparable. The increase in volume is attributed mainly to the immense growth of the neocortex, in particular the frontal lobe. This concerns brain areas involved in higher social cognition, behavioral and emotion regulation in social relationships, empathy, language, and understanding the feelings and intentions of others. Hence, from an evolutionary perspective, it has been proposed that the pressures leading to the development and evolution of the human brain can mainly be traced back to the intense and complex social life of early human primates. This may be corroborated by the empirical correlation between average cortex volume and average social group size across the various types of primates: the more complex and differentiated a social structure is, the higher the individuals' brain capacities need to be in order to orient within it.

> Following Dunbar (1993) and Ploog (1997), the medium group size for the gorilla is 34, and for the chimpanzee 65 individuals. If one relates this to the respective cortex volume, this would result in an expected average group size of around 150 for humans. This in fact corresponds to the range which is widely found in native village communities or clans. Accordingly, the "social brain hypothesis" (Dunbar 1998, Dunbar & Schultz 2007) attempts to explain the extraordinary development of the human brain by associating it with particular pressures that a species adapted to social interaction would have had to face, ranging from cooperation, communication, but also deception, to ways of jointly obtaining food, and caring for offspring (Adolphs 2009). These interactions imposed unprecedented needs for coordination and synchronization between members of larger groups, in turn based on increasingly complex brain structures related to social cognitions and emotions.

Regulation of social relations rests primarily on a differentiated repertoire of communicative routines: on the ability to express and recognize affective states in a differentiated manner, to imitate others, to cooperate with them, to communicate verbally, and on the ability of mutual perspective-taking. The increase of brain capacity required for this, however, applies primarily to *acquired abilities*. Thus, immediately after birth, brain size is only 25% of what it is in adulthood (Trevarthen 2001). By comparison, chimpanzee brains are nearly 50% of their adult size at birth, and macaque monkey brains about 70%. Here we can already see that the immense increase in genetically predisposed brain size in humans only develops postnatally. Since the human being is "physiologically premature," as Portman (1969) has shown, his biological development proceeds under the crucial influence of his sociocultural environment, which thus also becomes the relevant "ecological niche" for the development of the brain.

Our description of neuroplasticity (4.2.3) has already shown that growth and differentiation of the brain are not merely determined genetically but also proceed epigenetically. At the time of birth, only the most central neural networks in the phylogenetically earliest brain regions are developed. In the cortex, the growth of dendrites, axons, and the formation of synapses is in full swing. The developing microstructure of the higher brain regions primarily reflects the interactions with the environment, the patterns of which are anchored in neural networks through selection after initially excessive neuro- and synaptogenesis (see 4.2.3). Here genetically predisposed processes of maturation overlap with social ones. Whereas the selective processes are generally completed by the end of the third year (Markowitsch & Welzer 2009, 87), in the youngest brain region, the prefrontal cortex, the highest synaptic density is not reached before the age of 5. Its final structure is only completed around the 25th year of age (Sowell et al. 1999, Fuster 2001).

The human brain thus possesses a unique potentiality, which it cannot, however, realize by itself. The development of the embodied human mind does not only require interaction between brain, body, and environment, but essentially interaction with other humans. In the course of these biographically progressing interactions, the brain becomes a *social, cultural, and biographically shaped organ*. Of course, here we are not dealing with a "networking of brains," as neurobiologists like to formulate it, but with the interaction and shared practice of embodied creatures, that is, with *embodied intersubjectivity*. It is for this reason that current "social neuroscience" (Cacioppo et al. 2002, Decety & Ickes 2011, Cozolino 2014, and many others), which primarily attempts to identify the neural correlates of social cognition, largely remains within the conventional neurobiological paradigm. It applies the problematic notions of representation and simulation to the social other, who then, in turn, can only be construed or simulated as an internal model.

For these reasons, it is all the more important to include perspectives from developmental and social psychology as well as the cultural humanities in investigating the brain as a social organ—or as the "interactive brain" (Di Paolo & De Jaegher 2012, De Jaegher & Di Paolo 2016). The specifically human affective as well as cognitive capacities are certainly rooted in the organism, but they transcend the purely biological level. They originate in the shared cultural sphere and are only acquired in the context of social interaction. In the case of humans, society and culture have a much stronger impact on the development of emotional, cognitive, and social abilities than is the case with any other species. Infants' earliest interactive experiences have a sustained influence on their capacities of dealing with others, on trust and bonding, and thus on future relationship patterns. Over the further course of their development, children increasingly internalize the cultural symbol systems, the roles, and behavioral patterns of their society, which have a shaping influence on the brain.

In the ontogenesis of the human brain, biology and culture, as well as individuality and sociality, are intricately intertwined. Neither the development of the brain nor that of consciousness can be conceived of purely at the biological or individual level, even if they are based on genetic predispositions. Their realization is based on the embeddedness of the "social animal" (Aristotle's "*zôon politikón*") in the community, more specifically, on the unique human ability to recognize others *as their kin*, who can feel, think, and act like oneself. This enables the ability to put oneself in the shoes of others and thus "to learn not just from the other but through the other" (Tomasello 1999, 6). As we will see, the brain as an "organ in relations" is shaped by these processes equally as much as it enables them. The peculiar structure of human individuality as a "finding oneself through others," or the "eccentric position" also manifests itself in neural structures. Thus, the brain becomes the organ of *a human person*.

The following investigation into the social, cultural, and historical nature of the brain traces some basic tenets of this development without attempting to be complete in view of the sheer amount of new findings. In accordance with the dual aspect, the phenomenological level will, in each case, be presented first, followed by the level of neurobiological research insofar as it can be correlated with the former. Initially, the investigation focuses on the development of early intersubjectivity in the dyadic relationship between mother and child. Then it examines the neural foundations of embodied intersubjectivity, which are currently being heavily researched. Finally it turns to the development of "secondary intersubjectivity," especially language acquisition in early childhood. In conclusion, the course of further development is suggestively sketched.

5.1 Primary intersubjectivity

5.1.1 Prenatal development

Human development begins with conception and, accordingly, the development of the brain during the first month of pregnancy. From the beginning, it proceeds in close association with the motherly environment. Embryology and prenatal psychology have shown that the fetus already exists with multiple relations to its surroundings, be it at the sensorimotor or at the emotional level. From the early fetal period into the first year of life, the dyadic relationship between mother and child is the most important precondition for the mental development of the child, as well as for an adequate development of its brain.

Towards the end of the third month of pregnancy, the fetus for the first time shows intense motor activity, in the course of which it touches the umbilical cord and its own body or performs swallowing motions. Through the vestibular organ it perceives the mother's steps as cradling movements and from the fifth month on shows clear reactions to sounds. At 7 months, hearing is fully developed; the fetus now lives in an intrauterine space of touch, sound, and resonance, the impressions of which are already sedimented in implicit memory. Newborn babies verifiably show preferences for the voice of the mother and the sound of her heartbeat, as well as for melodies, rhythms, or verbal texts to which they were exposed during pregnancy (De Casper & Fifer 1980, De

Casper & Spence 1986). Even their crying melody at birth is shaped by the surrounding native language (Mampe et al. 2009). So they have already become familiar with the world before birth, and this facilitates the establishment of the early bond with the mother: what infants are familiar with, for instance, the mother's voice or heartbeat, has a calming effect on them (Salk 1962, Rosner & Doherty 1979).

The unborn child stands in relation to the mother also by means of affective processes. Her mental states can have a sustained influence on the development of the fetus and its brain. A severe stress reaction of the mother, for instance, leads to an increased permeability of the placental barrier for stress hormones, which can impair the functioning of the child's hippocampus and the development of the prefrontal cortex. Various studies confirm that children may later show cognitive and behavioral deviancies under such conditions (Huizinck et al. 2003, de Weerth et al. 2003, van den Bergh & Marcoen 2004). In the meantime, there have also been hints that stress can affect the temperament of the still unborn child (Huizinck et al. 2002). Thus, the influences of the maternal organism as the primary environment of the fetus have consequences for prenatal brain development and obviously also for the child's later development of personality.

5.1.2 Intercorporeality and interaffectivity

From birth onwards, the prenatal symbiosis of mother and child is transformed into a dialogical, intercorporeal relationship. By means of facial expressions, gestures, eye contact, and voicings, the baby actively contributes to the bond with the mother after birth. It tries to establish as much physical contact as possible in order to experience the sensations of warmth, smell, touch, and being held, which it perceives to be pleasant. Newborns also show significantly more interest in the facial expressions, gaze, and behavior of other persons than in inanimate physical objects (Valenza et al. 1996, Farroni et al. 2002, Turati et al. 2002). This "primary intersubjectivity" (Trevarthen 2001)[1] is mainly characterized by embodied, affective, and intuitive forms of relationships, which precede symbolic and verbally mediated communication. Let us take a closer look at these intercorporeal and interaffective relationships before we turn to their neural basis.

Research studies conducted during the last decades have mostly found that babies are able from birth onwards to imitate adults' gestures such as sticking

---

[1] This notion should of course not be taken as if the infant could already grasp the other-as-subject, but rather as the pre-reflective connection of two embodied subjects to a shared intercorporeality.

out their tongue, opening their mouth, frowning, and others (Meltzoff and Moore 1977, 1989).[2] They thus transform a perceived facial expression into the kinesthetic experience of their subject-body. Visual, proprioceptive, and motor modalities are integrated into a joint sensory space; there is an intermodal body schema that connects with the perception of others. That is, the newborn does not perceive its mother as a pure "image," but mimetically by emulating her expression within itself. This capacity for spontaneous imitation of others' expressions has been considered a crucial basis of early social development (Meltzoff and Brooks 2001, Meltzoff and Prinz 2002).

> Such observations corroborate Merleau-Ponty's phenomenological notion of an original sphere of communicative "intercorporeality" (Merleau-Ponty 1960): when two persons encounter one another in face-to-face contact, they are, from the beginning, involved in a systemic interaction that connects their bodies and establishes a pre-verbal and pre-reflective understanding. The feelings of others are empathically felt in their expression, because this causes an, usually unnoticed, bodily impression characterized by subtle sensations and precursors of actions and feelings (Fuchs 2017d). Now this bodily resonance caused in the one person in turn becomes an expression for the other; it will immediately affect his bodily state, and change his expression, however slightly. This creates a circular interplay of expressions and reactions running in split seconds and constantly modifying each person's bodily state, in a process that becomes highly autonomous. They have become parts of a dynamic sensorimotor and interaffective system that connects their bodies by reciprocal movements and reactions, that is, by *intercorporeal resonance*: the other is literally felt with one's own body (see Fuchs & De Jaegher 2009, Fuchs 2017d).

If we regard this connection of intercorporeality and interaffectivity as fundamental, we may assume that the infant gradually develops an emotional resonance with the mother via this, at first, merely bodily emulation. For as we have seen (4.1.4), bodily resonance is an integral component of emotions, which always feeds back into their experience. So if newborns begin to imitate emotional expressions such as smiling, furrowing one's brow, or surprise (Field et al. 1982), they already experience at least precursors of emotional exchange.

---

[2] Recent research with larger samples and a wider range of gestures presented to the infants challenges the results of an innate capacity of imitation, finding no significant excess of matching over non-matching reactions (Oostenbroek et al. 2016). But even if it turns out that imitation is not an innate capacity, but develops in the course of mutual exchanges and matching reactions during the first weeks, it still functions as a major component of primary intersubjectivity.

The mother, in turn, intuitively responds to the newborn's signals with suitable vocal and gestural reactions. In so doing, she uses unconsciously simplified, prototypical behavioral patterns (baby talk, exaggerated facial expressions, eye contact, etc.) which correspond to the still undeveloped repertoire of the child. Papousek and Papousek (1987, 1991) have described these forms of interaction as "intuitive parental competencies." According to this view, mothers and fathers possess an innate form of implicit knowledge that enables them to communicate with the infant by means of sounds, facial expressions, and gestures, to calm or stimulate it and, in so doing, be guided by the child's signals. Infant and caregiver also follow a turn-taking pattern, shifting the roles of agent and recipient in a non-random sequence (Jasnow and Feldstein 1986).

This "protoconversation" (Trevarthen 2001) of mother and child is characterized by baby talk and its musical qualities, the rhythm and dynamics of facial, vocal, and gestural interaction. The different sensory modalities and bodily sensations display a common cinematics which enables them to express the same affect. This may best be rendered by musical qualities (*"crescendo," "decrescendo," "accelerando,"* flowing, soft, explosive, etc.); Stern (1985) also speaks of "vitality affects." The bodily felt emotion of, for instance, joy and its visible expressive movement have an analogous intermodal dynamics, which allows babies already in the first months to recognize feelings such as joy, sadness, or surprise in the movements, gestures, or vocal intonations of others (Hobson 2002, 39–42).

This leads to "affective synchronization" and the "dyadic conscious states" of mother and child, as they have been emphasized in infancy research (Stern 1985, Tronick 2003). This is how the child's basic sense of living together with others in a shared emotional world and being connected with them develops. In the course of this early affective communication, the child gradually learns to associate a caregiver's emotional expressions with typical contexts and thus to differentiate their various shades of meaning. In turn, the child responds with increasing clarity of its own expression. At the age of 9 months, one can speak of fully developed interaffectivity (Stern 1985).

As it turns out, feelings are in no case merely subjective, but primarily intercorporeal phenomena. Bodily-affective communication is the universal language, with a basic grasp of which we are born and which becomes more and more differentiated. Embodiment is the basis of intersubjectivity, inasmuch as we do not ascribe abstract inner states to one another, but experience the facial features and the behavior of others as the expression of their affectivity. Contemporary social cognitive psychology is focused on the notions of "Theory of Mind," "mind-reading," or "mentalizing." According to such theories, it is only from certain stimulus cues that children learn to ascribe "mental states," that is, feelings, ideas, intentions, and goals to others (Carruthers & Smith 1996, Baron-Cohen et al. 2013). But neither children nor adults need theories with premises and conclusions nor inner models in order to understand one another on a basic level. Primary perception of others is not based on hypothetical inferences to an invisible inner sphere in their heads, but on intercorporeal communication and the mutual empathy of embodied subjects (Fuchs & De Jaegher 2009).

5.1.3 Intercorporeal memory

From the beginning, the bodily-affective communication between mother and infant is sedimented in the child's implicit memory. The motor, emotional, and social development do not follow separate paths but are linked in *interactive schemata* or *intercorporeal memory* (Fuchs 2012a, 2017c)

Implicit memory comprises virtually all forms of learning in early infancy. While the explicit, autobiographical memory system only develops from the second to the fourth year of life (Yim et al. 2013), implicit memory is already capable of extracting regularities from repeated interactive experiences from birth onwards (Amini et al. 1996). It comprises, as has already been outlined (see 4.2.3), procedural, that is, habitual motor learning, but also perceptual, cognitive, and affective abilities, which emerge without conscious attention or instruction. Among these count in particular the acquisition of prototypes or perceptual gestalts, object categorization (e.g., animate vs inanimate objects), as well as emotional reactions to similar situations, for example, by conditioning fear responses.

Implicit memory forms an excerpt of repeated, prototypical experiences with others and processes these into dyadic, later also into triadic, interaction schemata (e.g., "mummy and me changing nappies," "daddy, mummy, and me playing," etc.). Stern (1985) speaks of "schemes of being-with" which are structured equally in a sensorimotor, emotional, and temporal manner. Already in the first months of life, a memory for typical interaction sequences can be shown to exist, namely in the anticipation of motherly reactions. Babies learn quickly, for instance, which emotional expressions appeal to, activate, or turn away their mother. They clearly show expectations and, thus, also surprise and disappointment. In the well-known "still face" experiment (Tronick et al. 1978), in which mothers intentionally suppress all facial expression for 2 minutes, children at the age of 3–4 months at first react in an irritated manner and then with intense communicative signals (facial expressions, gestures, vocalizations). Obviously they expect to thus be able to stimulate their mothers to again participate in interaction.

This is how what Stern (1998b) calls "implicit relational knowing" develops: a pre-verbal, not symbolically encoded knowledge of how to deal with others—how to make contact, how to engage in enjoyable activity, how to express pleasure, how to get attention, and so forth. Children quickly learn how they can influence others through their expressions and thus experience self-efficacy. Relational knowing is an organized, so to speak, "musical" memory of the rhythm, dynamics, and emotional connotations that resonate in the interaction with others. Furthermore, it is procedural knowledge in the sense that it is only realized in the process of interacting with others in similar situations. This has far-reaching consequences: as the result of a process of learning that is similar to the acquisition of motor skills, in later life, humans create and stage their relationships according to the patterns they have extracted from their earliest experiences, mostly without being aware of this (Amini et al. 1996).

The traditional psychoanalytic notion of such experience formation has—according to the dualist dogma of an inner psychic sphere—been the "internalization" of attachment figures in the form of "object-representations." But it is not an object or an event as such that is sedimented in memory, but the *interaction of the organism* with this object or event. Meanwhile, in psychoanalysis, one speaks of the internalization of relational experience.[3] This, however, still presupposes the theorem of an inner psychic sphere. But children do not "internalize" anything if they develop in a social environment; rather, in being together with others and in joint practices they learn certain social forms of readiness-for-action that become part of their *implicit relational memory*.

In neural terms, this means that every interaction with others, by means of synaptic learning, leaves traces at the neural level; of course, not in the form of localizable, stored "memories," "images," or "representations" of the interactions or attachment figures, but in the form of *dispositions to perceive, feel, and behave in certain ways*. Such dispositions are based on widespread network connections that equally involve sensory, motor, and limbic-emotional centers. They resonate with current environmental situations or persons and activate the appropriate forms of behavior without the child having to explicitly remember earlier learning processes.

---

[3] An early example is given in Beebe's and Stern's article from 1977: "Hence, what is initially internalised is not an object *per se*, but an 'object relation': actions of self with reference to actions of object. The model of what is internalised, thus, includes mutually regulated sequences of maternal-infant actions with a particular temporal patterning" (Beebe & Stern 1977, 52).

5.2 **Neurobiological foundations**

Social neuroscience has identified several brain regions that are relevant for the processing of social situations. These include, in particular (Allison 2000, Adolphs 2001, Jacob & Jeannerod 2003, Amodio & Frith 2006):

- The superior temporal sulcus in the temporal lobe, which is specialized for the recognition of animate motion (e.g., faces, eye movement).
- The medial prefrontal cortex (mPFC), responsible for person perception and making inferences about other's thoughts ("mentalizing," Theory of Mind).
- The amygdala, which is the substrate of unconscious evaluation of the dangerousness of social situations.
- The orbitofrontal cortex as the correlate of the conscious evaluation of social processes.

Of course, the more complex social situations are, the more brain regions are involved in social understanding.[4] Furthermore, as we will see, there is an extensive neural system in the human brain for perceiving and communicating emotions. All these social capacities are not yet based on verbal communication, but on motor, sensory, affective, and motivational functions which involve the whole brain. If we furthermore take into account linguistic functions and their correlates, we can say: *the brain as a whole is a social organ*, not just individual "social brain regions."

Research in this field so far has mostly remained in the grip of the cognitivist paradigm. Instead of general stimulus approaches, now the reaction of the brain to "social stimuli" is measured. However, this leads to a methodical and conceptual short circuit between the brain and sociality: the social world and the other come into view only as input, corresponding to representations within the neural system.[5] From an enactive point of view, however, social cognition is crucially based on *interaction*: on the one hand, from a developmental perspective, the specific brain activities involved in social cognition are not innate but result from processes of interactive learning, where the brain provides the suitable *matrix* for the formation of social competencies. On the other hand, social

---

[4] Apart from the mentioned areas, these are, for example, the temporoparietal junction and the poles of the temporal lobe (Vogeley et al. 2001, Vollm et al. 2006, Decety & Lamm 2007).

[5] This short circuit is already visible in the title of many publications, for example: "How the brain understands intention" (Becchio et al. 2006); "The empathic brain" (De Vignemont & Singer 2006); or "Visualizing how one brain understands another: a PET study of theory of mind" (Calarge et al. 2003). To note, this mutual "understanding of brains" did not even concern a social situation, but a fictitious encounter which the test subjects should imagine—"Theory of Mind" in the literal sense, without even being in contact.

understanding during face-to-face contacts is bound to an ongoing feedback process of intercorporeal resonance which develops an autonomous dynamics (Fuchs & De Jaegher 2009). In other words, the interactive dyad forms a superordinate system which may not be decomposed into two bodies, even less two brains separately. In such embodied contacts, social interaction is not a mere external input or context, but itself *constitutive* of social cognition (De Jaegher et al. 2010, Di Paolo & De Jaegher 2012).

In what follows, we will take a look at two biological systems that can only be understood with a view to the circular relation between brain, organism, and social environment and which are of crucial importance for the development of primary intersubjectivity, namely the systems of *attachment* and "*mirror neurons*."

5.2.1 **The attachment system**

Currently the most important psychobiological model of the child's social development has been put forward by *attachment theory*. Developed by John Bowlby in the 1950s on a psychoanalytic basis, it has now, after having been at first neglected, become a significant line of research (Bowlby 1982, Hofer 2001, Cassidy & Shaver 2002, Fonagy & Target 2003). According to Bowlby, the social relationships in early childhood are regulated by a biologically grounded attachment system, which has the function of securing care and closeness of the caretakers. It comprises certain phylogenetically rooted signals such as searching, calling, looking at, crying, clinging, or protest in the case of separation. Not only are the child's most fundamental desires fulfilled by these means, but also in this manner the child gains the fundamental sense of trust and the secure basis, from which it can actively explore the world. Conversely, lacking attention, lacking security, or separation from the mother at first lead to stress reactions with heightened arousal, but then increasingly to resignation, apathy, or desperation.

Attachment theory gained substantial support from animal research, which showed that early separation from the mother regularly leads to sustained impairment of behavior as well as of physiological functioning. Among the latter are disturbances in the regulation of body temperature, heart rate, sleep, and hormone secretion (decrease in the level of growth hormones, increase in stress hormones). Ultimately it results in an increased susceptibility to stress and disturbed social behavior (Hofer 1994, Amini et al.1996, Insel & Young 2001, Levine 2002). It could also be shown that such disturbances were passed on transgenerationally: rats that are separated from their mother as newborns are not capable of later appropriately taking care of their own children (Meaney 2001).

Such research provides impressive evidence of how brain structures and functions are shaped and modified by early experience. Of course, this cannot be directly transferred to the case of humans, but there are extensive parallels in attachment research. In the case of the infant, too, the development of psychic and physiological regulatory mechanisms depends on a successful relationship with the mother, on her body warmth, her smell, her touch, her loving care, suitable stimulation, and calming. Such interactions play the same role for emotional and social development that visual stimuli play for the development of vision.

The early dyadic relationship between mother and child can—of course only to a certain degree[6]—be seen as a functional cycle, which encompasses two subjects, one of which represents the environment for the other. The attachment system is a central organizing principle of this system. The infant's still undeveloped organism including its brain consists, so to speak, of "open loops" which require external regulation and fine-tuning through the developed system of the mother in order to achieve homeostasis (Amini 1996). This not only holds for the vital desires for food, warmth, and shelter, but also the emotional desires for security and care. The attachment system supports the regulation of the child's homeostasis through the parents until its own psychobiological functions develop and become autonomous. Perinatal medicine has shown that such regulatory mechanisms can only be incompletely replaced by artificial clinical settings (Trevarthen 2001).

> Neurally the attachment system is mainly rooted in the phylogenetically older structures of the limbic system, for instance, the cingulum, but also involves the orbitofrontal cortex (Schore 1994, Amini 1996). A modulating role is also played by certain hormonal and neurotransmitter systems (oxytocin as an "attachment hormone," opiates, and monoamines). Conversely, Schore (1994, 2001, 2003) has extensively shown how early attachment relationships have an effect on the child's developing brain. He defines attachment as the "interactive regulation of biological synchronization between organisms" by means of "reciprocal processes of affective exchange" (Schore 2003, 162). In their course, cortical and subcortical activities during the first half year of life are integrated into cortico-limbic associative patterns. This primarily affects the development of the anterior cingulate, a region that is involved in play and separation behavior, laughing and crying, as well as facial expressions (Schore 2003, 170). Then, towards the end of the first year of life, the orbitofrontal

---

[6] From the outset, the relationship between mother and child is not only a biological, but a personal relationship, which becomes already visible in the intense mutual contact of gazes. If the mother did not treat the infant as a "little person," it could never develop its personal capacities.

system develops (see 5.3.3). In each case, these are *interactive* processes of growth, not endogenous ones.

In the dyadic regulatory processes of attachment relationships, we find the ecological or systemic parallel of phenomenal interaffectivity: an overarching self-regulating system of two organisms is formed as an extension of the child's still underdeveloped homeostatic system. Here we have a special case of the general relation between organism and environment: the child's organism already possesses the *pre-gestalt* or anticipation of complementary motherly functions. Conversely, the mother is biologically as well as psychologically oriented towards the needs of the developing child, up to the disposition towards intuitive competencies in dealing with the infant (see 5.1.2).

The relationship between mother and infant does not entail one system having an effect on the other, but a constant *reconfiguration of the dyadic system as a whole*. Here a pre-existing complementary relationship is continuously re-actualized— for instance, if a too great distance is balanced out again by re-approaching, or affective attunement is re-established after a misunderstanding (Stern 1985). If the complementation and regulation by the mother is successful, the infant increasingly becomes capable of regulating its own affective states. At the same time, its early relationship experiences are taken into implicit memory and anchored there as *secure attachments* which, as models, shape later relationships and regulation of social emotions up into adult life (Grossmann et al. 2002, Vrtička et al. 2009, 2012).

On the other hand, the reaction of the infant to a temporary disruption of the attachment relationship can already be impressively demonstrated in the above-mentioned "*still face*" experiment (5.1.3). Even more so, the complex pattern of separation behavior with its emotional and physiological components—crying, desolation, fear, despair, increase of stress hormones, decrease in heart rate and level of growth hormones (Hofer 2001)—is revealing of the close coupling of the dyad by means of the attachment system. With increasing frustration of its desires, however, the child's capacity for relating emotionally can be permanently damaged. In that case it cannot sufficiently regulate stress and intense affects. Not only the development of cognitive structures, but also the even more fundamental development of the emotional-limbic system is thus an experience-dependent process which can be subject to various disruptions.

The effects of deprivation in the first months of life are far-reaching. René Spitz (1965) performed well-known studies of orphans who, in the case of complete withdrawal of emotional care developed severe deprivation syndromes such as apathy, depression, or increased mortality rate. More recent investigations showed that Rumanian orphans who were adopted after having spent only a few months in an orphanage were physically, mentally, and socially less developed than a group of children who were raised in foster families from birth (O'Connor & Rutter 2000). Less severe relationship disruptions, for instance, due to postnatal depression of the mother, also have a negative effect on the cognitive and emotional development of the child (Murray & Cooper 2003).

Children who suffer early *traumatization* show lifelong abnormal physiological stress reactions, a reduction of size in the hippocampus, and impairments in the development of frontal brain regions in the right hemisphere (Hofer 1994, Insel & Young 2001). Such disturbances of brain development go hand in hand with a series of psychic impairments, for instance, of attention, of impulse control, or affective regulation, as is the case in borderline personality disorder (Cirulli et al. 2003, Schore 2003). The ability to adequately care for one's own children can be severely impaired if the mother herself experienced neglect or abuse. In this manner, early disruptions of attachment and development can be passed on transgenerationally (Ricks 1985, Chicetti & Carlson 1989).

The fundamental complementarity and mutual relatedness of brain, organism, and environment requires adjustments not only at the biological but also at the emotional and social levels for the brain to develop adequately. For this to happen, the innate language of embodied affective communication needs to be understood and responded to. Without this resonance, the homeostasis and the regulation of the child's organism are endangered. This can have serious consequences for the child's neurophysiological, emotional, and social development. The next section turns to another system that is of significant importance for this resonance.

5.2.2 The social resonance system ("mirror neurons")

A crucial step in investigating the social nature of the human brain was taken when the research group of Rizzolatti and Gallese discovered the system of the so-called mirror neurons.[7] This system may be taken as a confirmation of the phenomenological insight that social understanding and empathy are not primarily based on a "Theory of Mind," but on the embodied perception of the other, or on intercorporeality: we always already perceive others as being "of our own kind," because in each interaction our body is subliminally attuned to their facial and gestural expressions and their intentions of movement. The general interconnection of perception and movement thus becomes relevant for social perception, and the "mirror neurons" are ultimately nothing else than specialized components of the superordinate sensorimotor cycles which carry the interbodily exchange.

---

[7] Rizzolatti et al. 1998, 2001, Gallese 2001, 2002, 2005, Rizzolatti & Sinigaglia 2008.

It should be noted critically, however, that the term "mirror neurons" as such favors a representationalist conception of this system and therefore its choice seems unfortunate. It suggests, on the one hand, an internal imaging which is then re-projected onto the other ("mirrored"); on the other hand, it attributes nearly miraculous abilities to single neurons, as if they could accomplish social perception by themselves. In contrast to this, it should first be remembered that *neurons cannot mirror anything*. At the physical or physiological level, the representational relation of original and mirror image is nowhere to be found. A mirror, too, only reflects rays of light that hit it—to see this light *as a mirror image* is a feat that only a conscious being can achieve. Moreover, single neurons (even if their recording enabled the discovery) are certainly not able to respond to the complex sensorimotor situation in which consists social perception. For this reason, I prefer to speak of a "neural resonance system," but will also use the nearly unavoidable term "mirror neurons" in the following.

5.2.2.1 Foundations

"Mirror neurons" were first discovered in the premotor cortex of macaque monkeys, which organizes and regulates movement. In the meantime, they have been shown to exist, too, in human premotor, but also other regions of the brain.[8] Mirror neurons are activated both when a creature itself performs a specific action such as reaching for an apple or a cup, but also when it perceives the same action in a conspecific. Thus, the system intermodally connects interpersonal perception with one's own movement. This seems to happen in a specific manner; at least, in monkeys, mirror neurons react

1. Only to the movements of *animate* agents or conspecifics. If the movement is performed by a mechanical apparatus (e.g., pliers or virtual hands) instead of conspecifics, the mirror neurons remain silent.

---

[8] In the macaque monkeys studied by Rizzolatti and Gallese, mirror neurons are mainly located in the premotor area F5, in humans in the areas A44 and A45 which (this is of particular interest) are largely identical with Broca's area, an important substrate of speech. Whereas in monkeys they were identified by single-neuron recording, indications for their existence in humans were first found by EEG recording, fMRI scans (Buccino et al. 2004) as well as from subliminal activation of hand muscles during observation of another's hand movements (Fadiga et al. 1995, Gallese 2001). It was only in 2010 that a review of single-cell recordings in epileptic patients provided clear evidence of mirror neurons in the supplementary motor area and the medial temporal lobe (Mukamel et al. 2010), that is to say, also outside of the regions of the brain traditionally considered motor. It should be noted, however, that their function for action understanding is still a matter of debate (see, for example, Caramazza et al. 2014, Hickok 2014).

2. Especially to *goal-* or *object-directed* movements. This holds true even for incomplete goal-directed movements, that is, activation also occurs if the hand observed is trying to grip an object that was previously shown but now is hidden again (Umiltà et al. 2001). That is, the goal or the entire gestalt of the movement is completed or anticipated by the system.
3. The motor system is also activated if one only imagines performing a movement and, even more so, if one *imitates* an observed movement. Already while observing, there is subliminal activation of one's own corresponding muscles, that is, imitation is already being prepared (Fadiga et al. 1995).

A number of conclusions can be drawn from these results:

1. The social resonance system obviously contributes to perceiving movements of conspecifics as goal-directed actions by activating the observer's own motor system. The most probable interpretation is that this brings about a sensation of how an action "feels" for an agent and what purpose it serves.
2. The system establishes a resonance between homologous body parts of self and other. It furthermore enables the intermodal connection of visual perception, the motor body schema, and proprioception.
3. The resonance of mirror neurons also supports action readiness, in particular for imitating an action (Hari et al. 1998). Thus, the system could be a basis for model and imitation learning, a crucial human capacity for cultural development (Tomasello 1999, Meltzoff & Prinz 2002).

These results and conclusions correspond well with Merleau-Ponty's conception of intercorporeality as an intersubjective body schema: we understand the intentional actions of others, because our body "translates" them into its own actions. In other words, we use the "operative intentionality" of our body as a means to understand the intentional movements of others:

> The communication or comprehension of gestures comes about through the reciprocity of my intentions and the gestures of others, of my gestures and the intentions discernible in the conduct of other people. It is as if the other person's intentions inhabited my body and mine his. (Merleau-Ponty 1962, 215)

The sensorimotor neural resonance system is also in accordance with an enactive account of social cognition (De Jaegher & Di Paolo 2007, Fuchs & De Jaegher 2009, De Jaegher et al. 2010): perception and action are inherently connected, not only in the general relation of a living being to its environment, but also in social interactions. On the other hand, motor mirror neurons are not the only system that contributes to an understanding of others. They are complemented by resonance systems for the emotional expression of others.

Specialized regions, for instance, in the superior temporal sulcus of the temporal lobe, are involved in the interpretation of animate movement and in cases of "contagion" through the laughter, crying, or yawning of others. Regions of the sensory cortex in the parietal lobe, too, especially the inferior parietal region, play an essential role for the empathic sensations that are activated in perceiving others and which are required for full-fledged empathy (Jacob & Jeannerod 2004, Buccino et al. 2004). Of particular relevance are pain-specific areas in the frontal cingulate cortex, which react to one's own pain, but also to observed pain stimuli, and even if pain experience is only expected in someone else (Hutchison et al. 1999, Singer et al. 2004). Areas in the anterior cingulate and left anterior insula are also activated when observing an embarrassed person, that is to say, one experiences vicarious "social pain," as if it happened to oneself (Krach et al. 2011).

Furthermore, if one observes a disgust reaction in others due to an unpleasant olfactory perception, this activates a region in the frontal insula that is also involved in one's own disgust reactions (Wicker et al. 2003). If this region is damaged due to a stroke, the person is no longer capable of experiencing disgust himself, but also fails to recognize the expression of disgust in others (Calder et al. 2000). In all these cases, one's own embodied sensations become the medium of intercorporeal resonance, which can now be confirmed at the neurobiological level (see Fuchs & Koch 2014).

On the basis of such research results, one can now posit a complex social resonance system spread out over different brain regions. It is assumed to integrate the various functions of self and other perception as well as emotional sensation and thus to become a basis for intersubjective perception, imitation, and empathy.[9] Of course, this biologically grounded system does not itself produce human sociality. Much rather, it requires *real intercorporeality and interaffectivity* in order to develop. The actions registered by it and especially their emotional and intentional relations are not innate, but are based on typical, repeated experiences. Reaching for an apple can only activate the mirror system if the apple has acquired the *meaning* of a goal-of-reaching. Imitation, too, soon goes beyond simple neonatal mimicking and

---

[9] It should be noted, however, that empathy in its full sense comprises a whole complex of perceptions, feelings, preceding experiences, representations, and contextual relevances, that is, an encompassing affective-cognitive activity which certainly may no longer attributed to single (however specialized) neural systems (see Fuchs 2017e).

is directed towards the goals of the actions of others (Gergely et al. 2002). What is equally required for the neurally supported co-experiencing of emotional reactions is a contextual understanding, for instance, of the connection between smell and disgust. It should also be noted that the macaque monkeys with "mirror neurons" do not even imitate or appear to know about other minds, indicating that enculturation is crucially required for mere "mirroring" at the neural level to enable the understanding of others (Iriki 2006). Hence, the social resonance system can only fulfill its function if it is embedded in an *intersubjective space of shared action and meaning*.[10]

5.2.2.2 Simulation or resonance?

The theoretical interpretation of the mirroring system remains controversial. Gallese and Goldman (1998) have based a "simulation theory of understanding" on it and contrasted it to previously dominant "Theory of Mind" conceptions: internal simulation or "as-if"-imitation is taken to be the biologically grounded instrument of primates for understanding others. They achieve this by using their own bodies as mirrors of the others' actions and intentions: "our motor system becomes active as if we were executing that very same action that we are observing [...] action observation implies *action simulation*" (Gallese (2001).

Thus, simulation theory represents an advance over "Theory of Mind" insofar as it, instead of cognitive models or inferences, emphasizes the embodied perception of others. But for simulation theory, too, primary intersubjectivity essentially consists in the projection of one's own mental states onto others with the sole difference that the content of what is projected is a simulation rather than a cognition. The underlying model still is one of representation: mirror neurons give rise to "an internal representation of goals, emotions, body states and the like to map the same states in other individuals" (Goldmann & Gallese 2000, 255). In a sense, a person that perceives others does not directly interact with them, but with internally generated models or simulations of their actions.

---

[10] Accordingly, Catmur and colleagues (2007, 2011) could show that the sensorimotor social resonance can be reshaped in adults through experience. In a training phase, test subjects had to observe the movements of another person's little finger while moving their own pointer. After a few training units, a "counter-mirroring" occurred, that is, the observed movement of the little finger lead to activations in the motor cortex that were connected with the pointer. It may be assumed that not only such basal learning effects, but also social and cultural influences shape the neural resonance system—as a system of "open loops" formed and continuously modified through social interaction.

The complicated conception of an internal simulation, in which the perception of others is generated and then projected onto the outside world, can be significantly simplified by use of the suggested *notion of neural resonance*: in perceiving persons acting in a goal-directed manner, the external stimulus configuration obviously resonates not only with patterns of visual neural activity but also with the motor system in the brain, which is otherwise activated in oneself carrying out such an action (matters can be construed analogously in the case of emotional empathizing). This coupling is similar to the implicit auditory-motor coupling as we have described it in the case of a piano player: just listening to a melody activates the motor pattern of the corresponding movements (see 4.2.1).

But these coupled systems can, as components of the *entire neural resonance*, become part of the perception of the other without requiring any further "representation." In that case, there are no two separate processes, visual perception and internal motor simulation, which need to be pieced together, but the perceived movement is already understood as goal-directed and imbued with a sense of familiarity. In the context of this interpretation, mirror neurons do not generate internal representations, but serve only as specific carriers of *embodied social perception*.

Thus, the phenomenological notion of transparency (see 4.2.4) also applies to the sensorimotor perception of the other: one's own subject-body, by means of implicit coupling, is integrated in perception in a manner that we "see through it." Without us noticing, the sight of the other is tied together with subtle embodied sensations and creates a vivid overall impression. Social perception is always "colored" by empathic sensations, so to speak. This corresponds well with the phenomenological analyses by Merleau-Ponty: the resonance in one's own body is a transparent part of interpersonal perception, namely in the sense of an implicit coupling that can, in principle, be explicated and brought to awareness by reflective observation. Otherwise, these phenomenological analyses would not have been possible.

Let us summarize: "mirror neurons," which are part of the motor system but also have a perceptual function, can be well integrated into enactive concepts: the circular connection of perception and movement proves to also be fundamental for intersubjective relations. The understanding as well as the imitation of the actions of others are based on a form of "perception already containing movement." Empathy, too, that is, feeling emotions with others can be explained by the cycle of embodied affectivity (see 4.1.4, Figure 4.3). "E-motions," in a sense, also are movements and always sketch certain possible motor reactions, for instance, a sad, downcast gaze, a fist clenched with anger, or a joyously elated jump. If we furthermore take into account that social perception usually is not unilateral but, as in the case of the dyadic relationship between mother and child, is characterized by dynamic interplay proceeding within fragments of seconds, and in which both partners continuously modify their perception of each other, then we can say that in this case two organisms have been conjoined in one *dynamic resonance system* (Fuchs & De Jaegher 2009, Fuchs & Koch 2014):

> Between my consciousness and my body as I experience it, between this phenomenal body of mine and another as I see it from the outside, there exists a relation which causes the other to appear as the completion of the system. (Merleau-Ponty 1962, 410)

Even if this field of research is still in a state of flux, there are indications that at the biological level, too, there is an *intersubjective matrix* for the social perception of actions and emotions. At the phenomenological level, this corresponds to the primacy of intercorporeality as an encompassing sphere out of which self and other gradually emerge as independent poles and come face to face. We now turn to the further development of this secondary intersubjectivity.

5.3 Secondary intersubjectivity

Towards the end of the first year of life, the infant develops the ability of, together with a caretaker, jointly shifting the focus of attention to a third element, that is, an object or event. This marks a new stage of development, one which Trevarthen has termed "secondary intersubjectivity" (Trevarthen & Hubley 1978). The neurobiological foundations of this stage have only been rudimentarily researched and therefore can only be outlined here. In the present context, the aim is to characterize the development of higher cognitive structures of intersubjectivity as an *embodied and interpersonal* development (Fuchs 2013b): in particular, the human faculties of language, reflection, and perspective-taking emerge neither from genetically determined maturation nor by "social programming" of the brain, but primarily by means of *shared social practices* which are sedimented in the structures of the brain.

5.3.1 The 9-month revolution

At around 9–12 months of age, a series of developmental leaps occur, which Tomasello (1999) has called the "9-month revolution." Biologically, this development is supported by the fact that synaptogenesis, that is, the formation of new neural connections, peaks at this point before reducing in the second year of life. So here lies the phase of the most intense transformation of environmental experience, especially of social interactions into persisting network structures of the brain (Markowitsch & Welzer 2009, 150–151). On the other hand, the development of the hippocampus as well as of working memory structures in the parietal and frontal lobes reaches a point that enables keeping in mind complex information over a certain time span and thus makes possible the learning processes that follow.

From the age of 10 to 12 months, babies begin to jointly attend to objects with caretakers and, in so doing, reassure themselves of their attention by brief gazes—a phenomenon commonly referred to as "joint attention" (Tomasello 1999). Soon the babies go on to direct the attention of grown-up to objects by "deictic gestures," pointing towards objects that they want others to see, even when these objects are invisible or absent (Carpenter et al. 1998, Liszkowski et al. 2006). Conversely, they begin to understand the pointing gestures of grown-ups, that is, the "meaning" of the indicating pointer. This implies a shared relation to a third element that is jointly seen or handled. As Davidson has put it, "each is interacting simultaneously with the world and with the other agent" (Davidson 2001, 128), and each is also aware of the other's gaze which is often expressed by a shared affect such as "knowing smiles" (Carpenter et al. 1998). In short, sharing attention means being aware of this sharing.

Thus the circle of primary, dyadic intercorporeality is opened up and transformed into a *triangle*: its basis is the primary embodied relation between mother and infant, supported through mutual eye contact, and its sides are (a) the convergent directions of their gazes towards an object, and (b) the mother's or the infant's gesture of declarative pointing (Fuchs 2013b). With object triangulation, according to Tomasello, a specific form of human communication manifests itself: here lies the line of demarcation to the mental abilities of other primates, who cannot develop joint attention through declarative pointing. Even though great apes when growing up in human captivity may become capable of so-called imperative pointing ("Give me this!"), there is no declarative or cooperative meaning attached to it. This communicative and cooperative attitude has been highlighted by Tomasello as the crucial difference (Tomasello et al. 2005, 2007): only through the sharing of intentions, an actual *"we-intentionality"* is created ("Look at this!," "Now we are looking at this object together").

A further, connected phenomenon is *social referencing*: at around 9 months, infants learn the emotional evaluation of objects and situations, for instance, what is dangerous or harmless and so on, by looking at the reactions of their caregivers (Hornik et al. 1987, Hirshberg et al. 1990). According to the recent concept of *natural pedagogy* developed by Csibra and Gergely (2009, 2011), infants even show an innate tendency of learning from others: so-called ostensive cues of the caregiver such as eye contact, gestures, and vocalizations signal a shared learning context ("This is important!"), so that the infant understands the adult's subsequent object-directed action as meaningful and tries to imitate it.

In order to do this, the infant needs to grasp the goal, purpose, and means of the action. Thus it enters the "intentional space" of the other and attempts to understand for which purpose he or she uses the object. At the same time, infants also begin to realize that they are regarded as intentional agents by others. Thus, joint attention is not only related to objects, but also gives infants a new means of assessing how people relate to them. "In checking back and forth whether someone else is engaged jointly with them, infants begin also to check for their own existence in the mind of others" (Rochat 2009, 80).

In sum, joint attention, that is, being directed towards a shared context, grasping the perspective of the other as well as one's perspective on oneself, means a decisive step in human ontogenesis: here is the beginning of the "eccentric position," in which human personhood finds its specific expression. This step is taken long before the development of linguistic symbolization, before any theoretical knowledge of the mental states of others. Studies show that 10–11-month old children can already distinguish intentional sequences, that is, beginning, goal, and end in the continuous behavior of adults (Baldwin & Baird 2001). By 15–18 months, they understand what an adult intends to do and complete an unfinished action that the adult only attempted to accomplish (Meltzoff & Brooks 2001). Before they know of the significance of intentions at the theoretical or conceptual level, children already possess a pragmatic knowledge of them. The intentionality of others is no private mental state that first needs to be inferred or simulated, but is visible in the meaningful gestalts of their actions and expressed in the gestures of their subject bodies.

5.3.2 The embodied development of language

5.3.2.1 Language as social practice

In learning language, the child then acquires a fundamentally new, that is, representational or symbolic medium of communication, but also of knowledge of oneself and the world. This development proceeds via pre-symbolic forms of communication, which we have already dealt with: linguistic exchange presupposes intercorporeal and affective exchange (Fuchs 2016c). If we follow the socio-pragmatic approach to language development (Bruner 1983, Nelson 1996, Tomasello 2008), it does not proceed in a purely cognitive manner as though language were a symbol system to be learned abstractly; rather it is scaffolded by situations of intercorporeality, shared attention, joint practice, and ostensive cuing. The first words spoken by children are integrated into the actions in which they are involved and which are structured by the parents. The prerequisites are in particular:

1. The child's participation in an interactive framework that is already preverbally developed, in other words, verbal interaction presumes intercorporeal exchange.

2. Joint attention to a third entity, and specifically in the practical context that the speech refers to—that is, the triadic situation.
3. Understanding the communicative intentions of others as being based on their goal-directed movements, pointing, or expressive gestures.

Hence, social practice represents the reference point and at the same time the scaffolding context within which a symbolic language can be learned. In concrete terms, this means that the first words are connected with already comprehensible gestures, in particular, the pointing gesture. For example, the parents ostensibly look at or point to objects and name them ("Look! A ball!"). The child now must understand that the parent intends to share attention with her to some outside entity, or in other words, the communicative intention (Tomasello 2000, 2008). Of course, grasping the word as meaningful does not yet imply higher conceptual capacities, but rather a typification of proto-concepts according to similarities of shape and behavior ("ball" means "such round, rolling things"). In the sequence, this leads to a *reverse imitation:* Now the child uses the first words ("there!," "ball," etc.), often connected with a pointing gesture, to show the adult what she herself finds interesting and wants to share. The adult's understanding of the verbal gesture then acts as a reinforcement which stabilizes the new gestural meaning.[11]

A crucial question is how cognitively demanding this early communication should be conceived. Tomasello explains it in terms of Grice's (1989) complex theory of language and meaning:

> This is what a linguistic symbol is. It is a noise (or other behavior) that two or more individuals use with one another to direct one another's attention and thereby to share attention—and they both know this is what they are doing. (Tomasello 2000, 405)

This is already a high-level account of cognitive intentions, implying some kind of meta-perspective on the communication ("I know that you know what I mean"). It seems highly probable that this rather abstract level is only reached later on, whereas the early language use is based on situated and embodied interaction (Fuchs 2016c).

Thus, even if the verbal meanings can increasingly be detached from the concrete situation—at first, all of early speech acquisition is against the backdrop of interactive situations and short episodes: eating, washing, dressing, changing nappies, playing, building a tower out of blocks, feeding ducks, and so on. The

---

[11] Frequently, the interaction also selects wording from spontaneous sound production and the child's babbling, making them into meaningful signals: for example, when the child says "mummy" or "daddy," the parents presume her intention is to form these words and reinforce them accordingly. Recognizing the effect of her own sounds then leads the child to learn their meaning.

child always first learns co-involvement with the relevant practical situation and to form mutual goals, and then he orders the speech, which he has heard, into this context (Bruner 1983). He learns the word "ball" when playing ball, the word "there" in association with the pointing gesture, and the word "Ow!" in connection with an expression of pain and so on.

Hence, children's perception of the environment is synchronized with the corresponding verbal expressions that denote it and with the adult's visible attention and intentional behavior. They only adopt a word for a new object when his or her attention is actually directed towards this object. If the adult is looking in another direction or the voice is coming from a tape, the child doesn't connect word and object (Tomasello 2000; Dittmann 2002, 43). The capacity for speech therefore only develops within social scaffolding through an intercorporeal practice that is oriented towards a shared environment.

In fact, the word is a *vocal gesture* and initially only complements the pointing gesture as a first sign. But the voice also separates the sign from the physical movement and transports it into the invisible, no longer localizable medium of sound. Thereby, the possibilities of referencing multiply, and ultimately the sound signs can even be detached from the concrete situation. They are capable of pointing to absent objects, for example, to Mummy or Daddy when they are absent; they are even capable to pointing to "something like," that is, to similar, general, or abstract objects. The gestural-iconic representation is then increasingly replaced by propositional speech, and the remaining gestures accompanying verbal speech serve more visual aspects, for example, to illustrate forms, directions, and structures that are the topic of speech.

5.3.2.2 Neurobiological foundations

As we can see from this brief outline of speech acquisition, the body as the medium of all action and interaction plays a fundamental role in the process. How is this reflected in the *neuronal anchoring* of language?

Neuroplasticity is a crucial presupposition for language development; in the course of meaningful interactions with others, the brain also becomes the matrix of language. The overabundant formation of synaptic connections in the first years of life corresponds to the infant's universal endowment for vocalization and language, which by acquiring the mother language is then restricted to a culture-bound pattern.[12] This follows the general principle that a surplus of developmental potential is selected and specialized under the influence of sociocultural environment.

---

[12] Thus, it has been shown that in the first months, babies are still able to differentiate more phonemes than adults of their culture (Howe 2000, 5).

Two aspects are significant here. Firstly, EEG studies show that up to the second year of life the earlier developing right half of the brain, which is the dominant hemisphere for processing music, also manifests stronger activation while listening to language than the left half (Patel 2003, McCullen and Saffran 2004). This corresponds to the enhanced role of musical elements, namely, of speech melody, intonation, and rhythm for the perception of the toddler (Trevarthen 1998).

The more advanced the development of symbolic speech, the more areas in the left brain take over verbally relevant functions, in particular, Wernicke's and Broca's areas and other premotor areas as well as the basal ganglia (Kaan & Swaab 2002). However, even at a later stage in life, recent results suggest that the neuronal resources for processing speech and music still heavily overlap, in particular, in Broca's region and its counterpart in the right-half of the brain (Koelsch 2005, Koelsch et al. 2005). This suggests that, at least in infancy, the brain does not process music and speech as separate domains, but rather processes speech as a particular form of music, indeed *that the musical capacities of humans represent a decisive precondition for speech acquisition.*[13]

Both music and language are organized temporally, with the relevant structures unfolding in time, as patterns and sequences of rhythm, emphasis, intonation, phrasing, and contour (McMullen & Saffran 2004).[14] This is in correspondence with the central role of melodious-rhythmic interaction, vitality contours, and affective resonance in the early mother–child dyad, which was mentioned previously: the musicality of the interaction may be regarded as prefiguring the temporal dynamics in which language may then unfold. The theory of early "Communicative Musicality" is supported by acoustic analyses of the measures of rhythm, quality, and dynamics in the vocal interplay between infants and adults (Malloch 1999). Here, an emotional aspect of speech development is involved that is especially manifest in prosody. Accordingly, recent neuroimaging results indicate that responses to human vocal sounds are strongest in the right superior temporal area (Belin et al. 2002), near areas that have been implicated in processing of musical pitch (McCullen & Saffran 2004).

---

[13] The idea of singing being the ancestral origin of speech was first put forward by Giambattista Vico in his notion of "*Parlare cantando*" (see Trabant 1991).

[14] This correspondence of temporal structure has already been noted by Adam Smith in his essay *Of the imitative arts* ([1777] 1982): "Time and measure are to instrumental music what order and method are to discourse; they break it into proper parts and divisions, by which we are enabled both to remember better what has gone before, and frequently to foresee somewhat of what is to come after […] the enjoyment of Music arises partly from memory and partly from foresight" (quoted after Trevarthen 2012, 259).

This lends plausibility to accounts of musical and linguistic co-evolution that emphasize emotional communication through prosody as a primary root of both systems.

The second aspect is related to the embedding of speech acquisition in interactive contexts. Specialized brain systems are required for the neuronal connection of action, perception, and meaning through speech, and there is now evidence to suggest a crucial role for the sensorimotor resonance system. The localization of Broca's region in the inferior premotor cortex (responsible for speech production, but also for hand and mouth movement) and its coincidence with the main areas of the "mirror neuron" system suggests that language originally represented an *interpersonal resonance system for action schemes*: via the communication of the mirror neuron system, the voice was able to call up the idea of the intended actions and objects in both speaker and listener (Fuchs 2016c).

As mentioned earlier, mirror neurons are activated both when observing a conspecific reach for or grasp an object and when imagining oneself reaching or grasping without actually moving one's hand. Thus the system leads to matching an observed movement to the internally generated enactment of the same movement in the observer. Speculating on a connection to the evolution of language, Rizzolatti and Arbib (1998) have first assumed that the mirror neuron system also enables intentional meaning to be assigned to another's *vocal gesture*. The connection could be spelled out as follows (see Gallese 2008; Jirak et al. 2010):

Mirror neurons also react to suggested *goal-directed* movements, that is, they are activated when the hand of another individual reaches for an object that was already visible earlier, yet is now out of sight (Umiltá et al. 2001). This clearly corresponds to the pointing gesture which may be directed to a distant or even invisible object. Thus, the neural resonance system would be suitable to support the connection of pointing and the object, by evoking one's own experience of directed movement and gaze. The discovery of *audiomotor mirror neurons* in the Broca homologous area of monkeys also makes this plausible for *vocal gestures* (Kohler et al. 2002, Keysers et al. 2003). These neurons are activated (1) if the animal observes an action, which generates a sound—for example, knocking on a table or cracking a peanut; (2) if the animal performs the action itself; or also (3) if it *only hears* the knock or crack without seeing the movement. Transferring this to the voice, this would imply that the heard voice could potentially evoke the same action with an object that the listener could carry out himself.

Hence, in early speech acquisition when pointing and sound gestures are typically linked with each other, a neuronal coupling would be produced between

(1) the object being pointed to, (2) the related sound, and (3) one's own action with the object. As a result, *the originally only accompanying sound becomes capable of evoking the intended object and the object-related action scheme in the listener.*[15] At the same time, the gesticulating and pointing to objects become unnecessary and recede more and more into the background—as can also be observed in the development of infants.

In the acoustic medium, the word detaches itself from the speaker and is heard by him and the recipient together. The acoustic gesture is thus no longer subject-bound, but for both partners becomes a third entity, an *intersubjective symbol*. On a neurobiological level, this may be now understood as follows: verbal communication is grounded in the fact that via the medium of the resonance system, the word activates—in both speaker and listener—a congruence of neuronal patterns, and thus of ideas or action schemes. The concordant intention of both partners, which manifests itself in the word as an intersubjective symbol, would thus find its match in the resonance which forms between them on the neuronal level. Speech not only produces an intellectual connection among individuals, it additionally involves a biologically anchored *interbodily resonance system*. Thus, it is by virtue of our bodies acquiring, through social interaction, similar neurological structures and entering into resonance during conversation that we can share the meaning of words and sentences.

This is supported by recent fMRI research on coupled brain activity in speaker and listener during verbal communication: low-level auditory areas (processing the shared input), areas responsible for the production (e.g., Broca's area), and for the comprehension of language (e.g., Wernicke's area and temporoparietal junction) are simultaneously active in both partners, regardless of whether speaking or listening, and these areas exhibit joint, closely temporally coupled, response patterns (Stephens et al. 2010). Moreover, the stronger this neural coupling between interlocutors, the better the understanding (Stephens et al. 2010). This production/comprehension coupling is in accordance with the perception/action coupling mediated by "mirror neurons." Although it must be added that the precise functional relevance of these neural resonance systems for the evolution and ontogeny of language remains far from being clarified, they already offer strong empirical support for an embodied and enactive view of language.

---

[15] This connection is particularly supported by Aziz-Zadeh et al. (2006), who could demonstrate that the same cortical regions activated by *action observation* are also activated by the understanding of *action-related sentences*. On the linkage between word comprehension and related action tendencies observable in motor regions of the brain, see, for more detail, Pulvermüller (2005) and Fuchs (2016c).

Verbal communication is not a transfer of symbolic significances from one mind to another, but a "gesturing with words," co-enacting our actual and possible relations to the world, and scaffolded by our shared practical contexts. In particular, the pointing gesture, through uniting bodily movement and "we-intentionality," may be regarded as the lynchpin that leads from primary intercorporeality to the sharing of meanings through symbolic interaction. However, as Merleau-Ponty has argued, this transition never loses the gestural, enactive basis from which language first develops:

> The spoken word is a genuine gesture, and it contains its meaning in the same way as the gesture contains its. This is what makes communication possible. In order that I may understand the words of another person, it is clear that his vocabulary and syntax must be "already known" to me. But that does not mean that words do their work by arousing in me "representations" associated with them, and which in aggregate eventually reproduce in me the original "representation" of the speaker. What I communicate with primarily is not "representations" or thought, but a speaking subject, with a certain style of being and with the "world" at which he directs his aim. (Merleau-Ponty 1962, 213)

In other words, speech is primarily not a symbol system, but *transformed gesture*, enacted by the body and evoking possible actions in it. Speaking and understanding are lived acts in which our experiences as embodied agents are always present, both in the content and in the syntactical structure that expresses it.

Speech capacity therefore does not develop merely from a biological *Anlage* or genetic disposition, but unlike any other human capacity it requires embedding in a sphere of shared meaning structures and communicative practice in order to evolve. Verbal meanings only exist *between* individuals just as pointing with one's finger only attains its meaning from the jointly oriented gaze. Words are carriers of intersubjective meanings, which have formed within a culture and increasingly differentiated into a complex referential system. To learn these words, children must primarily be in intercorporeal, emotional, and practical contact with others. They must further develop the capacity to focus on the same object and to share this intention with them. Scaffolded by these triadic practical situations, the sound gestures may develop whereby we communicate with one another symbolically.

When in the embodied interaction with others the child learns their speech, then his brain functions as an organ of mediation that increasingly matches the heard words with neuronal patterns related to action, interaction, and object experiences. This matching only occurs if the child experiences the others as intentional actors who *intend to show him something through their speech* and whose goal is the intended object. In short, the child must experience himself as the *intended* participant of communication. Only then—and not by means of a mechanical-associative connection—can the new words become sedimented as neuronal patterns that are associated with experiences of acting and interacting. The coupling of language perception and motor activity, which is now demonstrated by numerous imaging studies of the brain,[16] shows that the meaning of words always remains connected to the interactive and embodied experiences in which they have been acquired.

The brain as such certainly does not become the location of meanings or the "symbol-processing organ," as it is sometimes referred to. The neuronal patterns, as correlates of speech, are only the *necessary* condition for the child to understand words as meaningful and thus participate in the shared world of the mind conveyed through symbols. Only such participation in the symbolic world is the sufficient condition for speech acquisition. Language, as I said before, is based on meanings, and meanings are ultimately based on embodied *relationships*. Correlates of these meanings are functionally and morphologically inscribed on the brain as neuronal patterns in the course of interaction. In this way, language becomes enmeshed in our organic life: we incorporate into our bodies a linguistic style of being. This is also the reason why "linguistic events have a direct route to even our physiology, why the complex sociocultural and interpersonal matrix disclosed by an insult or a compliment make our blood rush in quite different ways" (Cuffari et al. 2015, 1116). Language is nothing else than a manifestation of our embodied sociality.

---

[16] To give some examples of this connection (see Jirak et al. 2010 and Fuchs 2016c for an overview):

- Listening to the words "grasp," "go," or "shout" activates, alongside the receptive language areas, also the motor centers for the corresponding actions (Buccini et al. 2005, Jirak et al. 2010).
- Abstract usage of verbs such as "to give" or "to grasp" (to give a reason, to grasp a notion) activates the motor system no less than the concrete usage (Glenberg et al. 2008).
- Listening to the sentence "The alarm sounded and John jumped out of bed" will activate areas both in the auditory and motor cortex related to alarms and jumping out of bed (Kaschak et al. 2006, Winter & Bergen 2012);
- There is strong evidence for a somatotopy of language, that is to say, a differential activation of motor centers according to the limb or action involved in the sentence one listens to: Pulvermüller (2005) identified specific fMRI activity patterns in the premotor cortex for consonant verbs that refer to the mouth, arms, or legs, such as "lick," "pick," and "kick."

Of course, the same connections also apply when speaking these words. Thus, it can be concluded that at least an important part of language is embodied in the sense that its usage is associated with low-level sensorimotor processes.

5.3.3 Outlook: language, thought, and perspective-taking

Just as language gradually becomes detached from pointing and expressive gestures, it can also become detached from its primary carrier, namely the vocal gesture, and become the *idea* of spoken words. While the imagined words, at first, are still internally heard and articulated, their "disembodiment" finally reaches a point at which, as pure ideas, they are fully transparent for their content. Thus the spoken connections become *thought* connections. However, thinking always remains an activity of "operating with signs" (Wittgenstein 1958, 6), a virtual dealing with representations of objects and relations, or an "action in rehearsal," as Freud (1911) called it. As placeholders for complex behavioral procedures, thoughts substitute the real enactment of possibilities by trial and error with all involved risks. Nonetheless, thinking always remains an embodied performance, a form of inner speaking.[17]

Among the most important developmental steps that are connected to language acquisition are without a doubt the connected abilities of perspective-taking and reflexion. Understanding language involves taking into account the perspective of the other, which is enabled by the shared and mutual nature of linguistic symbols (Fuchs 2016c). Their use as such already implies a change in roles: the child uses the same signs vis-à-vis grown-ups that they have used in communication with the child (Tomasello 1999). According to Mead, it is the reflexivity of the linguistic gesture that encourages perspective-taking: the speaker also always speaks to themselves. "We are, especially through the use of the vocal gestures, continually arousing in ourselves those responses which we call out in other persons, so that we are taking the attitudes of the other persons into our own conduct" (Mead 1934, 69). Hence, the word as an intersubjective symbol already contains a reflexive element in itself. In particular, self-reference by use of their *proper names* gives children the opportunity to

---

[17] "Thought is no 'internal' thing, and does not exist independently of the world and of words. What misleads us in the connection, and causes us to believe in a thought that exists for itself prior to expression, is thought already constituted and expressed, which we can silently recall to ourselves, and through which we acquire the illusion of an inner life. But in reality [...] this inner life is an inner language" (Merleau-Ponty 1962, 213).

However, thought is not only related to speaking, but also to acting: the brain processes underlying abstract thought have been found to be most similar to those underlying action-planning and body movement. Thus, mental manipulations of objects in abstract reasoning involves schemes of virtual bodily interactions with objects, such as "putting together," "taking away," "putting inside," and so on (Schmahmann et al. 2001, Ito 2008).

encounter themselves in imagination and to see themselves as persons from the perspective of others.

Beyond joint attention, towards the end of the first year of life, language increasingly enables the child to take the perspective of others and to reconstruct their intentions—to assume the "eccentric position." In this case, too, it does, of course, not learn "theories" or "hypotheses" about the others—much rather, it learns to *understand* them as intentionally acting creatures with goals, intentions, and desires. It is also not the case that the child constructs a theory of its own experience when it learns to say "I know …," "I think …," or "I feel." The child is no external observer or scientist in the lifeworld, but it interacts with others as an embodied creature and, via this route, learns psychological understanding as a form of practical knowledge about how a person feels in this situation. The ability to recognize false beliefs which others erroneously hold—and thus to pass the "false belief" tests of Theory of Mind research (Wimmer & Perner 1983)—is something that the child does not learn before its fourth year of life. So before it is capable of explaining and predicting the behavior of others, to "read" or "simulate" their minds, it has long since understood them due to their gestures, expressions, and actions in the shared context of the lifeworld.

The internalization of linguistic interaction can then be identified with *reflexive or dialogical thought processes*. In this sense, Plato already understood thoughts as "speech that occurs without the voice, inside the soul in conversation with itself."[18] Reflective self-consciousness is the constantly given latent possibility or current realization of such an interior soliloquy (Fuchs 2013b). This also includes opposing inner voices, in which the child confronts itself in an evaluative manner. Then it observes its own behavior as though it were observing, commenting on or evaluating the behavior of another person. A central precondition for this is having a grasp of *negation*, the parental "No!" insofar as it gives the child cause to take a distance from its own impulses. This happens concretely by the child letting this "No!" internally resound or by imitating it. In the second year of life, one can, for example, observe how a playing child says "No, no!" to itself (or to a doll) or shakes its head (Bruner 1978, 79–80). Connected with negation, also the reflexive emotions of embarrassment, shame, or guilt arise, in which the child sees and evaluates itself through the eyes of others (Tangney & Fischer 1995, Fuchs 2013b).

On the basis of reflexion, negation, and experiences with a social environment that sets boundaries, in consequence, the special abilities of *inhibition, attention-direction and impulse control* develop as central functions of human

---

[18] Plato, *Sophist*, 263 E.

volition. We are dealing with inhibitory functions in the widest sense, which serve to choose between action possibilities and suppress disruptive impulses or distracting stimuli. In the positive sense, they serve goal-directed action planning and self-regulation of the individual in its environment. These capacities to inhibit and regulate are summed up under the term "executive control functions" in neuropsychology and are mainly ascribed to the specifically human development of the prefrontal lobe, the phylogenetically youngest brain region.

> The development of the prefrontal cortex begins fairly late in the third year of life—synaptic density does not peak before the fifth year of life—and is only completed around the 25th year of age (Sowell et al. 1999, Fuster 2001). The following regions are relevant for executive functions:
> 1. The *dorsolateral prefrontal cortex* is mainly seen as connected with working memory and is also central for mental problem-solving operations and weighing action-consequences. Damage to it leads to disruptions of long-term action planning and perspectival flexibility, with a tendency towards perseveration and sluggish thought.
> 2. The *orbitofrontal cortex* serves to regulate emotional processes and the neutralization of attractive stimuli or impulses by means of inhibitory paths to the limbic system. Here damage, on the one hand, can have a *disinhibitory* effect in the form of lacking impulse control or hyperactivity, and, on the other, as a *hyper-inhibitory* tendency towards depression, sluggishness, and lacking initiative.
> 3. The *frontopolar cortex* (anterior prefrontal cortex) obviously contributes to inhibitory functions as well as to perspective-taking—two functional areas that interestingly seem to be closely connected (Carlson & Moses 2001): consideration of others presupposes the inhibition of primary drives. In taking another's perspective or imagining the movement of others, this region is specifically activated (Ruby & Deety 2001, Decety & Sommerville 2003). Conversely, patients with damage to the frontopolar cortex show a pronouncedly *egocentric* perspective in moral decision tasks (Anderson et al. 1999).

As the development of the prefrontal cortex progresses, the structures of the limbic system increasingly come under cortical control. Its development requires a sufficient amount of socialization experiences—without learning and practicing volitional functions, self-control, persistence, and attention in social contexts, the prefrontal lobe cannot develop and fulfill its functions. These inhibitory functions are of particular importance for humans in order to relativize one's own primary egocentric perspective in favor of those of others and thus become a morally responsible agent. The capacity of personal freedom as a complex connection of the ability to postpone immediate impulses, of prudent consideration, and of taking a higher, intersubjective perspective may be outlined in the biological structures of the brain but requires a particularly long development in the cultural environment, to which it is ultimately due.

We have now come to the end of this discussion on the further development of human capacities after early infancy and, in conclusion, turn to the fundamental relation of brain and culture.

5.4 Summary: brain and culture

While genetic evolution progressed relatively slowly, humans have developed a much faster mode of "cultural inheritance," in the course of which abilities can be passed on and improved transgenerationally, once they have been developed. According to Tomasello (1999), this cumulative unfolding of human culture is essentially based on the identification with others as creatures acting and communicating intentionally. For the specific carriers of cultural development, namely tools and language, always point beyond themselves towards something else: tools to problems that they solve, words to situations and experiences that they represent. In order to learn the use of these cultural tools, children need to understand the directedness, goals, and intentions of others as well as to be able to "put themselves in their shoes." This, in turn, presupposes that they exist in an intercorporeal and emotional relationship with their parents from the beginning, in which—prior to any symbolically mediated communication—their crucial social abilities can develop.

Findings in both cultural anthropology and developmental psychology congruently show that the specific social and cognitive capacities of humans have developed through cultural evolution which became increasingly independent from its biological basis (Donald 2001). Therefore, humans, like no other creature, need their conspecifics in order to develop their dispositions into capacities. Nowhere else in the animal kingdom is progeny so dependent on support and teaching by the parents for such a long time. No other species comes into the world with as plastic and malleable a brain as do humans. To a significant degree, their neurological predispositions are "open loops" that need to be complemented by the emotional, social, and intellectual competences of caretakers in order to become stably fixed capacities.

The social and cultural environment with its shared patterned practices thus becomes the decisive "ontogenetic niche" for scaffolding individual human development and selecting appropriate neural structures (Tomasello 1999, Kendal 2011, Lende & Downey 2012). As we have seen, two stages can be distinguished here: at the pre-verbal, pre-reflective stage, *implicit* processes of learning predominate; with the development of joint attention and language they are increasingly replaced and expanded by *explicit*, verbally mediated processes of learning.[19]

- *Implicit learning*: the child's early interactive experiences already have a sustained effect on its abilities to corporeally and emotionally deal with others and thus on its future relationship patterns. They manifest themselves in affective coordination, embodied dialogues, empathy, trust, and attachment. Repeated sequences of shared social practice are incorporated into implicit body memory in the form of interactive schemata (Fuchs 2012, 2017c). This preverbal memory system forms an unconscious excerpt from prototypical experiences with others and processes them into *"schemes of being-with"* (Stern 1998a). But common forms of interaction and behavioral styles, from table manners to the style of dress and taste, are also consecutively integrated into body memory as *"schemes of interacting-with."* These collective patterns become second nature to the child, that is, they have an implicit, taken-for-granted effect and are usually not accessible as explicit knowledge ("knowing-how" instead of "knowing-that"; Fuchs 2016b, 2016c). The notion of "habitus" or "social sense" (Bourdieu 1977) designates such social practices that have become second nature and are ingrained in habitual reactions; here biology and culture overlap.

- *Explicit learning*: while the cultural environment, in which the child grows up, has a scaffolding function for processes of implicit learning, a new phase of socialization begins towards the end of the first year of life: by conceiving of others as creatures acting intentionally, children acquire the ability not only to learn through, but also *from them* (Tomasello 1999, 6). The place of automatic mimetic imitation in the first phase of life is increasingly taken by explicit imitation: children reproduce the intentional behavior of adults towards objects as well as their interpersonal behavior in a directed manner. They put themselves in the shoes of others, identify with them, and, often in playful imitation, take on their postures and roles. In this manner, they learn specific cultural techniques for dealing with objects as well as social

---

[19] In current cognitive neuroscience, this polarity is often expressed in terms of two sets of processes: those that are *automatic* and those that are *controlled*. The former are fast, spontaneous, dominant early in development, and often involve emotions. The latter are thought to be rather slow, effortful, and reflective; they arise late in evolution and individual development, and often involve language-based declarative reasoning and reflective thinking (Adolphs 2009).

competencies for dealing with others. Additionally, in acquiring language, they get a handle on symbolically mediated processes of learning by which they can go through cultural development in "fast forward," so to speak, that is, at a speed that could never be reached by imitation and model learning only.

If we take both forms of learning together, we can describe human socialization as an essentially *embodied socialization*. In its course, cultural techniques and life forms are acquired, on the one hand, by implicit intercorporeal learning, and on the other, by explicit identification, imitation, and verbal learning. These processes of learning can be seen as "social incorporation" insofar as specifically human capacities only develop in the context of shared, embodied practices, which are sedimented in brain development. In this enactive sense, culture is not merely a system of signs and meanings, but encompasses all formative processes of individuals and their capacities, which are anchored in their organisms, in particular in their brains. Only brains "socialized" in such an environment become carriers of the cumulative social learning processes, which mark a decisive advantage vis-à-vis natural evolution. Thus the individual gains access to the social and cultural memory of the community (Halbwachs 1939, Assmann 2011, Fuchs 2017c).

Mind and consciousness arise only in an overarching and dynamic interaction of organism, brain, and environment. Cognitive processes are not produced by an isolated neural apparatus which internally mirrors the world by means of representations. Much rather, they constantly transcend the boundaries of the brain as well as the body. The mind is based on *meanings*, and meanings on *relations*. They take root in the early experiences of joint attention, pointing, in the shared use of language, and in the intersubjective symbolism of words. Correlates of these overarching meaningful relations are functionally and morphologically inscribed in the brain throughout the course of socialization as neural patterns. Thus, the brain becomes the organ of the mind—but the mind is not "in the brain," for *it is the overarching manifestation, the gestalt, and the ordered patterns of all relations that we have to our environment as animate beings, and as humans to our fellow humans.*

This extended mind refers to information that is not only accessible on the basis of neural correlates in the brain, but is available in the manifold structures of the environment: in the structures of the human body, in human relationships, in written and spoken language, in collective myths and practices, in art, literature, technology, and science. All these cultural products are not phenomena produced by the brain, but, conversely, phenomena shaping the brain. Human cognitive agents actively structure their world, and those structures play a crucial role in their cognitive processes, which in turn guides further structuring. The human mind is thus not only extended into its respective environment, but also socially scaffolded: human beings construct cultural niches which influence the cognitive development of their offspring and exert their influence over generations (Sterelny 2010, Krueger 2013, Sutton 2015).

This points to the inherent connectedness of embodiment, enaction, and culture (Durt et al. 2017): humans use the items in their environment not only as objects, but also as carriers of shared symbols and information which in turn influence their embodied memory. We invent writing, books, calendars, or computers in order to use them as "external memories." Language itself arguably is the most important memory that humanity has developed, and which is incorporated in early development. By means of the senses and limbs, the brain, in each case, connects to those carriers and circuits of information in the human environment, just as the subject-body incorporates tools, instruments, and words in skilled usage. This is only possible, because the brain, as the organ of mind, already has become part of the overarching system of cultural meanings, resources, and processes. It has become an organ in relations.

Chapter 6
The concept of dual aspectivity

> **Overview**
> Following the description of the brain with reference to its embodied and ecological aspects, Chapter 6 progresses with an examination of the concept of "personal dual aspectivity." The unity of the living organism and its enactments of life provides an alternative to the separation of the mental and physical in philosophy of mind (6.1). A critical consideration of identity theory further develops this conceptual approach (6.2). The concept of integral causality is then differentiated in the light of emergence theories, emphasizing the primacy of holistic functions over their components, and the reciprocity of downward and upward causation. The role and function of consciousness as the integral of the organism–environment interaction is discussed in detail. Here, the brain assimilates biographical experiences as a "matrix" that serves as a basis for a person's mental capacities, thus enabling their integral causality (6.3). This gives rise to several conclusions regarding the intentional determination of neuronal processes, particularly an embodied notion of free will, as well as an explanation of psychophysical interrelations (6.4).

A survey of culture completed the description of the brain in its biological and social environment. In this chapter, I re-examine previous results and elaborate upon the central concept of personal dual aspectivity, in particular, with a further examination of this concept in the light of current mind–body theories.

6.1 The mental, the physical, and the living

I have repeatedly argued that a characteristic shortcoming of contemporary theories lies in the exclusion of an autonomous concept of life. This forces a short circuit of the mental and physical, or of consciousness and the brain. Peter Bieri already drew attention to this aporetic situation with his classic description of the trilemma of philosophies of mind (Bieri 1981, 5):

1. Mental phenomena are non-physical (*nicht-physische*) phenomena.

2. Mental phenomena are causally effective in the sphere of physical phenomena.
3. The sphere of physical phenomena is causally closed, that means, for each physical event p_1 there is a sufficient physical cause p_2.

Although it is tempting to accept each of these statements, further inspection shows that they are based on a non sequitur:

> In each case, two of the three sentences imply the falsity of the third: if mental phenomena are non-physical phenomena and if mental causation exists, then the sphere of physical phenomena cannot be causally closed. However, if it is causally closed and mental phenomena are non-physical, then mental causation cannot exist—contrary to all appearances. And if mental causation exists despite the physical world being causally closed, then the sentence of mental phenomena being non-physical must be false. (Bieri 1981, 6; own translation)

Different mind–body theories now attempt to solve the trilemma, generally by relinquishing one or two of these three sentences.[1]

Of course, the formulation of the trilemma in itself pre-structures the problem through the dualism of two distinct spheres: "mental" phenomena belonging to the subjective inner world and "physical phenomena" being part of an objective, physically describable (and therefore causally closed) external world. This generates the irresolvable dilemma of how something non-bodily and internal can be in contact with something bodily and external. The very definition of the problem, however, already excludes an entire class of *animate, bodily and interbodily* phenomena. Their character is neither purely inner nor purely outer, not purely mental nor purely physical, and yet this is exactly why they form the essential fabric of our everyday lifeworld: laughter or crying, suffering pain, talking, writing, playing the piano, mutual greeting, dancing together, and so forth. All these acts of life can be experienced both from the inner perspective of the first person as well as perceived from the outer perspective of the second person (see 3.1.3). Yet Bieri's definition splits the unity of these life acts from the very start.

The basically dualistic framework of the majority of contemporary mind–body theories was contrasted in Chapter 3 by the conception of a grounding unity of the living being. This may be regarded under two different aspects: firstly, as displaying *integral acts of life*, like the ones just mentioned, generally speaking, conscious experiences and acts; and secondly, as consisting of *physiological processes*, which manifest a complex, organically

---

[1] Identity theory thus negates sentence (1), while functionalism and epiphenomenalism negate sentence (2). Eliminative materialism negates both sentences (1) and (2), while interactive dualism finally negates sentence (3).

structured order. Both complementary aspects are equally connected to a *physical being* which appears as precisely *one* unitary and animate substance on the one hand, and as an organized totality of all its material parts and processes on the other.

The mind–body problem is thus recast as a "body–body problem" (namely body-as-subject vs body-as-object or *Leib–Körper* problem). In one case, the body appears as the living center and medium for the enactment of life, and in the other case as an observable and materially composed object among other objects. The inherent split of the Bieri trilemma is thus overcome and, in particular, the reduction of the *physical* to the domain of physics (see 3.1.2). For although integral acts of a living organism cannot be decomposed into separate "physical" (*physikalische*) particles, they are nevertheless "physical" (*physische*), bodily processes (both in the sense of body-as-subject and body-as-object)—and certainly in all these acts a more or less pronounced change occurs to the configuration of the body's physiological conditions.

One objection here might be that a class of phenomena has once again been overlooked: that is, *genuine* "mental" phenomena such as ideas, fantasies, or thoughts, which are not or not necessarily expressed through a person's bodily behavior. Such phenomena seem to be capable of grounding an autonomous mental domain, and they could consequently deprive the other acts of life of their "mental components" so as to incorporate them within their domain. Hence, the unity of life processes would be split up again into a subjective inner sphere and outer behavior, which could only be observed in behavioristic terms. It is all the more important to highlight the concept of *embodied subjectivity* which constitutes *all* conscious processes as permanently enacted on the basis of the lived body's overall state, and thus being always embedded in this subjective state (see 4.1.2). Given that this bodily background experience participates in the basic coextensivity of subjective body and physical body, this also applies for all apparently purely "mental" conditions of human consciousness. The subject of all mental activities is always *a bodily subject*, "incarnated" in the whole physical body, and it is thus a *spatial* subject as well.

The conventional wisdom of the "non-spatial" quality of the mental sphere—a legacy of the Cartesian dichotomy of *res cogitans* and *res extensa*—must therefore be strongly disputed. While this idea naturally applies for the intentional *content* of thought, it is not valid for its *realization*. As a conscious activity, thinking is not possible without a latent, albeit often diffuse and variably extended bodily background experience. In section 1.2.2, I showed that this spatial dimension of experience is not simply illusionary, and thus should not be overlooked when characterizing mental phenomena. To assume a non-spatial mental sphere is only valid given the entirely unfounded condition of space being something only described by physics and comprised of separable elements. On this condition, the merely secondary objectivized space is falsely considered as primary, indeed as the only space. But neither the living organism nor the undisputed extended spatial experience, for example, of hunger, diffuse pain, fatigue, feeling of illness, and so forth, can be divided into single parts. Both the living organism and the subjective body are unitary, non-divisible, and yet extended. Hence, they are *neither* "res cogitans" *nor* "res extensa."

The presumed "pure thought" is also part of the life process insofar as it always engages the whole person, for example, in the form of volitional directedness, of concentration and the related effort, of inner speech and its corresponding motor tendencies (see 5.3.2.2), and so on. In no sense is this enactment of life a disembodied act. We can certainly see that a thinking person is alive and conscious. Indeed, we can even *see him thinking* if his behavior is in the mode of Rodin's *Le Penseur* [*The Thinker*], though we do not perceive the contents of his thoughts. Thinking—like emotions—is admittedly not *entirely* observable from the outside. But that does not mean thinking, as an enactment of life, can be displaced to an independent mental world.[2]

A fortiori, no "purely mental" components could in any way be detached from other enactments of life such as speaking. Speech is not an independent motor activity occurring alongside thinking. Rather, it is the expression of a person's thoughts, for thinking itself is "inner language" (see 5.3.3), and conversely the act of speaking as such already prescribes and prompts the path of thought. Thus verbal "expression" is the example par excellence for a process, which jettisons the dualism of inside and outside, even more so since its origin and meaning is derived only from being embedded in a context of social interaction.

---

[2] Wittgenstein's late philosophical writings about a private language provide us with some evidence for this: our talk about the "inner" or emotional world, as with all language, ultimately originates from a shared social practice of discussing our experience, desires, and ambitions with others (see 5.3.2.1). Unlike conversation on the topic of external objects, the nouns which are formed in this practice, such as "pain," "feelings," "thoughts," and so forth, do not represent specific objects, processes, and conditions that are encountered in the material world. Hence, in each case they refer not to "a specific something" that "there is" like stones, apples, or birds, with the only difference that it remains invisible and therefore confined to an autonomous inner world. Here, formation of the nouns suggests an incorrect objectification that ultimately must lead to the dualistic idea of a "ghost in the machine" (Ryle 1949). In fact, in accordance with their social derivation, expressions for emotional experiences always remain *a person's* descriptions for experiences and, thus, for his or her enactments of life. (See Wittgenstein, *Philosophical* Investigations, I, 304–307.)

In general terms, the concept of embodied subjectivity implies that all forms of experiencing and acting, whether given in the first- or second-person perspective, are primarily understood as integral enactments of life, thus as *activities of a living organism in relation to its environment*. The lived body and life itself therefore become the bridge between the "mental" and the "physical." Nevertheless, this conception still implies a duality, namely as the dual aspect which the living being shows. However, such duality corresponds not to two essentially distinct domains of reality, but rather to two opposing *perspectives* and *attitudes*, which we can adopt towards life, and which are not mutually transferable.

In the first or *personalistic* attitude of the lifeworld, speech is naturally a bodily expression of my own intentions or of another person's intentions when I am listening. I perceive myself not as a mental being motivating the body to utter certain linguistic phrases to carry information from within my consciousness to the outside world. Nor do I suppose a consciousness concealed inside the other's body, on whose putative content I could only make inferences from certain linguistic and other messages. The other person rather appears to me directly "in the flesh" and through his articulation of words, and thus as a unitary being.

Only in the second, *naturalistic* attitude can I begin to examine the processes at work during speech as physical processes. This entails increasingly detailed analysis of the acoustic and physiological conditions of utterances and tracking these processes ever deeper within the living organism. Even so, they evidently remain processes *within a living being*, although they lose their aliveness which is connected to spatially and temporally superordinate forms of movement, expression, and meaning. Insofar as we aim to understand physiological processes in system theory as complex, hierarchically structured self-organizing processes, this type of description approaches the original perception of the unity of the living being, yet without being able fully to reconstruct this. Taking a third-person perspective, the naturalistic attitude is not in a position to do so per se. 'Even to know another life requires a certain sympathetic engagement, for life is noticed only by the living' (Spaemann 2006, 183). In a similar vein, it is also true that: "One must participate in life in order to research living beings" (von Weizsäcker 1986).

Even if we can appropriate this second, naturalistic attitude, the primacy of the first- and second-person perspective or the participant perspective still holds true. If we were not always to perceive and interact with others as beings *of our own kind*, that is, as embodied subjects, we could not even identify the phenomena whose physiological or neuronal correlates we seek to find. The first, integral aspect therefore corresponds not to a naive ontology that the naturalistic perspective might overcome. On the contrary, the point of psychophysiological research is that there is precisely no escape from the participant perspective, and thus from the personalistic attitude towards ourselves and other human beings. In no way can phenomena of consciousness be investigated by merely varying environmental stimuli and measuring corresponding neuronal activities.

> At some future date, we might become convinced that artificial intelligent machines truly possess consciousness, that is, can sense pain or pleasure and not merely simulate them. This could not be concluded, however, from specific configurations of their internal processes—regardless of how closely such processes resemble mechanisms we observe in human organisms and their brains. For in a naturalistic sense, we "know" just as little whether human beings actually feel pain, and no examination of the brain can ever prove this. Rather, we would have *to perceive* those machines *as suffering beings*, thus *acknowledging* them as participants of a shared life form, or as "fellow creatures."
>
> This perception would correspond to a fundamentally new attitude towards those machines: they, too, would now appear under the dual aspect of life. Consequently, dual aspectivity always corresponds to an intrinsic freedom of our attitude. We *can* certainly also view human beings as intelligent machines with integrated "self-modeling" circuits. In this case, however, we also refuse them recognition of their "being-with," or the personal or *you-relationship*. "That Lise is a person, then, is not something that we first suspect and then, as our suspicions grow stronger, reach a judgement on, so that finally we recognize her personal status. Only as we recognize Lise as a person can we conceive of her being one" (Spaemann 2006, 182).

Thus, even a complete naturalistic description would not suspend a person's dual aspectivity. But the opposition between both aspects is not as stark as for the "mental" and the "physical." The duality is not ontological in this sense, but epistemological, that is to say, it is dependent on our chosen attitude. Both characterizations refer to one and the same living organism: although their *intension* is different, their *extension* is identical. Both aspects also manifest various *correlations, isomorphisms, and structural similarities*.

The existence of correlations or regular connections between, for example, physical stimulation and perceptual impression is taken for granted as the basis for all psychophysiological research and hence requires no further examination here. However, closer attention should be paid to the more fundamental isomorphisms and congruencies between both aspects. A first example was *coextensivity* (see 1.2.2): lived bodily and physical bodily space overlap, thereby indicating the living being's intrinsic unity. The forms of *integrative synthesis* comprise another parallel. In experiencing and acting, holistic forms of perception and movement evolve by means of implicit coupling. Likewise, corresponding links are also formed on a physiological level between specialized neuronal assemblies and brain areas. Accordingly, it is possible to identify the substratum of a person's *incorporation of experience* and thus to understand the intertwinement of *process and structure* (see 4.2.3, Figure 4.7). Both aspects are connected or interlinked via the history of the living organism, that is, via developmental biology. The formation of capacities and dispositions can be described both phenomenologically as implicit coupling and biologically as neuronal coupling.

The central coincidence, however, which is also the leading principle informing our entire investigation, involves both aspects of the living organism as fully describable *only in relation to the environment*—on the one hand, phenomenologically, and on the other hand, ecologically. Living subjectivity is bodily "being-towards-the-world" and cannot be reduced to a purely "mental" consciousness that is cut off from the world. The living organism, in turn, only exists and reproduces itself through permanent interaction with the environment. The life process as a whole constantly transgresses the body's physical limits, that is, the living organism and its complementary environment form a superordinate system. Hence, as I stated in Chapter 4, the *phenomenology of bodily being-towards-the-world* and the *ecology of the organism-in-its-environment* correspond to one another, without being identifiable with one another.

We treated *language* as a specifically human paradigm for this interconnection of both aspects with the environment. Learning and using a language are first and foremost dependent on interpersonal, intercorporeal resonance and participation in a superordinate context of meaning. By the same token, the emergence of language in human societies seems to depend on a biologically inherent and linguistically appropriated neuronal resonance system connecting the individuals to each other at the organismic level. The human being is thus uniquely distinguished as "*zōon politikón*," namely as "*zōon logón échon*"—an animal that has language, and whose biological structures are co-influenced by his social and linguistic environment.

Such concurrences show that the aspect of integral acts of life and the aspect of physiological–organismic processes are not externally opposed, but rather intertwined within the individual person as an animate and historical being. The idea put forward here can therefore be summarized as "biological aspect duality," albeit without any intentional link to "biologism." Neither the independence of the material world nor of the cultural world is questioned here. But humans only participate in both realms to the extent that they are *living organisms*. "Persons, therefore, are living human beings. There is no other way of being a person—for example, thinking or having one or another type of conscious state—than that of being a human" (Spaemann 2006, 69). Persons are the particular type of living organisms gifted with speech and intellect, who perceive each other as a unity of inner and outer, as subjective bodily and as intentionally directed beings, and who also share the capacity to focus their attention together on the same objects.[3] Hence, the "biological" becomes the *personal aspect duality.* In what follows, a comparison with viewpoints frequently under discussion today, namely, identity and emergence theory, will shed more light on this concept.

6.2 Differentiation from identity theories

The suggested concept of a dual aspect of lived and living body is clearly differentiated from another aspect dualism, which is particularly represented by the various versions of *identity theory*. First suggested by Place, Feigl, and Smart in the late 1950s, it found its most pointed formulation by Armstrong (1968): all mental states are identical with brain states. Accordingly, the mind and the brain (or rather parts of the brain) are considered as two properties or aspects of the same entity. A particular and circumscribed form of the brain's neuronal processes can be both observed externally and described from the internal perspective as phenomenal consciousness.[4] Although this idea has been heavily criticized in analytic philosophy of mind,[5] it is still a common background assumption of many neuroscientists and psychiatrists.

Granted, this seems to correspond to the extensional congruence despite intensional difference, which I have also ascribed to an animate being's dual aspect (see 4.2.2). However, while in that case the living organism or person as

---

[3] "[Persons] are encountered only in the context of a world that is common to them and to us. They are encountered in such a way that we understand them only when we 'look in the same direction,' i.e. enter into their carrying out of their intentions" (Spaemann 2006, 587).

[4] "Instead of conceiving of two realms or two concomitant types of events, we have only one reality which is represented in two different conceptual systems—on the one hand, that of physics, and on the other hand, where applicable (and in my opinion only to an extremely small part of the world) that of phenomenological psychology" (Feigl 1960/1981, 349). Spinoza was the predecessor of this type of double aspect theory, advocating in his *Ethics* (1677) the view that "the mind and body, are one and the same individual, which is conceived now under the attribute of thought and now under the attribute of extension" (*Ethics III*, Prop. 2, Note; Spinoza 2000, 58). Admittedly, this doctrine is already based on a Cartesian precondition which excludes life.

[5] Mainly for the problem of the multiple realizability of mental states in various substrates; see Putnam (1967).

a whole represented the common referent (or common extension of characterizations), here it is only a specific brain condition. This restriction of the referent now has a series of grave consequences.

> In Chapter 2, we already referred to the aporias of reductionism, which also apply to identifying conscious states with brain states. They primarily relate to the individual subject's perspectivity and intentionality which are not captured by any objectivizing description of neuronal processes. It is not possible to demonstrate how *my* experience as a subjective, perspectival, and qualitative event can be identical with the condition of a physically describable process. Similarly, no equivalent exists for intentionality in neurophysiological conditions, since these reveal neither directedness nor meaningfulness. Both properties of human consciousness have proven as extraordinarily resistant to all naturalization attempts. Finally, I have pointed to the temporal integration of consciousness as being irreducible to brain processes. However, I have no intention of reiterating these arguments again, but now turn to consider other viewpoints.

The first consequence to be examined is mainly methodological. Since human persons cannot report anything about their brain conditions and brains give nothing away about their possible internal conditions, ultimately identity theory can only ever be based on *correlations* between both areas. Nevertheless, the "gold standard" for such correlations remains the statements made by human subjects, in other words, by *conscious animate beings*. On the one hand, the supposition of identity is thus related to neuronal micro-phenomena in circumscribed brain regions. On the other hand, it is related to macro-phenomena, namely the experiences of human persons, who can only undergo and report these experiences as whole animate beings. *"Consciousness" or "conscious events" can neither be found nor researched anywhere in the world*—there are only conscious animate beings.

The absence of the living being as a shared referent thus leads to a *fundamental asymmetry* of both links of the identity statement. But the equal ranking of two phenomena or their appearance "under the same aspect" (like the morning star and the evening star) would be the precondition for their identification. This equality is often suggested by speaking of "mental states" and "brain states" as though the same "state" were applicable in both cases. However, a "mental state" is always *somebody's state*, that is, it is experienced by an embodied subject. If a volunteer in a tomograph explains that he or she feels anxious, then neither the anxiety nor the statement about it are the activity of a pure, mental subject. To sense anxiety (the shortness of breath, the tension in the body) and speak about it, one must be an embodied person.

On the other hand, brain states represent *nobody's* states, but only specific configurations of organized matter. No brain process exhibits any measurable consciousness. Neither the anxiety nor the volunteer's statements are observable objects such as a falling stone or firing neuron. On the contrary, *the section of the world, which the third-person perspective grasps, excludes the second- as well as the first-person perspective*. With this, however, the preconditions of identification no longer apply. A neuroscientist's assumption that what the volunteer currently experiences and articulates is *identical* with changes of brain activity displayed by the scanner is not supported by any substantial evidence. Purely functional correlation is no evidence of identity. Identity theory—no matter how many and how exact the correlations to be found—is ultimately grounded on an act of faith.

One could raise the objection to the concept of biological or personal aspect duality that it suggests the same premise of both aspects being based on extensional identity. However, the common referent—that is, the living being or the person—in this case has an entirely different character. For the unity of embodied conscious experiences from an inner perspective corresponds here to the *unity of the living organism* from the outer perspective. Insofar as it delimitates and sustains itself as against the environment, a living being in all its acts appears to us as a "self" and thus corresponds to the unitary subject of experience. Even in the observer perspective, on the proviso of its holistic characterization as a living system, the organism is assumed to be a unitary entity and thus emerges as a suitable and plausible candidate for bringing forth integral and conscious acts of life. Drawing on Merleau-Ponty's notion of the bodily subject as primary subjectivity, we can indeed say that *one's conscious self is nothing else but the human organism that one is*, namely insofar as it is considered as the source and the agent of conscious life. Self-awareness is not an additional byproduct of the brain, but a living organism's self-awareness.

On the contrary, one can hardly assume that a circumscribed brain condition—such as, for instance, Edelman's suggestion of a constantly changing thalamocortical "dynamic core" of reentrant neural processes as a substratum of consciousness[6]—is suitable as such to represent a "self" or "a whole." Accordingly, the localization and demarcation of such a substratum of consciousness remains still an unresolved issue, as does the question of how the partial systems involved are integrated. There is no central module or processor in the brain where everything is gathered to form the conscious self, no Cartesian

---

[6] Edelman & Tononi 2000, 143–145; see 2.2.2.

theatre or a director in charge of the play.[7] The phenomenal self, presently the focus of the intensive search for neuronal correlates,[8] is not be located inside the brain at all, for its unity only has a correlation *with the organism as a whole*. The nervous system and brain undoubtedly achieve an integration, which is indispensable for the phenomenal self. *What* the brain integrates, however, are not merely its internal processes but also the entire vegetative, emotional, and sensorimotor processes within the system of organism and environment. Consequently, self-awareness is not a specific "mental state" in the brain but *the integral above the entire system of bodily "being-towards-the-world"* (see 4.2.2).

In contrast, the restriction of the common referent to circumscribed brain conditions leads to the aporias of the "brain as subject" which I have highlighted in Chapter 2. If a mental process were identical to a specific brain process, strictly speaking we would have to accept that neuron associations are *subjects* of thoughts, perceptions, or actions. Yet this surely imprisons the subject into an inner world. When we listen to a person speaking it is then only a custom of "*folk psychology*" to perceive him as a speaking, embodied person. In truth, however, "he" is a state localized inside the brain and behind the visible body. His speech is merely an external movement in which he does not appear or express himself. Identity theory therefore breaks up the unity of acts of life into certain neuronal processes (with consciousness) and other physical processes (without consciousness). But neuronal processes as such are always only *fragments* of the superordinate system processes of organism and environment, which form the basis of conscious experience, even if such fragments—as we may assume—are in regular correlation with the qualified conscious acts of life.

6.3 Emergence

6.3.1 The primacy of function

In describing the living organism's basic structure, the circular causality of life (Chapter 3), and the neuronal foundations of the emergence of *gestalts*

---

[7] The demarcation of "consciousness-bearing" from other cortical processes leads to numerous diverse suggestions, with most attempts being based on fluctuating neural associations, even varying from person to person (Edelman & Tononi 2000, 144). Thus, at the neurobiological level it is not at all clear what should actually be identified with what. Even if it were possible to isolate cortical activities necessary for consciousness in a relatively satisfactory manner, their connection with sub-cortical and overall organismic (vegetative as well as sensorimotor) processes would still be required for the emergence of consciousness (see 4.1).

[8] See, for example, Kircher & David 2003, Vogeley et al. 2004, Northoff & Bermpohl 2004, Ochsner et al. 2005, and Rameson et al. 2010.

(Chapter 4), I repeatedly referred to a hierarchical structure of processes, which are intrinsic to the organism, and to their vertical–circular interaction. This structure evokes another concept of the mind–brain problem, namely, *emergence theory*. However, this concept comes in multiple versions, and we have to examine carefully how it fits with an embodied and enactive view of the organism under the double aspect of subject- and object-body.

The idea of emergence is based on the assumption that sufficiently complex systems may display novel properties—properties not possessed by their parts. This idea has been spelled out in various ways. Following Stephan (1999a, 1999b) and Chalmers (2006), we can distinguish between weak and strong emergence. Weakly emergent phenomena are merely unexpected, given our knowledge of the domain from which they arise. Strongly emergent phenomena are not just unexpected; they cannot (not even in principle) be deduced from the domain from which they arise. One could also describe the first type as *epistemological,* the second type as *ontological* emergence.

> In a way, the philosophical morals of strong and weak emergence are diametrically opposed. Strong emergence, if it exists, can be used to reject the physicalist picture of the world as fundamentally incomplete. By contrast, weak emergence can be used to support the physicalist picture of the world, by showing how all sorts of phenomena that might seem novel and irreducible at first sight can nevertheless be grounded in underlying simple laws. (Chalmers 2006, 146)

Weak emergentism has been advocated by a number of philosophers and neuroscientists (e.g., Bunge 1980, 2003, Searle 1992, Swaab 2014). Here, consciousness is usually described as a *higher-level system property* of the brain or of specific cortical networks, analogous to many macro-properties of physical systems such as liquidity, elasticity, and others. For example, Searle proposes:

> The brain causes certain "mental" phenomena, such as conscious mental states, and these conscious states are simply higher-level features of the brain. Consciousness is a higher level or emergent property of the brain in the utterly harmless sense of "higher-level" or "emergent" in which solidity is a higher level, emergent property of H_2O molecules when they are in a lattice structure (ice). (Searle 1992, 14–15)

Although Searle equates emergence with causation and rejects any identity of brain states and conscious states, a weak concept of emergence can even be linked with positions of identity theory.[9] As Chalmers has rightly argued,

---

[9] For instance, in Bunge's theory of emergence, which he defines as follows: "(i) all mental states, events, and processes are states of, or events and processes in, brains of higher vertebrates; (ii) theses states, events, and processes are emergent relative to those of the cellular components of the brain" (Bunge 1980, 21). This is well compatible with what he calls "psychoneural identity theory": "Every fact experienced introspectively as mental is

emergent properties *without* downward causation result in an epiphenomenalist picture: there is a fundamentally new quality, but it plays no causal role with respect to the basic level (Chalmers 2006, 249). This weak form of emergence comes close to supervenience theories which assume that higher-order properties simply supervene over lower levels (i.e., they are determined by the configuration of components) without causal impact (Davidson 1980, Kim 1993). Here the primacy of the physical level is undisputed.

As I have already shown in section 3.2.1, an embodied and enactive concept should be based on a *strong version of emergence*. For this, two requirements have to be fulfilled:

1. Primacy of the whole or of holistic functions vis-à-vis their components.
2. Reciprocity of global-to-local (downward) and local-to-global (upward) influence, or circular causality.

We can now further specify these requirements.

(1) Primacy of function. Undoubtedly, the hierarchy of basal, intermediary, and integral levels presents a fundamental structure not only of higher organisms, but also of their nervous systems. In particular, as we have seen, brain activity depends, on the one hand, on a high degree of distributed parallel processing, and, on the other hand, on increasing integration in higher centers, that is, on a vertical–circular interaction, which is required for the emergence of consciousness. However, circular causality between the whole and its parts not only implies the emergence of higher function levels from individual components, but also the converse. *The conditions for all life processes are always the respective functions,* and even if these must first develop through ontogenesis, they nonetheless only evolve through increasing differentiation of the already existing organism as a whole. The substratum level is not organized of its own accord, rather taken into service, transformed, and organized by the superordinate function. This assigns to the living whole not the role of a product but rather of a producer:

> Whole organisms are not assembled by bringing together disparate parts but by having their parts differentiate from one another. Organisms are not built or assembled. Although they grow by the multiplication of cells, these divide and differentiate from

---

identical with some brain activity" (p. 73). Similar concepts are found in many neuroscientific accounts of consciousness, for example, in Swaab (2014, 170): "Consciousness can be seen as an emergent characteristic generated by the joint functioning of specific areas of the huge network of neurons in our heads. Brain cells and areas have their own separate functions, but their functional links with one another jointly endow them with a new 'emergent' function."

> prior, less differentiated precursors. Both in development and in phylogeny, wholes precede parts. (Deacon 2006, 116)
>
> Emergent wholes have contemporaneous parts, but these parts cannot be characterised independently from their respective wholes. (Kronz & Tiehen 2002, 345)

This reciprocal relationship of the whole to its constituent parts is best captured by the term "*dynamic co-emergence*" (Thompson 2007, 60–62), implying a two-way enabling relation. Whether we look at organic molecules, at the cell organelles, the cells, or the organs, in all cases the next higher organizational level is the condition of its components which wouldn't be produced or sustained outside of the organism. There is no base level of elementary entities to serve as the ultimate "emergence base" on which to ground everything. At the level of the brain organization too, "the distinction between pre-existing parts and supervening wholes has no clear application: One might as well say that the components (local neural activities) emerge from the whole as much as the whole (dynamic patterns of large-scale integration) emerges from the components" (Thompson 2007, 423).

Of course, the primacy of function or of the whole in relation to the parts does not imply a new vitalism. Neither a vital force nor any other transcendent principle is instilled in the substance comprising the living organism. It is *one living material substance* that due to its complex organic structure is in a position to bring forth highly differentiated acts of life, such as feeling pain, perceiving trees, or writing books. Matter is not the basis that produces or forms the living being of its own accord. On the contrary, the living organism transforms the material in a suitable way, assimilating and transforming this into its constituent parts. The form of life emerges not from the material, but rather *organizes* this and turns it into *its own* material. Thus, the components display new, emergent properties or behaviors, such as the previously mentioned example of iron in hemoglobin (see 3.2.1).

The concept of self-organization is often taken to suggest that complex structures emerge spontaneously from the physical components like eddies in a river. However, we always only encounter organisms already representing a vital whole with varying degrees of complexity and differentiation. At the same time, there is no disputing the fact that entirely new properties and functions emerged during the path of evolution. This subject of *diachronic emergence* merits a study in its own right and is not our chief concern here.[10] In any case, in

---

[10] A short remark may be in order: we may assume that in the course of evolution spontaneously developing novel life forms "inserted" themselves into pre-existing and suitable ecological niches or "empty spaces" offered by the overarching system of life and environment, thus favoring the emergence and preservation of higher functions and organisms. This would be analogous to the top-down influence of an attractor within a living organism.

the course of its development, the organism always precedes the formation and differentiation of the means for its self-preservation. Hence, it can be claimed that the organism's *functions bring forth the organs* enabling them rather than the converse. This factor is particularly demonstrated by the brain's plasticity, enabling it to even substitute for lost functions to a certain extent by the use of its other hemisphere (see 4.2.3): "*the function creates its cerebral organ*" (Brodmann 1909).

Therefore neither life nor consciousness is only a "higher-level property of material systems." Searle's comparison of consciousness with the solidity of H_2O molecules in a lattice structure overlooks that this structure does not fulfill any function, nor does it reproduce itself over time. A living system is not assembled under specific environmental conditions like water molecules to ice only to disintegrate again under different conditions. Rather, the living form and function precedes the parts which do not "organize themselves" as a system. Instead, the organic system is the form in which the living being organizes and maintains its own material basis.

(2) Circular causality. The second requirement of strong emergence is captured in the following definition of emergence (Thompson & Varela 2001):

> A system or network, N, of interrelated components exhibits an *emergent process, E,* with emergent properties, P, if and only if:
> 1. E arises from the non-linear dynamics of the interactions of N's components, and
> 2. E has a global-to-local (downward) determinative influence on the dynamics of the components.

Thus, in circular causality, the superordinate process results from the interaction of the components while in turn it determines their behavior and dynamics. As already pointed out in section 3.2.1, this downward determination does not imply efficient causation or an external force that acts on something. Instead, the superordinate dynamics of the system *constrain* the behavior of the components so that they no longer have the same behavioral alternatives open to them as they would have outside the system (Haken & Stadler 1990, Kelso 1995). "Constraint" is not an efficient, but a formal or topological notion: the form, configuration, or topology of a system narrows and limits the *range of possibilities* in the system's phase space (as I have shown by the example of iron integrated into hemoglobin; see also Thompson 2007, 427). The superordinate

---

Novel and expanding forms of life in turn changed the environment, creating new niches with new selection pressures that influenced the further development of species (e.g., plants, through the development of photosynthesis, changed the earth's atmosphere which could then be used by air-breathing land animals). In sum, the processes of niche formation result in a *co-evolution and co-determination of life forms and environment.* On this, see also Thompson (2007, 201–218).

configuration functions as a *global-order parameter* modifying the probability of events within the system, even though these higher-order regularities are realized by lower-order interactions. One could also say that it acts as an "empty space" or attractor which does not enforce the behavior of the parts but *draws* them into its dynamics.

Deacon (2006) has further distinguished "morphodynamic emergence" in non-organic self-organizing systems (as observable in the formation of a snow crystal or in the so-called Bénard cells forming in a heated dish of fluid) from the "teleodynamic emergence" in living systems: Here, a *memory of past states* crucially constrains and orders future processes. This leads to a particular form of downward causation established in repeated organism–environment interactions, which turn a superordinate *process* into a modified *structure*, thus furthering the self-sustainment of the organism under changing conditions (see 4.2.3, Figure 4.7). Teleodynamic emergence enables *adaptivity* (Di Paolo 2009), meaning that the living being is capable of regulating its own interaction with the environment, not only by currently selecting between suitable reactions, but to a certain degree, even *by changing its own structure and dispositions over time*. This is the basic function of implicit or body memory.

Now the crucial function of *consciousness* in this context is to establish an integrated superordinate process—conscious experience or *Erleben*—which enables a specific type of adaptivity, namely *learning*.[11] This means an organism's change of structure under conditions of conscious awareness. To illustrate this, let us once more take the example of Pavlov's conditioning which I have already described as a coupling of meaning (see 4.2.4; Murphy 2006). Here the higher-order system, as seen at first from an objectifying point of view, is the dog's living body in the context of the particular situation. The superordinate regularity or order parameter is the pairing of the bell with the smell of the meat. This produces a repeated simultaneous wiring of the neural networks within the dog's brain related to the sound, the smell, and to saliva secretion. After a sufficient number of repetitions, a neuronal coupling is established at the micro-level which changes the future behavior of the dog with regard to similar situations. Each time such situations occur, a modified resonance will now arise between brain, body, and environment. In sum, downward or global-to-local causation has occurred by selecting and constraining the synaptic weights and connections within the dog's brain, resulting in suitable patterns of future activation.

---

[11] Of course, the usage of this term *sensu strictu* forbids calling a neuronal network or some other artificial system a "learning system." Learning is bound to conscious adaptivity, in the case of humans augmented by purposeful or *self-induced* activity, such as learning a language.

However, it is decisive that all this has taken place *with the dog's conscious awareness*. The conditioning could not have been established in an unconscious state (e.g., if sound and smell were only presented during sleep). The reason for this is that *only the conscious state of the dog is in a position to establish the superordinate spatiotemporal unity, or the integral of the situation*, within which the sound, the smell, the taste, and the dog's desire for anticipated meat may be combined *into a unitary experience*.[12] Only as a result of this spatially and temporally extended integration are the processes in the dog's brain contextualized in a new way: they become components of a specifically extended system, namely the dog's lived body related to its perceived environment (including intermodally paired sound and smell), with its anticipation of imminent saturation, and with its specific history (sedimented in the dog's body memory). This overarching context as enabled by conscious integration explains the reciprocity of process and structure, which we have considered in section 4.2.3, in terms of circular causality: *conscious experience, as enactment of life, is the superordinate process, which shapes the participating structures at the microlevel, and is thereby incorporated in form of lasting dispositions.*

6.3.2 Downward causality and dual aspectivity

"Teleodynamic emergence" thus enables conscious learning processes. However, we should not call consciousness itself an emergent process, but only the higher-order system regularities that *correspond* to conscious experience (i.e., the biological and ecological aspect, seen from a third-person perspective). Any reference to "emergence" is possible only within a *single* aspect or from *one* methodological point of view. Similarly, the concept of the "co-emergence" of the whole and parts should only be used with respect to the biological–organismic aspect and not to characterize a link of one aspect with the other.

This is particularly important if we want to consider the problem of 'mental causation' in the appropriate way. How can the mind exert its causal powers in a material world? When theorists of emergence aim at solutions to this problem, we often find a short circuit of conscious and biological processes, for example in neuroscientist Roger Sperry's account:

> [Conscious] phenomena as emergent functional properties of brain processing exert an active control role as causal determinants in shaping the flow patterns of cerebral excitation. Once generated from neural events, the higher order mental patterns and programs have their own subjective qualities and progress, operate and interact by their

---

[12] As we have seen in section 4.2.4, the intermodal binding enabled by conscious awareness most probably corresponds to a specific synchronization of responses in the various neuronal networks that process the attended stimuli (Singer 2009, 193).

own causal laws and principles which are different from and cannot be reduced to neurophysiology. (Sperry 1980, 201)

Similarly, according to neurophysiologist Walter Freeman, consciousness may itself be seen as an order parameter and "state variable-operator" that "mediates relations among neurons" in the brain (Freeman 1999, 132). But how should conscious processes or subjective qualities be in the position to control or order neuronal activities? The category mistake results from the short circuit of conscious and neural processes: rather than understanding consciousness as the *integral activity of living beings*, it should adhere to specific physiological processes while simultaneously affecting them, as if from the outside. *But the routes and actions of a human being are never directed by consciousness as such.* They are directed and enacted by the conscious human being *as a whole,* including all its physiological and brain processes.

Based on the conception of the dual aspect of the living person, we have to characterize those actions in two entirely different ways. Though being complementary, they must not be confused. The following example highlights this:

A. A hungry person sees a delicious apple and reaches out to grasp it to satisfy his hunger.

B. Between an organism in a homeostatic imbalance ("hunger") and his surroundings, a resonance arises, connecting visual neuronal patterns stored in the organism's brain with current patterns in the environment (in particular, the "apple"). Due to the organism's current deficiency, this external pattern has become relevant and salient for it, corresponding to a particular attractor landscape in the neuronal phase space. Based on neuronal couplings formed in the organism's pre-history, a sequence of other state changes now ensues (such as the activation of evaluation and reward-related systems in the brain, the resort to taste-related neuronal patterns, the emission of saliva, etc.). This leads to a new overall condition of the organism–environment system, particularly including the activation of motor patterns in the brain's premotor and motor areas. The system's resulting instability ("tension") is finally released in appropriate neuro–motor impulses, which constrain the muscles' excitation preparedness according to an ordered pattern and end up in the muscular movement of grasping the apple. Now, a new overall state has formed between the organism and the environment, including a remaining instability which is immediately transformed into the next state leading to "taking a bite" of the apple.

At no point in the processes (B) just described (obviously in still a very simplified form) did consciousness exert any type of *additional* influence on neuronal and other organic partial processes. Indeed, in this objectivizing account it consequently failed to emerge at all. Is there any sign of its whereabouts? Certainly not in the hungry person's brain, as a mere byproduct. Consciousness is rather the integral of all the relations of organism and environment described in (B), and this integral constantly changes, in accordance with the reconfiguration of the overall physiological constellation of the organism in its situation. Both descriptions render complementary aspects of the life process, and there is no interaction between them—after all, they are only aspects of *one and the same process*. One might be tempted to somehow single out from the flow of consciousness an "impulse" or a "decision" to grasp the apple, and to ascribe a special efficient power to it. But this dualistic intuition is misleading. It would suspend the complementarity of the aspects, or metaphorically speaking, allow one side of the coin to impact on the other.[13] While it is certainly true that preceding impulses and decisions have a role to play in our actions, nonetheless, they too can always be accounted for under both aspects in parallel and in a continuum. There is no way to escape the duality of aspects or to render it "permeable" at certain points.

Does this mean then that the description (A), that is, hunger, seeing, and grasping the apple is only a naive "folk psychological" account of this event, while case (B) illustrates what "actually" happens? Are we not sliding into a new "biological epiphenomenalism," basically turning conscious experience into a dispensable varnish over the organism–environment relationship systemically described? And could the organism not achieve the same results without implicating consciousness?

The answer is no: for conscious processes enable the living being to grasp its own state in a far more complex, yet at the same time integral way and to choose potential actions with much more flexibility than unconscious steering routines would allow for. As I have already argued in section 2.3.3 ("The role of consciousness"), this is mainly achieved by the capacities of consciousness:

1. *To produce a unified intermodal action space with integral gestalt units* ("apple," "grasping").
2. To be *intentionally and affectively directed* towards relevant objects ("perceiving the apple," "hunger," "desire").

---

[13] This misleading intuition also formed the basis of the aforementioned experiments by Libet (1985) who started from the premise of a mental "impulse" as triggering a motor activity, with consciousness exerting an intermittent influence on neuronal processes (see 2.3).

3. *To transcend the momentary present,* either anticipating what is about to come ("reaching for the apple"), or retaining what has just been experienced.[14]
4. To provide a sense of *self-awareness* which integrates the organism's current state with regard to its own self-preservation as well as in relation to external objects ("satisfying my hunger by eating the apple").

In these ways, consciousness obviously multiplies a living being's possibilities of action and adaptation to various situational conditions, while at the same time reducing the complexity of myriads of micro-processes by transforming them into integral experience (see 4.2.4). In contrast, an organism without consciousness, perception, feeling, and desire would not only behave quite differently; it would also have *an entirely different structure*. Most of the processes outlined in scenario (B) would not emerge in the first place. The living organism would interact with the world in a much more one-dimensional way and only with restricted patterns. Hence, not only the events described in (A), *but also those in (B)* would not advance at all.

One might object that consciousness could still be a byproduct of the systemic processes occurring within and outside the organism. Have we really achieved anything with this whole investigation apart from widening the basis for the reduction of consciousness, that is, from the brain to the organism or to the system of organism and environment?

However, here we just need to turn the tables: the crucial insight that we have won by integrating life is the *primacy of function*, including, above all, the function of consciousness itself. The experiential processes (A) do not proceed in this manner *because* the organic processes (B) proceed as they do (and could do so even without consciousness). On the contrary, the processes (B) advance in exactly this manner *because human beings have consciousness, sense hunger, perceive, and initiate motor activity.* For consciousness is the crucial function of higher living beings, enabling them to have feelings, perceptions, volitions, and to perform actions. It is only in order to realize these integral functions that the necessary central nervous structures have developed at all. Hence description (A) does not state a mere epiphenomenon, but *the actual meaning and biological purpose* of the processes described in (B).

Granted, there would be no human consciousness without a brain; but equally, *without consciousness there would be no human brain*. If, during the course of evolution, the functions of consciousness had proved superfluous

---

[14] As I have argued, all brain and body states, as such, are always strictly present (physiological mechanisms and control loops cannot "anticipate" anything), whereas the overarching temporal integration of consciousness allows grasping the possible future as such.

or disadvantageous, the brain structures necessary for them would not have developed further. And if consciousness did not exist, they would not have developed at all. Hence, feeling hunger and seeing the apple are derived from a *phylogenetic history*, in the course of which such functions have shaped the organic structures of living beings.[15] Now, it is only because the human person possesses conscious capacities by virtue of his suitable organic structures that he behaves as he does: he reaches for the apple *because he is hungry* and *because the apple looks delicious*. Neither of these qualities nor the subject of the experience, his volition and action can be retrieved in the complex description (B). And yet, all these properties are what human life actually consists in—in the biological sense too.

This does not create a substance or entity of its own named "consciousness." As a function of the organism, consciousness is necessarily immersed in organic processes, as embodied subjectivity. This also makes a complete account of its carrier processes possible in principle—on the proviso of sufficiently extending the basis to the *entire* system of organism and the environment—without stumbling upon consciousness anywhere, as it was suggested in case (B). And yet this description of the process would not actually be complete—indeed, to believe it is would be a fundamental mistake. For account (B) refers to the organic processes, which *precisely serve the purpose* of supplying the human being with the capacity for conscious activities, and which are *structured and characterized in exactly the right way* for a person to realize this capacity. Only the complementary account (A) supplies the biological meaning and purpose and, so too, the sufficient explanation of those processes.

Thus, particularly from a biological viewpoint, it would be entirely inappropriate to eliminate consciousness or to treat it as an epiphenomenon due to its no longer being observable in the naturalistic attitude (B), or not interposing

---

[15] Of course, this is an abbreviated formulation. More specifically, the evolution of specialized brain structures may be seen as the result of a matching between spontaneously developing life forms and ecological niches as "spaces of possibility" (see 6.3.1, footnote 10). The formation of novel affective or cognitive functions may then be explained by a combination of vertical and horizontal circular causality: spontaneously developing genetic variants (upward causality) are integrated by novel functions (downward causality), provided that these functions or capacities prove effective in coping with the environment (horizontal circularity). This is only the case, however, insofar as they fit into pre-existing ecological spaces of possibility. In this sense, one can say that in evolution, too, the superordinate function shapes the developing structure of organisms, or in other words, the functions create the conditions of their own realization. Of course, the evolution of human culture crucially changed the ecological niches in which the brain further developed, resulting in a co-evolution of brain and culture (see Menary 2013).

itself in the physical causal chains. This would be synonymous with the claim that precisely the same physiological processes could also have emerged and now proceed without consciousness, thus turning matters upside down. It would mean to conceive of a merely physical or object body without a lived body.[16] However, the functions of consciousness outlined above could by no means be realized by blind physical processes. And consciousness itself, as a process that integrates space, time, and the self, could by no means be a mere epiphenomenal product of physical processes. Rather, it is the *necessary complement and purpose* of certain organismic, in particular neuronal processes connected to the environment at a sufficiently high level of complexity. *Both aspects are irreducible, and yet ontologically inseparable.*

Hence, the realization of conscious perception, feeling, and action requires specific physiological structures and processes. *In turn, however, these would not have emerged at all without those integral functions.* All this points to the description of experience and action (A) as not a mere varnish glossing over "real" processes (B), for these only exist as the means to the end of being able to experience and act. Hence, (A) also states the *actual cause for the apple being grasped in the person's hand.* Someone grasps the apple because he is hungry—not merely because certain homeostatic and neuronal regulatory processes are at work within his organism. Integral causality of life cannot be reduced to physiological processes, wherever they are located and however they are extended.

To give a full account of the cause of the process, however, we still have to expand the functional–phylogenetic perspective and include the *individual historic dimension*. For if the person in question had neither seen nor tasted the apple beforehand (an overarching conscious act of life that included

---

[16] In analytical philosophy of the mind, this would correspond to the hotly debated "zombie" argument: if physicalism were true, then one could imagine a sufficiently complex organism showing precisely the same behavior as a human person, yet without possessing consciousness—a zombie. But since we are not zombies, physicalism must be false (see Harnard 1995, Chalmers 1996, Kirk 2008, Bailey 2009, among others). This argument is supposed to provide evidence for the irreducibility of qualia to physical processes, yet with the unwanted side-effect that there is no longer any causal role to be assigned to consciousness. Indeed it ultimately amounts to a Cartesian dualism: if it is at least conceivable that an organism such as ours could function completely without being conscious, then consciousness would seem to be something non-physical.

For obvious reasons, in contrast to Descartes's time, the opposite of a zombie is rarely discussed today, namely an *angel*: a disembodied immaterial consciousness. However, both ideas are equally mistaken conclusions from the duality of aspects, which hypostasize one of the aspects to a reality of its own. One could argue that as long as we are inclined to consider one of these possibilities even to be *theoretically* possible, we still do not conceive the unity of embodiment, life, and mind radically enough.

corresponding neuronal processes), he would not now be in a position to recognize the apple. He would be entirely unaware of the apple's nourishing properties as food and have no idea about its taste. But his knowledge of apples was not constructed by brain processes. It was established thanks to embodied and also social interaction—other people taught him about apples and their significance, and so he tasted them. Without such conscious, holistic pre-experiences, however, neither process (A) *nor* process (B) could be fully explained now. If a person had only ever been instilled with grated apple in his sleep since childhood without ever seeing them, this would not have left the neuronal patterns that now enable him to recognize the apple as such. Conversely, if we assume that later on an acquired fructose intolerance had led to an incompatibility of apples, then the person will only *avoid* the apple next time if she has *consciously experienced* the nausea or abdominal pain and thus remembers them vividly when seeing the apple.

Hence, without an integral pre-history or real-life experiences, a current integral enactment of life—such as seeing and grasping (or avoiding) an apple—is neither possible nor explicable. Account (B) is therefore incomplete also in this respect, even though we may assume that it fully describes all *physiological* occurrences in cross-section. Even so, it only offers a snapshot of the superordinate course of events and therefore only a *fragment* of the life process. The full reasons why somebody grasped the apple only rest with the individual person himself, namely, as a conscious living being, with his personal learning history, in his present overall state of being, and in this particular situation.[17]

In conclusion, we can return to the concept of downward causality or macro-determination. As I have shown, the idea of vertical circular causality developed in section 3.3 can only be sensibly applied if the organism as a whole or the respective function is already presupposed. An integral level emerges not through self-organization at the substratum level. The living being with its individual learning history is the cause of its acts of life, not the complex conditions of matter within its body. Conversely, downward causality does not mean that integral functions such as, for example, conscious pursuits actively intervene

---

[17] This does not yet resolve the question about man's free will. The relevant concept of free actions naturally requires additional qualification, such as activities including imagination and deliberation which have only played a minor role in the example of grasping the apple. Here, the key point was to trace such everyday actions back to *the integral causality of life*. Nonetheless, this paves the way for a non-dualistic idea of embodied freedom (see 6.4.2). For integral causality does not assign a "mental sphere" its own power of influence, but embeds conscious reflection, decision-making, and action within a sequence of enactments of life. These are always physical in nature as well and therefore may equally have causal implications in the physical world—such as the hand grasping the apple.

in neuronal processes. Rather, the latter are structured and proceed in such a way *because* the living organism has developed the capacity for specific conscious activities both phylo- and ontogenetically and now puts this capacity into practice.

Hence, these activities are neither in opposition to the neuronal processes nor do they steer them. No laws of physical energy conservation are contravened by an "additional intervention." This does not mean, however, that conscious acts of life are explicable purely by physics simply because they are *also* physical occurrences. In isolation, this level is grossly under-determined—what it can contribute as an explanation is restricted to certain physical–chemical conditions for cellular micro-processes. But even the complex biological system processes in account (B) offer no full explanation of what occurs. We *can* naturally describe things on this level, and no gap will emerge in the description at any point that *forces* a change of our attitude. However, to return to an earlier comparison, it would be as if a country's rail transport—with all its procedures for transport, energy supply, circulation, and control—were completely accounted for and described in every detail, while crucially overlooking the fact that people are sitting on the trains and that this entire exercise is only intended to provide them with a service and means of transport.

6.4 Consequences for psychophysical relations

Our discussion of identity and emergence theories further refined the concept of personal aspect duality. Following on from this analysis, we can now revisit the "Bieri trilemma" (see 6.1) and recast its statements in accordance with the concept. First, we should set aside the "mental," since this notion is virtually impossible to separate from the dualistic implication of a disembodied, inaccessible inner world. But even the significance of the "physical," as previously outlined in section 3.1.2, is markedly altered: this implies no exclusive describability by physics, but also refers to living processes, including the integral activities of animate beings. Hence our reformulation goes like this:

1. Phenomena of consciousness consist in the conscious activities and life acts of living beings within their surroundings. As holistic processes, *they are also physical in nature*.
2. As life acts of the organism as a whole, phenomena of consciousness are also *causally effective* in the physical sphere.
3. Within the domain of living organisms, the physical sphere is *causally under-determined* in terms of the laws of physics.

Statement (3) is grounded in the fact that physical processes in organisms are incorporated into their integral life activities. To this extent, while being subject to the natural laws of causal relation, they are simultaneously constrained by the superordinate influence (downward determination) of these integrated activities. In particular, they are subject to the bodily, emotional, and intellectual capacities, which an individual has appropriated throughout his learning history. Their superordinate determination is manifested in each concrete realization of such capacities. In other words, *only living organisms as a whole, and not their constituent parts, are the sufficient causes of some physical occurrences in the world*. From this follows that the agent-causal statement: "Tom raised his arm" may not be fully transferred into an event-causal statement which goes: "Some event or some process *within* Tom caused a raising of his arm," although such events or processes certainly act as proximate causes for Tom's movement. They are necessary, but not sufficient causes.[18]

What now follows is a discussion of some implications for psychophysical interrelations arising in this scenario. However, on the basis of the concept of dual aspectivity, the notion of psychophysical or psychosomatic "interactions" is no longer tenable. *No side of the coin impacts the other*. Thus, research into psycho-physiological interrelations has to be essentially restricted to the search for correlations, co-variations, and structural links. Yet at least we can differentiate various procedures at work within the organism, deciding in each case which aspect is "leading" and co-determines or constrains the other—namely, the integral or the (micro-)physiological aspect.

6.4.1 Intentional and psychological determination of physiological processes

Intentionality determines physiological, in particular, neuronal processes. As I momentarily reflect and write these sentences, I do so in harmony with certain physical laws, at least to the extent that this activity can certainly be described as a process of neuronal activations inside my brain, of finger muscle contractions,

---

[18] As can be seen, this contradicts the principle of causal closure in Bieri's trilemma, namely the assumption of a *sufficient* physical cause p_2 to exist for every physical event p_1 (Kim 2006). However, this principle is not even usually claimed by physicists, in particular in view of quantum physics. For example, it is *not determined* by a preceding physical event at which time an unstable atomic nucleus such as radium decays. In that sense, it is an uncaused event. Hence, the principle of causal closure, though intuitively plausible in terms of Newtonian physics, should be abandoned. Nevertheless, the reformulation (3) is not necessarily based on an indeterminacy at quantum level; it only presupposes that the strict premise of causal closure is generally unfounded. On the other hand, it should have become obvious that this is not meant to imply a "mental causation" by a non-physical mind.

and so on. At no point does any type of non-physical energy intervene here. Nevertheless, those laws are neither sufficient to determine nor to explain what I write—this is the product of my thought processes. My conscious activity of thinking and writing is therefore the *superordinate cause* of the physical changes, which make this written text appear in the world. The integral causality of my life activity *inherently comprises and orders* the micro-causal links of neuronal and muscular processes which realize my writing.

In the course of my reflections, the confusing idea can occur to me that my thought patterns are not free at all. Instead, I should be thinking and writing in such a way "as prescribed by my neurons." The physical mechanism of my neuronal processes would proceed relentlessly as prescribed by laws of nature, and there would be no escape from the iron cage of my brain. Yet again, we must turn the tables at this point. Firstly, my thoughts—as agile as they may be—are never absolutely free. Rather they follow various paths which my experiences pre-draft as potential routes, such as logical paths, which have formed in the refinement of my capacity for thought, associative paths (ideas, images), which correspond to my individual experiences, or emotional paths, which link my thoughts with specific values, wishes, and aspirations. *However, the neuronal processes must also orient themselves along such pathways.* In other words, the neuronal system conditions $N_1, N_2, N_3 \ldots$ which run their course while I think, are linked in this way *only because* my corresponding psychological conditions $P_1, P_2, P_3 \ldots$ are linked with each other via the laws of logic, semantics, form similarity, and so on. My individual pre-experiences of such meaningful relations have been incorporated in my plastic brain structures as "open loops" which I am now able to actualize in my intentionally directed thoughts. Hence, my fingers move the way they do not merely because of the participating proximate or micro-causes (depolarization of neuron membranes in my brain, acetylcholine release in the motor endplates of my finger muscles, etc), but *because of the superordinate activity of my thinking and writing* (downward or global-to-local causality).

These reflections can be summarized as follows: *human beings transform intentional, semantic, and other meaningful relations into causally relevant dispositions of their organic basis.* Hence the brain is no cage, but an *organ of potentialities*. It is not the mind that must do what the neurons prescribe, but conversely the neurons facilitate everything unfolding in the mind. The inherent meaningful, that is to say logical or associative structure of my thoughts becomes an organizational structure for neuronal processes which enable these thoughts to be realized. In this sense, *reasons or motives take on a causal effect*: my thoughts can gain influence in the world as my life activities. So this written text may result from them without infringing the law of energy conservation or any other physical laws.

However, the same applies for all reason- or motive-based human action. Its *actual* explanation is related to psychological and teleological, not to physiological or even physical statements. The intrinsic links of experience, dispositions, motives, and behavior can only be understood psychologically, even if we increasingly know more about the neurophysiological processes involved. Why am I offended by an insult someone has directed at me? My sense of self-worth and the possibilities of feeling hurt were not autochthonally manufactured inside my brain, but trace back to complex social experiences. The insult presupposes that as a child I learned the *meaning* of linguistic signals in their social and situational context. Back then I also developed a reflective understanding of my personal status in the social group. Hence, I connected related value *judgements* with my own emotional responses. Today, this implicit relational knowledge and value system operates "without thinking," due to corresponding neuronal links that were formed in those processes. But its effect is still based on understandable connections.

The explanation of the neural processes involved in my reaction in the various brain areas associated with cognition and emotion may progress as far as possible. But if a neurobiologist were to try and describe these processes to me as 'my feeling slighted', he would after all need to resort to a psychological description, based on a highly problematic assignment of experiential to neuronal conditions. In the final analysis, I would not be any the wiser about why I was insulted by the comment, let alone what an insult actually involves. On the contrary, the inherent meaningful connection of my experience, once objectivized to an external, quasi-mechanical process, undergoes an alienation. It is as if I have been dispossessed of my own reaction as it is turned into an impersonal process. This is not an argument against the neurobiological investigation of its correlates; but it does contradict the view that my experience can be described through its materialization in brain processes in a clearer, "more realistic," or even "actual" sense. The opposite is the case.[19]

The two descriptive systems do not rank on an equal footing here. Rather, primacy is attached to the psychological–hermeneutical aspect both *genetically* and *methodologically*. The cause of a mourning reaction, for example, is not the activation of the cingulate cortex which may be observed in its course, but certainly is the painfully experienced loss. And it is not the activation of the amygdala which causes fear but primarily the subjective perception

---

[19] The criticism leveled by Freud, a former neurologist, against neurological explanations of states of anxiety is similar: "Today, however, I must remark that I know nothing that could be of less interest to me for the psychological understanding of anxiety than a knowledge of the path of the nerves along which its excitations pass" (Freud 1917/1963, 393).

and evaluation of a threatening situation—and this superordinate perception cannot be found in the amygdala, however necessary it is for the experience of fear. Hence, a lesion of both amygdalae leads to a loss of fear reaction (Feinstein et al. 2011), but the overreaction of the amygdala in panic disorders is only a result of its physiological adaption to repeated experiences of threat (LeDoux 1998). Moreover, since the relevant neurophysiological processes only occur as they do *because* previous experiential and motivational connections have sedimented in organic links and dispositions, we initially learn nothing whatsoever about the *causes* of the current psychic process merely by examining brain conditions. First, these have to be brought into a regular relationship with specific types of experiences. For this, a most fine-grained and differentiated communication between the subjects is called for in terms of their experiences and motives. In other words, the theoretical caliber of phenomenological and hermeneutical methodology is required to describe these experiences.

Only in a second step can neurobiologists proceed to identify the correlates associated with these primary subjective and intersubjectively accessible experiences. Assuming that the correlates can be identified with sufficient validity, they only ever represent components of the overall functional cycles of embodied experience. Only by embedding them within the pre-history and the superordinate dynamics of the current situation can neuronal process patterns be given a function and purpose that they do not possess themselves. Here, "explanation" and "understanding" cannot be strictly kept apart. Rather, any explanation that includes neurobiological processes is not possible without understanding the experiential and motivational connections.[20]

6.4.2 Embodied freedom

The question of free will undoubtedly represents one of the cardinal problems of psychophysical interrelations. The topics dealt with in sections 2.3.1 and 2.3.2 are therefore revisited and considered here in greater depth. Briefly, these sections yielded the following insights.

---

[20] To give a final example: Hardcastle and Stewart (2009) analyzed the overall gain of numerous fMRI studies on pain, with the result that in spite of many detailed findings, "most of these studies are not telling scientists anything that they did not already know from traditional psychological and clinical investigations [...] In short, what we have are largely replications of previous psychological experiments, but now in color" (p. 191). The authors conclude: "Thus far, it appears that the imaging technology has not improved our theoretical understanding of cognition; it has merely given us vivid illustrations of the cognitive processes that psychology had already surmised were there" (p. 192).

Neuroscientific enquiry into free will takes as its point of departure a dualistically defined alternative. Decisions are either to be seen as products of brains, or arrived at by a fictitious, bodiless "Ego" or mind. The mind's autonomous, rationally grounded power of decision-making is then disputed in a second step (with reference to experiments by Libet and others). However, ascribing decisions to the brain not only implies a mereological fallacy, but also creates an aporia: it is no longer possible to assert any function for the activity of conscious decision-making. Subjectivity thus remains an equally inconsequential as well as inexplicable epiphenomenon of neuronal processes.

This conception ignores the fact that decisions always belong to an *enactment of life*. Making a decision does not present an isolated and lightning-like act of volition. Instead, it is a "maturing" process that occurs over a period of time. However, this process by no means involves purely conscious reflections and rational justifications. Rather, it includes pre- and unconscious motives, feelings, experiences, and expectations—ultimately the entire life history of the actor which he embodies as the individual person. To take a decision and act upon this therefore signifies a type of self-determination that can only be attributed to the organism or person as a whole. In place of the dualistic alternative of the rational subject or brain, I therefore propose the idea of *embodied freedom* as a particular variant of a libertarian concept of free will.[21] I will elaborate on this in what follows.

6.4.2.1 A phenomenology of decision-making

Neurobiological experiments taken as exemplary descriptions in the analysis of free will, especially those conducted by Libet and followers, focus on instructions for motor activity or pre-selected reactions that happen in fractions of a second. But by looking too closely at the minutiae, this misses an adequate assessment of the phenomenon. A relevant interpretation of freedom is linked to a time-spanning *process of decision-making* that cannot be broken down into arbitrarily short time episodes. The decision as the outcome of this process is inseparable from it and does not arise through an arbitrary "lightning volition" at the end. The unique human capacity of free will therefore rests primarily on a particular qualification and inherent structure of the decision-making process. This warrants closer analysis.

---

[21] Libertarian accounts assume an incompatibility of free will with determinism; compatibilist theories take the opposite stance. The question of determinism will be discussed further later in this chapter. Of course, it is impossible to take the literature on the problem of free will into account here even on an approximate basis. For an overview, I refer in particular to Kane (2011) and Pereboom (2009).

The ability to take a decision is first based on bracketing what is merely factual and thinking through the available alternatives *as* possibilities—I could or could not do this, or I could do something different. The precondition for freedom is thus initially a space for thinking and imagining, where I am free to move among possibilities without factual constraints. This space of possibilities arises from an *inhibition*: we possess the capacity *to suspend* our own impulses and desires, to pause for thought, and test whether and in what way we convert them into actions.[22] Frequently, this pause for reflection results from an ambiguous situation in which various conflicting possibilities emerge, thus leading to a temporary disorientation. Should I keep my promise to go on a weekend trip with a friend, or change to another, more attractive alternative? The inhibition or interruption of an unconsidered life trajectory opens up a *moratorium*, a more or less extended time period for the process of deliberation, keeping one's own counsel, and voicing and clarifying motives and reasons. In virtual trials, a person anticipates future possibilities, weighing up advantages, risks, or hurdles in order to find a new *coherence* and thus reorientation of the personal life trajectory.

Arguably, this is not a strictly systematic, but rather a dynamic and creative process in which conscious and unconscious components, feelings, desires, ideas, expectations, and reasons mutually influence each other, modifying and spurring each other on. Hence, the result is not derived from pre-existing psychological determinants. We are neither concerned with a vector addition of independent psychic motives, nor with a rational calculation or algorithm of reasons. Rather, the various components flow into the process's open "horizon of possibility." In this sense, a free play ensues and takes on the form of inner dialogue and *a relationship to oneself*. This is the central condition of freedom: by engaging with our motives, desires, and reasons, they do not remain the same, but parade on an inner stage so to speak, where we can weigh up, evaluate, and modify them. By identifying with the perceived possibilities on a trial basis, as if we were "trying them on" to see how they fit us, we can gain the freedom to take a stance and make our choice.

Hence, the relationship to oneself fundamentally transforms the pattern of events. If considering and deciding were only a linear sequence of conscious

---

[22] John Locke described this as follows: "For the mind having in most cases, as is evident in experience, a power to suspend the execution and satisfaction of any of its desires, and so all, one after another; is at liberty to consider the objects of them, examine them on all sides, and weigh them with others. In this lies the liberty man has [...] we have the power to suspend the prosecution of this or that desire, as every one may daily experiment in himself" (Locke 1689/1825, book 2, chapter 21, § 47; p. 169).

events, we would have no influence over the outcome. The freedom of the decision depends on keeping counsel with oneself, the personal self-relationship where the implicit balance of all previous experiences holds sway. Cognitive, emotional, and intuitive components are linked and permeate each other in backward and forward movements. This forms a progressive clarifying process, so that the person's motives become more transparent and he is finally in a position to identify with his choice.

The experience of "being in agreement" does not emerge through a purely rational assessment of intelligible reasons, however. Successful decisions equally require an emotional and bodily intuition in which former experiences are implicit present—an *emotional body memory*.[23] Were a person to ignore such feelings, this would indeed be at the risk of self-alienation. A person will only arrive at a sustainable decision if a sufficient degree of *congruency* ensues between tentative identification and emotional-bodily intuition. Decisions are more freely taken, the more aspects and layers of a personality are incorporated into the dynamic process of evaluating and self-questioning. As Bergson (1950, 172) put this, we are free "when our acts spring from our whole personality, when they express it."

Arriving at a decision means an active closure of the horizon of possibilities which opened up by suspending preceding impulses. But again, a decision cannot be understood as an isolated act of an independent authoritative "Ego" intervening at some point and in a haphazard way. Rather, we should regard the entire process of evaluating and deciding as a dynamic and open movement, in which the person taking the decision is involved and also changes in it, at least to a minimal extent. Hence, this process involves a hermeneutical circle: the personal subject cannot stand by and look on indifferently, but rather experiences and articulates himself in forming his will. This process not only expresses, but also shapes and modifies the primary, pre-reflective self-feelings; it thus implies both aspects of *self-articulation* and *self-creation*. At the same time, however, the subject is also the authority presiding over the process, driving it forwards and taking the lead.

The freedom to choose an alternative in the decision-making process is therefore not founded on a higher-level reason outside this process. On the other hand, the choice does not happen in an arbitrary or purely decisionistic manner. Rather, a person identifies with an alternative, embracing it and thus making the corresponding reasons effective. The relationship to oneself culminates

---

[23] From a neurobiological point of view, a similar concept has been proposed by Damasio, namely his "somatic marker" theory (Damasio 1995; see 4.1.4).

in a *self-agreement* with the choice, and thus with the type of future a person is willing to accept on the basis of this identification. Hence, we experience a decision as an enactment of self that is also expressed in the consequent course of action, namely, as an accompanying and guiding intentionality. For this reason, and not merely due to a social attribution, we later encounter ourselves as also responsible for our actions.

This brief synopsis of the decision-making process suggests an idea of freedom as a person's ability:

- to suspend his primary impulses
- to clarify his intentions in a moratorium that ensures these are in accord with personal motives, feelings, and attitudes
- initially to identify with possibilities in a tentative way and in the light of the personal relationship with oneself
- to arrive at an inner coherence or felt agreement in taking a decision, and finally
- to transfer the decision into action or a series of actions.

The triggering of an action, which is the primary focus of neurobiology, is thus only one small part in the entire intentional arc of human freedom, which was opened up by the suspension and interruption of the unproblematic enactment of life.

6.4.2.2 Free will and integral causality

This concept of embodied freedom is already prefigured in our account of integral causality (see 3.3.3 and 6.4). If Tom raises his arm, it is not some event or process *within* Tom which gives the sufficient cause for this action—neither an act of a "will" or an "Ego," nor some process in the brain. Neural processes only function as *proximate causes*, as do the physiological events in the motor endplates of Tom's muscles. The sufficient cause of raising his arm is Tom himself as a conscious living being including his mental and bodily capacities. This corresponds to Aristotle's notion of the "self-movement" of living beings.

The integral causality of living beings in general is raised to a higher level in human decision-making. For here, the options of action are not just selected immediately depending on the organism's current needs and the requirements of the situation—one might say, more or less on "autopilot." Rather, they are chosen only after Tom detaches himself from the situation and anticipates and evaluates the possible alternatives *as such* (see 2.3.1). Deciding and acting are no longer only acts of self-movement, but also acts of *self-determination*. This comes close to the concept of "agent causation" defended by some libertarian philosophers,[24] assuming that for free will to be possible, an agent or a person should be able to start a new chain of events by a decision that is not fully conditioned. It is indeed a person that performs acts of free will, rather than events in his body or in his history causally determining them. Event causation is not sufficient for integral causality.

However, embodied freedom assumes no "unmoved mover" or independent initiator of a novel chain of events, an idea that would admittedly approximate agent causation to substance dualism. The deliberating and deciding subject is embodied at any time; it thus contains and integrates its history, its emotional dispositions, motives, and intuitive evaluations, which all enter into the dynamical process of decision-making. This process is not a linear sequence of events, at the end of which the decision occurs as the last link of the causal chain. Rather, it forms an overarching temporal unity of components mutually permeating each other. This unity is based, on the one hand, on the moratorium of deliberation which suspends the impulses, motives, and options such that they may be anticipated and compared in "virtualized" form, and, on the other hand, on the reflective self-relation of the subject, which relates these components to himself and evaluates them.

The fact that prior to reaching a decision we genuinely have a choice between alternatives is thus based on the *horizon of possibility* that is created by this suspension, and on the *reflective relationship*, which we can adopt towards our possible options. Hence, a decision or a willed action is not fully determined by a causal chain of preceding events; there remains an ultimate spontaneity of will. Willing means choosing a course of action, in other words, *a bifurcation of the course of the world actually exists*: this becomes possible precisely by our being aware of the alternatives as such. Spontaneity does not mean random choice, however—the process of decision-making clarifies and increasingly restricts the remaining options, and there are always "good reasons" for the final choice, which the person makes effective through his decision, even if they do not suffice to *cause* it.

As we have seen, taking a decision is not the intervention of an autonomous self, but the activity of an embodied subject who must have learned and incorporated the capacities for inhibition and reflexion in the course of his biography. Free will is thus a complex capacity of human agents whose components can only be acquired and practiced through a *self-cultivation* in the course of social interactions. Section 5.3.3 provided a short account of how these experiences are deposited in the structures of the prefrontal cortex, thus making the brain

---

[24] See Taylor 1973, Chisholm 1976, Clarke 2003, and Lowe 2003.

an "organ of freedom." Embodied freedom is based on the formation of organic structures that facilitate the creation of spaces of possibility. But one could ask once more at this point: does this realization of decision-making through neuronal processes not suggest a determinist standpoint? Is our account of embodied freedom nothing more than an illusion after all?

The objections raised against this sketch of freedom, as *actually* being able to do otherwise—that is, a libertarian concept—are in the end based on a general or domain-specific determinism. However, this assumption is ultimately not founded on empirical evidence, but rather on a "worldview" or doctrine. There are convincing arguments against universal determinism, in particular, the fact that physical natural laws cannot determine how the world turns in every last detail (they are no "pushy explainers"). Rather, the laws that physicists find should be treated as regularities describing what occurs in a systematic order. Thus, they have no *prescriptive*, but only *descriptive* validity. The actual course of the world moves through manifold overlaps of lawful regularities, singularities, as well as chaotic processes. Last not least, probabilistic quantum physics dealt a body blow to the doctrine of universal determinism. Indeterminacy no longer seems to be an exception to the rule, but rather a basic element of all natural processes.[25]

Special or neuronal determinism represents another unproven assumption. So far, there are no determinist neurobiological laws that would even allow a certain prediction of a person's actions within the next few seconds or minutes. A prognosis is far more likely on the basis of genuine psychological insights. This is not only due to the brain's complexity, but also to its dependence on a person's pre-history, to its plasticity, and ongoing reshaping during every interaction with the environment. On these terms, it would be a meaningless task to search for determinist laws for brain processes, and thus, for a person's actions. Even Libet and his followers, who modeled their experiments on free will using fragments of seconds, were at best able to calculate statistical probabilities for a subject's ensuing actions. But no matter whether the outcome is 30%, 50%, 70%, or 95%, nothing suffices to demonstrate the brain's determinism. There cannot be "a bit of determinism"—it either equals 100%, or not at all.

Correlations between experience or action and simultaneous neuronal processes also provide no evidence for determinism. Although they are usually expressed in terms such as brain processes being "at the basis of" actions, or "conditioning," "controlling," "causing" them, and so on, such phrases only convey a

---

[25] See also 6.4. This cannot be described here in full; on arguments against universal determinism, see especially Cartwright (1999), Dupré (2001), and Keil (2007).

deterministic impression, but imply no strong physical determination according to laws of nature. Upon closer inspection, there is no proof of the brain's determinism, which neuroscientists often claim to be the case. Today, nobody can say for definite whether microphysical indeterminism or reinforcement mechanisms, such as those familiar in chaos theory, do not break through to the macrophysical level in the brain–mind relationship. In any case, stochastic processes can be found both at the molecular and cellular level of the brain: fluctuations of membrane potential, dumping of transmitter packages, or activations of individual neurons cannot be precisely predicted in advance. Thus, an incoming action potential leads to a neurotransmitter release at the synapses only in 10–20% of cases (Craver 2007, 22–25). The synaptic signal transmission thus proceeds in a non-deterministic way and allows only for a probabilistic prediction of the output of a neural network.

In sum, given a non-linear dynamic as we find it in the brain, it cannot be excluded that minimal deviations on the micro-level may lead to huge changes on the system level. Of course, this is no reason for a chance neuronal happening being in charge of the decision-making process: all that is needed is the presupposition that the course of the world or of brain processes, respectively, is *not completely determined* for all future. The *positive* determination of freedom emerges not from the micro-level, but from the integral aspect, that is, the intentional enactment of life of the conscious and embodied subject, as described in the previous section.

Anyone who believes that freely made decisions are incompatible with the physical worldview has not fully appreciated the status of science. This so-called "scientific worldview" has little to do with scientific practice. It is more apt to refer to a *scientistic* worldview that is not empirically grounded, but assumes the character of a metaphysical doctrine. In fact, in the natural sciences (both physics and neurobiology), no empirical evidence irrevocably contradicts the experience of free choice. The reservations about a libertarian conception of human freedom are ultimately always rooted in latent dualistic intuitions, assuming that this type of freedom was based on an immaterial mind steering the activities of neurons. In contrast, the concept of embodied freedom treats decisions as superordinate, intentionally directed enactments of life performed by an embodied person—enactments that are facilitated, but not determined by the neuronal processes involved.

6.4.3 "Psychosomatic" and "somatopsychic" interrelations

In the context of dual aspectivity, we now consider three groups of psychophysical phenomena that are particularly relevant in medicine. Specifically, these are

(1) "psychosomatic" and (2) "somatopsychic" interrelations, and (3) functional failures. As we will see, they can be explained by various aspects of *vertical circular causality*.

1. Integral acts of life, especially if they are linked to emotions, not only incorporate neuronal processes, but also other somatic (neuroendocrine, autonomous, muscular, etc.) processes. The entire organism thus becomes a "resonance body" for experience (see 4.1.4). These "psychosomatic" interrelations often favor dualistic intuitions of an "interaction" between the psychic and physical. However, a feeling of shame is not an external *cause* of blushing, just as little as fear *causes* palpitations or cold sweat. Rather, feeling ashamed and afraid are integral acts of life that equally *involve* intentional, emotional, and bodily components. Thus, the relation of emotions and bodily resonance is one of *implication*, not of succession or effect as it is the case in linear causality.

   True, the integral aspect takes the lead here, since shame or fear emerge from a superordinate experience of the current situation and not from any localized physiological event. Under the biological-systemic aspect, the situation can then be described as vertical circular causality: the organismic overall condition, corresponding to the experience of the shameful or threatening situation, is transformed via the brain into physiological partial reactions, such as blushing or trembling (*downward* or *global-to-local influence*). These partial systems and processes, conversely, contribute to the overall emotional state, in the way described as a cycle of embodied affectivity (4.1.3, Figure 4.3). As we have seen, a modified body resonance may also favor a corresponding emotional state (e.g., holding a hot cup of coffee may elicit a warmer feeling towards target persons, as shown by Williams and Bargh (2008); see 4.1.3). However, this does not mean an "interaction" between somatic and psychic aspects either, but corresponds to an *upward* or *local-to-global influence* of bodily conditions.

   As we already saw in the case of the insult, such integral reactions by the organism refer back to corresponding *couplings of meaning*, which we find at the most basic level in Pavlov's "conditioned reflexes" (see 6.3.1). The meaning of specific situations is linked to physiological processes that usually serve to provide suitable functional loops for coping with the situation. Regarding emotional reactions, such implicit links primarily form in social contexts in early childhood, conveying to a baby or toddler the meaning and interpretation of situations experienced in others' company (e.g., through social referencing, see 5.3.1). But they can also be newly established during the course of a lifetime. Here, a crucial point is that the connections, once formed, generally withdraw from conscious awareness. Similar situations can then provoke an autonomous organismic reaction without the person in question being aware of its cause.[26] On the other hand, physiological (autonomous and muscular) reactions may also withdraw from their former integration in superordinate emotions and become independent or *particularized* as permanent organic dysfunctions (such as hypertension, irritable bowel syndrome, lumbago, etc.). Having lost their expressive or activating function for coping with certain situations, they are now removed from superordinate control and regulatory feedback.

2. We now turn to the opposite type of "somatopsychic" interrelations that are particularly apparent when chemical or other agencies have an immediate effect in the brain. The examples are familiar: consumption of alcohol leads to a light-hearted mood, analgesics alleviate pain, hallucinogens produce hallucinations, electrical stimulation of the temporal lobe triggers remembered images, and so forth. Despite their being so obvious, it is not the case that holistic acts of life like moods, pain, hallucinations, or memories are directly produced by chemical or physical agents; rather, they are *instigated* or *called forth* by the agents through upward or local-to-global causation. The changed integral aspect (mood, memory, etc.) is the reaction or *answer* of the living organism to the effect on the physiological micro-level.

Consider for a moment the "mood-enhancing" effect of an antidepressant. How can we describe this process with respect to the dual aspect? Initially, the organism absorbs, metabolizes, and transforms the chemical agent, so that it is "recognized" as a substance. Now it can initiate certain neurobiochemical modifications in the brain, for instance, altered transmitter concentrations in the synapses of the limbic system, and after some time also reactive changes in postsynaptic receptor density, and so on. As a result of this bottom-up influence, the organism is able to enter into a changed relationship with its current environment (e.g., by readjusting the stress hormone balance, neurovegetative activation, etc.), in which it can fulfill certain requirements in an improved manner. This adaptive reconfiguration of the entire organism–environment system on the macro-level corresponds to a changed *mood state* that the person in question experiences in relationship to his surroundings.

---

[26] This is the case, for instance, with post-traumatic stress disorders when particular environmental factors, which resemble the original traumatic situation, may later trigger sudden stress or panic reactions (van der Kolk 1994).

In abbreviated form, one could say that the drug "caused" a mood enhancement. But there is no direct effect of chemical agents on acts of life or consciousness. Once again, both sides of the coin remain inaccessible to each other. Nor is it possible to refer to a linear connection of cause and effect at the physiological level. Rather, the agent gives the organism an *occasion* or stimulus to reorganize itself and its relationship to the environment.[27] Here, we can again refer to the vertical circular causality or transformation performed by the organism and especially the brain. The mood change, which manifests itself at the subjective level, is the result of this self-activated transformation and not the direct effect of the drug.

3. The last type of psychophysical interrelations focuses on the *restriction or loss of acts of life*. These are disorders of mental functions that can be traced back to dysfunctions or functional failures at the physiological level. Here, the organic aspect is leading in the sense that it no longer carries the integral acts of life in an unrestricted way. As already implied under (1), this can relate to particularization of physiological processes that were originally incorporated into the living being's integral reactions. A good example is a hyperactive amygdala that is no longer incorporated into acts of life in keeping with the situation—for instance, the adequate fear of an approaching predator. Instead, with the amygdala having become independent, even harmless stimuli may provoke a massive anxiety attack. Similarly, recurrent depressive episodes may still be triggered by specific experiences, but are now also rooted in permanent dysfunctional neuronal circuits that have formed in the course of earlier episodes (Moylan et al. 2013). Finally, macroscopically traceable *lesions* of specific brain areas or generalized damage to the neuronal structures (such as in Alzheimer's disease) can restrict certain conscious acts in a more or less serious manner, or even disable them completely.

Interpreting these interrelations presents no major problem, if we set aside the fact they are one of the most common causes for the localization fallacy highlighted in section 2.2.2. If a function fails, this always only ever points to a specific region being an important or necessary condition, albeit not a *sufficient* organic condition for a conscious act of life. Consciousness, in whatever form it may manifest itself, is not assembled from partial functions or modules. Instead, like the organism itself, it presents a *primary*

---

[27] The biological effect of medication is to this extent far less "mechanistic" than its critics or advocates often suppose. Furthermore, it is also important to take into account the interpersonal level of treatment, that is, the *meaning* of the medication for the patient (Chapter 7 deals with this in more detail).

unity, which is differentiated in the course of the individual's development in specific forms and capacities, and which actualizes these capacities as each situation demands it.

6.5 Summary

This chapter drew together the various strands of the account of "personal aspect duality" first highlighted in Chapter 3, combining it with an ecological description of the brain as an organ of a social living being. In summary, the key outcomes are as follows.

In place of an unmediated opposition of the mental and physical, the personal dual aspect assumes a fundamental unity of the living person. This appears, on the one hand—in the personalistic attitude of the first and second person—in the form of integral, conscious and embodied enactments of life. On the other hand, in the naturalistic attitude of the third person, the human being appears as a physical body that can be basically analyzed and broken down into physical processes, yet also reveals a complex physiological order. It can be understood from the perspective of an ecological biology and dynamical systems theory as the living being's hierarchically structured self-organization or *autopoiesis,* yet without fully reconstructing the person's conscious experience as such. Life as *Erleben* or experiencing oneself surpasses the system perspective.

In terms of each aspect, the human person can only be fully accounted for in the *relationship to the natural and social environment.* Living subjectivity is bodily and inter-bodily "being-in-the-world" and "being-with" (*Mitsein*). The living organism, in turn, only exists and endures through permanent exchange with the environment. The life process as a whole therefore constantly transgresses the body's physical limits and forms a superordinate system with its complementary surroundings. Since we have thus also gained an integral viewpoint on the biological side, we could also highlight *structural links* between both aspects in the course of the analysis. These are:
1. With regard to the *potentiality* of life, the relation of integral *capacities* to enabling organic *structures*, in particular, neuronal couplings or "open loops."
2. With regard to the *actuality* of life, the relation of integral *acts of life* to the corresponding superordinate processes of the organism–environment system, in particular, the production of coherence through *neuronal resonance.*

However, both aspects are also interconnected *diachronically* through the history of the living person or via developmental biology. The formation of capacities and habits can be recorded phenomenally as an *implicit coupling* in body memory and biologically as a *neuronal coupling*. In this way, it become possible to characterize a person's integral capacities, formed throughout their life history via the matrix of neuronal plasticity, as also physically causing present action. Hence, conscious enactments of life become causes of physical occurrences in the world.

Further clarification of these interrelations resulted from engaging with the currently popular *identity* theories. The identification of activities of consciousness with specific brain processes extracts them from the person's bodily, living unity, with the result that the subject can only be localized in certain partial processes of the organism. But the full biological basis for consciousness as well as a human person consists in the organism as a whole in its relation to the complementary surroundings.

In examining conceptions of *emergence theory*, I rejected what is mostly an underlying assumption of the primacy of the material basis. Physically describable material processes do not themselves bring forth life functions and acts of life. They can only (a) *facilitate* or *realize* them as a substratum, (b) initiate or *trigger* them as a stimulus, and (c) disrupt or *render them impossible* as a harmful influence. Physical matter is not the foundation that produces the living being through self-organization. On the contrary, the living organism transforms the material in a suitable way, assimilating and transforming this into its constituent parts. The form of life emerges not from matter, but rather *organizes* and turns it into *its own* material. Likewise, living functions—unlike those of machines—are not assembled from partial structures or processes. On the contrary, they are the condition for the development of organs and sub-systems through which they are realized.

An embodied and enactive concept should thus be based on a *strong version of emergence*, or on a *co-emergence* of the whole and its constituent parts. This includes both a two-way *enabling relation* and a reciprocity of global-to-local (downward) and local-to-global (upward) influence, or *circular vertical causality*. The respective function or higher-level organization forms and constrains the activity of partial components such that they are integrated and allocated a place within the whole (see also 3.3.1). Moreover, this relation implies a developmental aspect: as I have shown by the example of classical conditioning, a crucial function of consciousness consists in establishing a superordinate context which enables learning processes as an anchoring of novel experiences in an organism's neurobiological structures. However, it was also emphasized that consciousness should not itself be considered an "emergent process," since any reference to "emergence" is possible only within a *single* aspect or from *one* methodological point of view.

The possible objection of ending up with a "biological epiphenomenalism," which could only describe consciousness as an entity uselessly suspended above the organism, was rejected on two grounds:

1. Firstly, the primacy of function implies that the central nervous processes are not self-sufficient, but serve precisely to realize a living organism's integral capacities that bring it into a holistic contact with its environment—that is, particularly feeling, perception, and self-movement. Conscious activities are therefore not secondary accompaniments of neurophysiological processes, but rather their evolutionary meaning and purpose.
2. Secondly, a person's current acts of life and their unique content can only be explained on the grounds of an individual learning history. The specific knowledge and capacities, which are required for perception or action in the present, were not constructed by brain processes. Rather, they were only founded through integral embodied interactions with the natural and social environment.

From the combination of both viewpoints, it results that only the *person as a whole* causes integral acts of life, even though this causation is mediated by organic structures and patterns formed on the basis of a life history. Thus, if someone practices logical thought sequences or mathematical calculations, his neuronal processes are subject to the laws of logic or mathematics, not the other way round. The decisive condition for this lies in the individual's prehistory: thinking in logical terms or in accordance with mathematical laws has to be trained. The brain's plastic microstructures serve as a matrix for the logical, semantic, and equally for the associative and motivational connections that an individual has appropriated in the course of his learning history. They have become structural couplings and in this sense, the condition for realizing those capacities. Hence, the brain is not the organ of determination, but the *organ of potentialities*. It is not the producer, but the *mediator* of a person's activities.

Along these lines, I finally drew several other conclusions about psychophysical interrelations which seem to suggest a dualistic interaction of the psychic and the physical. The concept of integral acts of life comprising enabling physiological processes helps avoid such erroneous conclusions: the concept of *mutual implication replaces interaction*. In this context, particular attention was given once more to the issue of free will, which was defended as embodied freedom against determinist challenges. Other psychosomatic relations were primarily of medical relevance and conveniently lead to the next chapter.

Chapter 7

Implications for psychiatry and psychological medicine

Overview

Chapter 7 examines the conception presented here with regard to implications for psychiatry and psychological medicine. After an introduction on current neuroreductionist trends in psychiatry (7.1), the next section develops a concept of mental illness as a fundamentally circular process with a pivotal impact on a person's self-experience and interpersonal relationships (7.2). This dimension is traced as far as etiology (7.3). Somatic therapy and psychotherapy are then contrasted from the standpoint of dual aspectivity; here, the principle of transformation is particularly significant (7.4). In summary, an orientation towards subjectivity is shown to be indispensable for psychopathology and psychological medicine (7.5).

The approach outlined in previous chapters is applicable to various scientific and practical contexts. In conclusion, and as broadly representative of other disciplines, in this chapter I investigate the consequences of dual aspectivity and the ecological conception of the brain for psychological medicine—that is, for psychiatry, psychosomatics, and psychotherapy. This also links up with distinctions already drawn in the preceding chapter.

7.1 Neurobiological reductionism in psychiatry

Since its development around 1800, psychiatry has been moving between the poles of the sciences and the humanities, being directed towards subjective experience on the one hand and towards the neural substrate on the other hand. Dualistic opposition of "psychic" (or "moral") and "somatic" explanations prevailed during the nineteenth century (Verwey 1985). Wilhelm Griesinger's dictum that "we ought to regard mental illnesses as diseases of the brain" (Griesinger 1845) was far ahead of his time.[1] The poor attempts towards the end

---

[1] It should be remarked, however, that this was not primarily intended as a reductionist claim, but as a statement against the two prevailing approaches of that time: the moral

of the nineteenth century, for instance by Carl Wernicke and Theodor Meynert (1884), to subsume mental illnesses under the "diseases of the forebrain" were still derided by Jaspers as "brain mythologies" (Jaspers 1913/1997, 18, Fuchs 2013c). Apart from the neuropathological study of brain lesions, approaches towards localization of mental disorders still lacked an appropriate technical basis.

Today, however, the traditional dualism seems to be overcome by a naturalism, which identifies subjective experience with neural processes: "*Mental diseases are brain disorders*" (Insel & Quirion 2005, White et al. 2012). Two developments contributed, above all, to the dominance of the biological paradigm: first, progress in molecular neurosciences led to a more complex view of neurophysiological processes including receptor regulation, neuron interaction, neuroplasticity, and the like.[2] Second, the rise of cognitive neuroscience and neuroimaging, in particular functional magnetic resonance imaging (fMRI), increasingly enabled a localization of correlates of mental functions and dysfunctions. Moreover, it crucially contributed to a popular "neuroculture" by (although wrongly) suggesting that we can literally watch "the brain at work."

Since the first "decade of the brain," inaugurated in 1990, great hopes were placed on this biological turn of psychiatry. As a natural science discipline, psychiatry would soon be able to explain mental disorders as malfunctioning brain circuits and make objective diagnoses by means of neuroimaging and other biomarkers. On this basis, highly specific drugs could then be developed, and even persons at risk for mental disorders could be identified by genetic screenings for preventive treatment (Charney et al. 2002, Hyman 2003, Haag 2007). Psychiatrists should be renamed "clinical neuroscientists," for this would also accelerate psychiatry's integration into medicine in general and contribute to a destigmatization of the patients (Insel & Quirion 2005).[3]

---

approach on the one hand, and the somatic approach on the other hand, inasmuch as the latter linked mental illness to body processes in the lung, liver, or other organs.

[2] The journal *Molecular Psychiatry*, founded in 1997, is now one of the most prestigious and most cited journals in the field.

[3] This hope has proven elusive, however. Meta-analyses of numerous studies (Read et al. 2009, Schomerus et al. 2012, Kvaale et al. 2013) have demonstrated that the biomedical concept of brain diseases has not at all led to a destigmatization, although having become widespread in the public in the last 20 years. On the contrary, the majority of people experience a mental disorder as alien, abnormal, or even threatening if it is based on a genetic or brain disorder rather than if it may be attributed to psychosocial causes.

The predominant neuropsychiatric view of mental illness may be characterized by the terms of (1) reductionism, (2) reification, and (3) isolation.

1. *Reductionism:* neuropsychiatry regards subjectivity as a product or epiphenomenon of the brain's activity. All mental processes take place in brain tissue, therefore mental disorders must be brain disorders.
2. *Reification*: mental states seem to be *localizable* in the brain; consequently, a mental disorder must be more or less equivalent to either too high circuit activation, reduced metabolic activity, or some other dysfunction in certain areas of the brain.
3. *Isolation:* as a further consequence, this view tends to isolate the individual patient and to consider his disorder separate from the current interconnections with his environment—even if it is conceded that the brain is epigenetically influenced by certain conditions such as early life trauma or disturbed attachment relations.

It is obvious that the neuroreductionist view is in contrast to almost everything I have elaborated in the previous chapters: psychic processes may not be reduced to the brain or to localized neural activities; they are embodied, inherently intentional, and context related; and they are inseparable from the intersubjective world of shared meanings and interactions. As we will see, this applies to dysfunctional or disordered mental processes as well. Nevertheless, the progress made since the take-off of the neuropsychiatric paradigm in the first "decade of the brain" seems impressive at first. Regardless of whether one thinks of the identification of brain structures involved in numerous disorders such as anxiety, obsessive–compulsive, or traumatic disorder, of the epigenetic connection of gene variants, life events, and vulnerability (Meyer-Lindenberg & Tost 2012), or also of research on neuroplasticity and the influence of early socialization on brain development—without doubt our knowledge of the brain, its interactions with the environment, and also its dysfunctions has grown considerably.

And yet after three decades, the outcome of neuroscience for psychiatry is more than sobering. Despite all promises and billions of invested research funds, scarcely any clinically relevant findings could be brought to light. Apart from Alzheimer's disease, there is no possibility to reliably diagnose any mental illness by instrumental means or biomarkers, or to attribute it to specific gene variants. Nor did psychiatric therapies change as a result of neuroscientific findings. Even major parts of the pharmaceutical industry have withdrawn from research in psychiatry due to a lack of prospects of success (Abbott

2010, Miller 2010). All this is now even granted by leading representatives of neuropsychiatry—to quote only some of them:

> Despite obvious and rapid scientific advances, there is widespread frustration with the overall pace of progress in understanding and treating serious psychiatric illness. (Krystal & State 2014, 201)

> Unfortunately, there have been no major breakthroughs in the treatment of schizophrenia in the last 50 years and no major breakthroughs in the treatment of depression in the last 20 years. (Akil et al. 2010, 1580)

> Despite decades of research, the neurobiology of MDD [Major Depression] is largely unknown, and treatments are no more effective today than they were 50–70 years ago. (Holtzheimer & Mayberg 2011, 1)

However, the usual consequence is not a reconsideration of the underlying reductionist paradigm. On the contrary, it is the traditional, fuzzy nosology and the outdated, subject-oriented psychopathology which are identified as the causes preventing the success of biological psychiatry (Cuthbert & Insel 2013, Krystal & State 2014). As a consequence, the American National Institute of Mental Health (NIMH) has demanded a radically new classification of mental disorders according to biobehavioral dimensions (e.g., reward, attention, arousal, approach, anxiety), which should better comply with the molecular and imaging techniques—the so-called *Research Domain Criteria* (RDoC; Cuthbert & Insel 2013, Carpenter 2016). This follows a well-known principle: what can be measured and grasped by technical means determines what is considered significant and finally regarded as the actual reality. Such a research program may be perpetuated until the last convolution of the brain has been measured under special conditions of activity.

Of course, classical, subject-oriented psychopathology is not cast in stone. But already 10 years ago, Andreasen and other leading psychiatrists had deplored the decline of psychopathological expertise as a result of the criteriological diagnostic systems (Andreasen 2007, Mezzich 2007). The question arises whether psychiatry should run the risk of further losing this expertise, with the result that future psychiatrists know everything about reward systems in the brain, however, without being still able to distinguish schizophrenia from hysteria. Moreover, it is by no means clear whether mental disorders may be analyzed into the neat modular functions postulated by the RDoC system. It seems much more probable that we are dealing with highly complex, mixed, context-related, and therefore inherently fuzzy processes (Sprevak 2011, Wakefield 2014).

Despite all promises for "translational research," there is a risk that academic psychiatry along this path increasingly separates from clinical care and therapeutic practice—even though ever new decades of the brain, even the "century of the brain," are proclaimed (Blakemore 2000, Editorial 2010) and therapeutic breakthroughs are announced ("There is great promise for development of more effective treatments in the upcoming decade"; Insel 2014). In view of the impasse of the neuroreductionist paradigm, it seems time to question the underlying assumption that mental disorders are brain disorders. What should be searched for instead is an overarching paradigm that is able to found psychiatry as a relational medicine in an encompassing sense: *as a science and practice of biological, psychological, and social relations and their disorders.* This is what I want to develop in the following.

7.2 Mental disorders as circular processes

Since its early beginnings, psychological medicine shadowed a debate about the biological, psychological, and social explanatory models, which each in turn took precedence as the main paradigm. In recent decades, the widely advocated "biopsychosocial model" (Engel 1977 vs Uexküll & Wesiack 1996) offered a form of compromise between different approaches, albeit merely resulting in an eclecticism of factors. How the biological, psychological, and sociological factors should be integrated is only poorly understood (for a critique see McLaren 1998, Ghaemi 2010, Hatala, 2012).

An ecological approach, based on concepts of embodied and enactive cognition, offers an alternative paradigm that conceives brain, organism, and environment in their dynamical unity.[4] The neuronal processes are then regarded as components of superordinate processes, which may be regarded on different levels: (1) on the *macro-level* of psychosocial processes or interactions of persons; (2) on the *meso-level* of interactions between the individual brain, organism, and environment; and (3) on the *micro-level* of neuronal and molecular processes within the brain. Descending to the next level, the chosen focus or section of the process narrows each time (Figure 7.1).

As we have seen in the preceding chapters, these interactions should be considered as circular processes, including horizontal as well as vertical causality. Horizontal circularity characterizes in particular the macro- and meso-levels of social and organism–environment interactions, whereas vertical causality is effective between higher and lower levels within the organism. Nevertheless, there are also circular (top-down and bottom-up) relations between the macro- and micro-levels. Thus, as we will see, a psychotherapeutic treatment, as an interactive, intentional process on the macro-level, involves neuronal processes on the micro-level, which result in a modification of the patient's brain

[4] For an excellent account of enactive psychiatry, see also De Haan (forthcoming).

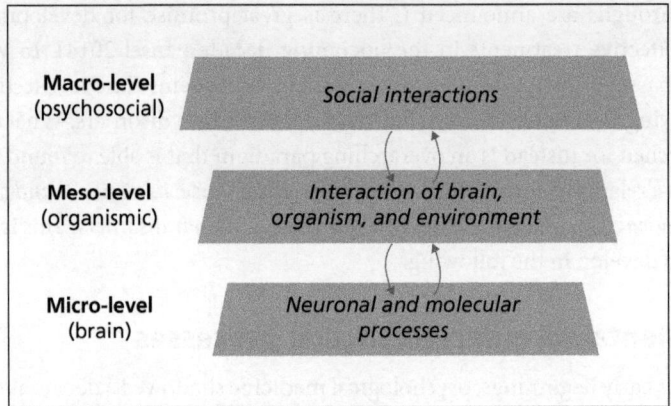

Figure 7.1 Levels of embodied interactions with top-down and bottom-up relations (↓↑).

structures—*top-down*. The modified neuronal structure, in turn, enables the patient to change his interactions with the environment—*bottom-up*, and so on. Over time, this leads to a mutual influence of superordinate psychosocial interactions and neuronal substrate, or of process and structure (see 4.2.3, Figure 4.7).

On this basis, the concept developed here may be outlined as follows: mental disorders are marked, on the one hand, by a disruption of *vertical circular causality*, that is, of the interplay between lower-level processes and higher capacities of the organism. As we will see, this primarily affects a patient's relation to him- or herself, which continually influences the course of the illness including the neural processes on the micro-level. On the other hand, mental disorders are characterized by a disruption of *horizontal circular causality*, that is, of social relationships and the ability to adequately respond to the demands and expectations of others. This leads to negative feedback loops in socio-functional cycles, which also have a crucial influence on the course of the illness. Both kinds of circular causal processes are tied to mediation by the brain, but cannot be located within it. For this reason, reduction of mental disorders to brain disorders is in principle not possible.

7.2.1 Vertical circularity

Human illness as the basic object of medicine is an ambivalent state in which the dual aspect of lived body and physical body emerges as a special case, that is, as a state of *disorder*. In a healthy state, there is a fluid relation between both aspects, with the body continuously shifting from the foreground to the background of awareness, and vice versa (see 3.1.2). In feeling ill, however, something alien, irritating, or painful makes itself noticeable in the body. To some extent, the object-body "particularizes itself," withdrawing from one's control and becoming an obstacle to the enactment of life. Instead of mediating one's being-towards-the-world (Merleau-Ponty), the body emerges as a person's adversary.[5] Medicine now describes and explains this particularization and estrangement of being unhealthy as a localized defect of the physical body, thus allowing the patient to take a certain distance from the disconcerting experience. The feeling of *being ill* now transpires as a *state of disease*.

However, mental disorders are not manifested in the same way. Initially, they may be recorded as "bodily" disorders as well, in which the lived body loses its mediality and becomes conspicuous or resistant—for instance, through panic attacks in anxiety disorders, or as heaviness, numbness, and lack of drive in the case of depression, or as disorders of perception, agency, and loss of "natural self-evidence" in schizophrenia.[6] Nevertheless, a mentally ill patient, and for that matter also a psychiatrist, does not succeed in attributing the disorder merely to the body, thus putting it at a distance. The disorder primarily affects a person's *self-experience*, be it in emotions, perceptions, thought, or behavior, and its subjective aspect involves not just a secondary reaction to physiological dysfunctions.[7] To a certain extent, a kind of "self-division" or self-alienation is always implied. Something *inside me* confronts me, and yet is removed from my influence or otherwise manipulates me, be it a panic attack, a depressive mood, a compulsion, or a hallucination, while I vainly attempt to reinstate my sovereign control.

Mental disorders therefore affect the core of one's self-relationship and self-understanding. The central equilibration and integration of mental functions fails. Different from somatic and even from neurological disturbances, the experience of illness may therefore not be regarded as an epiphenomenon of "actual" physiological processes.[8] Rather, *the patient's changed self-experience*

---

[5] On this, see also Gadamer (1996).

[6] See my analysis of depression or schizophrenia as the lived body's loss of mediality and transparency, in Fuchs (2005, 2010a, 2015). On the "loss of natural self-evidence," see Blankenburg (1970).

[7] This is not to say that serious or even fatal physical conditions may not also affect a person's relationship with the self. In this case, however, there is always the character of a *reaction* to the condition which does not capture the person as such.

[8] According to Graham (2013), mental disorders should be categorized as necessarily involving conscious and intentional states, or in other words, the mental *qua* mental. As such, they can be distinguished from neurological disorders, which arise from brute brain affections (such as stroke, neurodegeneration, or infection) and only present with secondary mental symptoms that are usually not sensitive to psychological treatment.

and self-relationship is the "substance" of the disorder itself. As such, it also forms a continual and permanently effective component of its trajectory. However, this means that regardless of the type of initial etiology, *vertical circular causality* always plays a substantial role for explaining the condition.

As an example, consider the case of a depressive illness: no matter how the various (genetic, neurobiological, biographical, interpersonal) causal conditions interact in the respective case, from the moment a depression emerges, it is per se *a personal affliction*. It manifests itself in a fundamental change, namely a narrowing or constriction of the lived body, including psychomotor inhibition, oppression, heaviness, and lack of drive. Scarcely any other kind of mental illness affects a person's bodily subjectivity in the same way (Fuchs 2005, 2013d). At the same time, it also gives rise to negative self-perceptions and evaluations (self-reproaches, guilt feelings) as well as typical depressive thought patterns. In turn, these negative self-assessments become self-fulfilling prophecies, thus increasing the likelihood of further failures and reinforcing the depressive condition. Similar vicious circles are well known, for example, in anxiety disorders. They have the following pattern: occurrence of physiological features of stress (activation of sympathicus, arousal, faster pulse rate, shortness of breath) → perception of physical symptoms as "threatening" → catastrophizing cognitions and evaluations → increased physiological stress, and so on. The subjectivity of experience, as a *relation to oneself*, thus becomes an important component affecting the course of the illness.

Every psychopathological experience is characterized by a personal meaning that the patients attribute to it, and a certain stance that they take towards it— suffering passively, giving in, acting out, interpreting it in a certain way, fighting against it, detaching oneself from it, and so on. This position-taking is a relevant clinical feature in itself. Of course, these subjective modes of experience and behavior are enabled by neuronal processes. Otherwise, they could not be effective within the organism. The brain here functions as a transforming organ that converts peripheral and central, lower- and higher-level components of the mentioned vicious circles into one another.

However, the phenomena of subjective ascription of meaning, assessment of a situation, and relation to oneself cannot be equated with processes in the neuronal substrate, as these lack acts of meaning-making or intentionality. That all thought is realized in neuronal activity does not make it the case that it is identical with brain processes. Intentional content and directedness, as we have seen, is inseparable from a subject's relation to the world. If neuronal processes function as vehicles of intentional acts, they can do so only as part of over-arching life processes that include the organism as a whole and its environment.

In a similar vein, it is not possible to reduce mental illness to circumscribed neurobiological dysfunctions—no matter how reliably such correlated dysfunctions can be identified. For, on the one hand, the subjective experience of the illness, in its specific quality and its intentional contents, is not reducible to physiological descriptions. No imaging of brain activities can provide a psychiatrist with an understanding of what it is like to be depressive, to experience a panic attack, or to hear voices. A description of biological markers of anxiety or depression, however detailed it may be, will tell him nothing about whether the patient in question is worried about a failure in the past, a threatening loss of her job, a public speech she has to give, or a current illness of her child. Obviously the biological data will be of very limited value as long as they remain isolated from their experiential context. In fact, they do not even provide criteria for what counts as a pathological or as an ordinary physiological process—this can only be known from clinical practice, that is, from the patient's experience and behavior. Neural or genetic data only yield statistic deviations, not diagnoses. This means that the definition of mental disorders crucially depends on subjective and cultural factors that fall outside the domain of natural science.

An even more crucial reason for this irreducibility is given by the patient's *self-relationship*, which is continually involved in the illness process, influences it positively or negatively, and, as such, bars us from seeing mental disorders as purely biological processes. The perception and assessment of one's own condition are genuinely personal phenomena that also limit the transfer of animal models to circumscribed components of the illness. They give rise to a unique, specifically human kind of vertical circular causality, namely the feedback from subjective perceptions and evaluations into more basic processes of the illness. Not least the possibility of suicide—only an option for the human individual—evidences the fact that the relation to oneself can decisively influence the course of an illness, though, in this case, fatally.[9]

7.2.2 Horizontal circularity

Just as mental illnesses cannot be entirely detached from the person and be ascribed exclusively to the neuronal substrate, it is also not possible to see them as purely individual dysfunctions, that is, detached from their interpersonal aspects. Irrespective of their causes, mental illnesses are always disturbances of

---

[9] This is not to characterize suicide as a freely chosen action, since this is almost always based on a seriously restricted perception of the situation. However, it presupposes that a patient adopts a personal standpoint to his or her situation—no matter how distorted or restrictive this may be. As such it cannot be regarded merely as a manifestation of a neurobiochemical dysfunction.

the patient's interactions and relationships on the macro-level. They are accompanied by various curtailments of freedom to respond to situations, offers and demands of the social world in a flexible and autonomous manner. As such, one can characterize them as impairments of a person's responsivity (Fuchs 2007b): certain abilities of the patient to shape social relationships according to their needs are either inhibited due to the illness or have not been developed in the first place. Thus, a significant part of psychopathology cannot be assessed in isolated patients, let alone their brains, but only as interactional or *horizontal dysfunctions*.

As soon as social responsivity is impaired, feedback effects necessarily occur in the socio-functional cycles and, from the very beginning, influence or even determine the progression of the illness. The functional cycle of social perception and action is impaired or interrupted; the patients lose the usual resonance with their environment. Therefore, one can also characterize mental disorders as *communicative disturbances* in the broadest sense. Symptoms of the illness evoke these disturbances, but they, in turn, are sustained, promoted, or even generated by the communicative impairments.

In the case of depression, for instance, a loss of emotional and intercorporeal resonance occurs, that is, a severe dysfunction of the responsivity and exchange with the environment (Fuchs 2001, 2013a). This dysfunction in turn increases the patients' depressive self-perception, yet it also has an impact on their social system. Family and friends at first react by giving support, but in the further course become increasingly helpless, feel guilty, and experience latent or open anger. Their generally inconsistent behavior and the patient's depressive state mutually reinforce each other in a vicious circle. The crucial influence of partnership interaction and social support on depression has been repeatedly confirmed (George et al. 1989, Sherbourne et al. 1995, Mundt et al. 1998, Backenstrass et al. 2007). Further factors aggravating the illness are detrimental consequences in the workplace, the feared or actual stigmatization of the patient, but also a possible secondary gain. All of these influences on the macro-level of social interactions are certainly not generated by the brain, but are continuously taken in and transformed top-down into altered dispositions of experience and behavior.

7.2.3 Synopsis

In summary, we can state: mental disorders may firstly be comprehended as *vertical circular disorders*. The central integration of partial functions or impulses of the organism fails. These functions become independent and evade the person's control, for instance, in the form of neurotic symptoms, compulsions, panic attacks, impulse control disorders, self-disorders, hallucinations, and so on. Such particularized processes affect the person's self-relationship, thus leading to various attempts at reintegration and coping. On the other hand, secondary reactions and symptoms also emerge (e.g., "fear of fear," guilt feelings, etc.) that further reinforce the illness.

On a biological level, such particularized dysfunctions can frequently be described as local hyperactivities of specific brain centers, though this does not allow to draw conclusions about causal connections. Undoubtedly, it is interesting that, for instance, acoustic hallucinations also activate areas of the primary auditory cortex in the temporal lobes (Dierks et al. 1999). But this does not mean that hallucinations as such can be localized in this area—as a patient's enactments of life, they cannot be localized in any form. Nor does this suggest that the activation was the cause of the hallucinations, as they may equally result from a generalized disintegration of the neuronal system leading to a disinhibition of particular processes.[10] Similarly, in the case of obsessive–compulsive disorders (OCD), the observable hyperactivity of the caudate nucleus (a basal ganglia core) is rather a consequence than a cause of the disorder. Though genetic and neurobiological factors such as abnormalities in orbitofronto-striatal circuits and serotonin metabolism are certainly involved in OCD (Abramowitz et al. 2009), the development of the disorder is crucially based on the person's avoidance reaction to his own rejected impulses or fears, that means again, on his self-relation (Doron & Kyrios 2005).

Local deviations of neural activity and brain metabolism as such do not determine the cause of a disorder—they may as well be an accompaniment or result. The cause of a severe grief reaction is not the activation of the cingulate cortex which may be observed as a correlate (Gündel et al. 2003), but an experienced painful loss. Similarly, it is not the amygdala as such which causes fear, but the subjective perception and evaluation of a threatening situation—and this is not to be found in the amygdala, though it is necessary to experience fear. Granted, brain images suggest their own reality and may well tempt one to confuse correlate and cause. But with the linear causality of nineteenth-century physics—brain state A causes disorder B—it is impossible to grasp the complex causal connections involved in mental disorders, even less so without the patient's subjective experience.

As I have further shown, mental disorders should also be considered as processes of *horizontal circularity*. This is because they are linked to more or less

---

[10] At the end of the nineteenth century, J. Hughlings Jackson already proposed this explanation for acoustic and visual hallucinations with his influential concept of "disinhibition" and "release": According to him, such positive symptoms are attributable to a release of lower-level activities due to a failure of inhibitory control by higher centers (Jackson 1958).

pronounced interferences with responsivity towards the social environment. The lack or loss of a person's "responded effectiveness" (Willi 1999) within the surrounding world essentially jeopardizes the ecological foundation of psychic stability: human beings are intrinsically dependent on the resonance of their actions within the social context. Furthermore, in relationships with significant others, negative feedback often emerges along with vicious circles that sustain or further intensify the symptoms of the illness.

From both aspects of circularity, and prior to any etiological analysis, it follows that a simple linear–causal description and explanation of mental disorders based on neurophysiological conditions is not adequate for the level of complexity involved. No psychiatric condition may be diagnosed, described, or explained without reference to a patient's subjectivity and interpersonal relationships. *Mental disorders always afflict the human person in relationships with other persons.*

7.3 Circular causality in pathogenesis

If we now turn to the etiology of psychic disorders, the circular structure of psychophysical interrelations is principally repeated, albeit the basic structure takes on different forms. First, I again refer to the example of depressive illness.

7.3.1 Etiology of depression

The manifestation of a depression is usually preceded by a personal situation which the individual concerned finds threatening, while assuming that the necessary coping measures are not available ("learned helplessness," Seligman 1974). Thus, the overarching situation and its subjective perception, including negative expectation and self-fulfilling prophecies, constitute the crucial triggering constellation. At the neural level, mediated by linking of prefrontal and limbic centers, and with significant involvement of the amygdala, this is linked to a physiological stress reaction. Initially, this corresponds to the organism's functional short-term reaction of fear and readiness (*"fight-or-flight response,"* McEwen 1999). However, under the conditions of a perceived impasse and feelings of helplessness, it leads to massive dysfunctions of the organismic functional cycles. The activation of the corticotropin-releasing hormone–adrenocorticotropic hormone–cortisol and sympathicus system as well as the disturbed serotonin-transmitter regulation in the limbic system places the organism in a permanent state of stress, corresponding to the aforementioned experience of bodily depletion and constriction (Glannon 2002, LeDoux 2003).

The self-perception of this altered organismic state, as a negative feedback loop, intensifies the physiological symptoms of stress. As a result, the initially functional organismic reaction is decoupled from its integration in superordinate feedback cycles and eludes the person's control (see 6.4.3). Negative horizontal feedback loops connecting to the social environment then influence the further course as described earlier. The desynchronization of biological rhythms (hormonal, sleep cycles, etc.) and the desynchronization from the social environment mutually reinforce each other (Fuchs 2001).

Social desynchronization may also be a decisive cause for the incidence of depression on the macro-level. The typical triggering situations are mostly characterized by a disruption or at least a serious endangerment of relations and bonds: a loss of relevant others (bereavement, divorce, or marital crisis) or of important social roles (loss of a job, moving to another city), furthermore experiences of backlog or defeat, poverty, social exclusion, or isolation resulting in a desynchronization from others (Brown & Harris 1978, Burns et al. 1994, Monroe et al. 2009). Such situations, under the condition of an individual vulnerability (see 7.3.2), are perceived as threatening and unsolvable by the patients: they do not feel equal to the pace of changes or are unable to cope with losses. They surrender in the face of painful processes of detachment or grief, or they refrain from necessary role changes.

It thus emerges that, for the manifestation of depressive disorders, subjective and intersubjective experience in no way merely plays an epiphenomenal role. Rather, the illness originates in a specific perception of the situation, that is, in an *individual act of meaning-ascription* that is not, as an intentional relation to the environment, reducible to neuronal processes. Depression results from a perceived loss of meaning and social resonance, not from a lack of serotonin. Moreover, it is not the objective features of the situation, but their subjective evaluation as insurmountable, which is decisive for the depressive reaction. As a result, biographically acquired dispositions such as lack of self-worth or self-efficacy become highly influential factors in pathogenesis. Only as a consequence do the physiological reactions take on a life of their own as a sustained regulatory dysfunction affecting the entire organism. Granted, in later stages depressive episodes may result from minor events or somatic triggers, since the first episodes obviously lead to a neurosystemic alteration facilitating further disorders (Moylan et al. 2013). But even then the organismic dysfunction always remains circularly connected to the patient's subjective perceptions as well as to their illness-related behavior in interpersonal relations.

7.3.2 **The development of vulnerability**

The anchoring of dysfunctional dispositions of perception and behavior, which can lead to the manifestation of illness in appropriate situations, is a biographical process dating as far back as early childhood. Essentially, this occurs in accordance with the principles of social learning set out in Chapter 5. On the basis of genetic disposition, individual temperament, and constitution, and through early social interactions influencing the epigenetic maturation of the brain, implicit emotional and interactive schemes develop, and especially attachment patterns as a foundation of relationships formed later on. Attention was already drawn to the substantial role played by unfavorable early attachment experiences for the emergence of mental disorders (Schore 2003; see 5.2.1). Dysfunctional interactions and harmful environmental influences lead to a deficient development of relational and coping capacities (Braun & Bogerts 2001). This is a crucial component of individual *vulnerability* or susceptibility to mental disorders.

The concept of vulnerability (Zubin & Spring 1977, Nuechterlein & Dawson 1984) collates different factors, primarily, genetic, temperament, and personality factors within a multicausal model, yet treats such "vulnerability markers" as relatively stable, biological, and individually determinable properties. However, vulnerability can be also described intentionally or interpersonally and as a result of early interactions, although it naturally yields corresponding neurobiological correlates. In particular, this interpersonal description is reasonable when it comes to altering such dispositions through psychotherapy (Stamm & Bühler 2001).

> As regards depressive disorders, genetic polymorphisms have been detected that increase a person's sensitivity towards stress (De Kloet et al. 2005). However, genetic factors cannot be directly effective, but only as *components* of vertical circular causality, namely via *subjectively experienced* stress reactions. In the context of corresponding interactions with the social environment, genetic factors may also contribute to personality traits such as dependence, introversion, or neuroticism. Such properties, in turn, impair the capacity to establish favorable relationships, thus taking on a pivotal significance for later depressive illness (Kendler & Kessler 1993, Kendler et al. 1995).
>
> The clustering of depressions in families is thus not only explicable genetically, but at least also via interpersonal circular causality: "depressive families" develop their own social dynamic (Cumming 1995). Postnatal depressions in motherhood and also later interactions with a depressive parent may influence childhood development in the sense of deficient training of social competencies, negative self-evaluations, and basic assumptions about the world that favour depression (Harris & Brown 1996, Bedi 1999, Murray & Cooper 2003). All this contributes to a dependent and conformist, strongly norm-oriented and risk-averse life conduct, the so-called Melancholic Type (Mundt et al. 1997, Kronmüller et al. 2005). This personality structure, however, is particularly vulnerable

to unavoidable losses, disappointments, failures, or other forms of social desynchronization which sooner or later lead to decompensation and depressive illness.

In the last decade, social neuroscience has made significant progress in exploring the connections between environmental influences, gene expression, brain structure, and vulnerability (Akil et al. 2010, Heim et al. 2012, Meyer-Lindenberg & Tost 2012). Reductionist tendencies may not be overlooked, however, since in typical descriptions, experience and relationships scarcely play any role, and the mechanisms explored in animal models are simply transferred to humans:

> Exploring the mechanisms of gene-environment interactions for depression is not substantially different from understanding how environmental toxins contribute to cancer or how diet influences cardiovascular disease. (Insel & Quirion 2005, 2222)

Here it remains completely out of the account that it is crucially dependent on the *subjective experience and evaluation* whether a stressor becomes a trauma for the one person or favours resilience in the other; or how unemployment, separations, and social exclusion impact on a person's psyche, to mention just a few examples. *There is no direct impact of environmental factors on the brain*—a brain concussion left aside. What changes brain structures enduringly are *the experiences a person makes in her social environment*. However, these experiences may not be described from a third-person perspective, for they are bound to consciousness, communication, and relationships.

Vulnerability is thus not a sum of single biological properties. Rather, it emerges in a complex biological–psychosocial feedback process, in turn affecting a patient's later conduct of life and relationships. Cultural factors also play a key role in the development of psychic vulnerability: the rise of narcissistic forms of depression in Western societies and their increasingly earlier age of initial manifestation[11] should primarily be traced back to higher individual expectations regarding performance and success, faster work processes, and therefore a more frequent experience of failure. Ultimately, this depends on the growing discrepancy between a culturally influenced self-image and potential self-realization (Ehrenberg 2004). To the extent that such self-images incorporated in the form of neuronal networks become effective triggers for depression, here, too, the brain emerges as a culturally shaped organ.

Though the weighting of the factors involved differs in other disorders, we can generalize the paradigm of depression insofar as we always find in mental illnesses a *complex interplay of circular processes* both at the vertical, organismic

---

[11] See Sartorius et al. (1989) and the studies of the Cross-national Collaborative Group (1992).

level and at the horizontal, interpersonal level. In each of these internal and external circularities, the brain functions as an organ of transformation or mediation, that is, as the carrier of the biological component of pathogenesis. In turn, however, its structure is continually shaped and modified by psychosocial interactions. In this way subjective experience, as a significant component of the interaction of organism and environment, has a structuring or top-down influence on the neuronal substrate—an insight that is of no little relevance for psychotherapeutic practice.

7.3.3 Summary

Let us return to the question: are mental disorders "really" brain disorders? Should anxiety, compulsion, depression, or schizophrenia be ultimately regarded as neurobiochemical dysfunctions? Such statements are already untenable on grounds of the dual aspect. As the experience and behavior of a mentally ill person can only be recorded in the first- and second-person perspective, the analysis of neuronal substrate processes only reflects the complementary aspect. It does not render superfluous the careful psychopathological description of the subjective experience of illness—on the contrary, it presupposes this description.

Yet also with respect to etiology and given the complexity of causal interrelations, such statements mean untenable reductions. Anxiety disorder, as we have seen, is not caused by the amygdala, nor is OCD caused by the caudate nucleus. A change in local metabolism as demonstrated by functional neuroimaging is only one, albeit a key *component* in the circular processes of the illness. Without embedding in the system of organism and environment, such images only reveal fragments of a superordinate process. Even if neurosystemic maturation disorders in schizophrenia or the massive hyperactivity of the amygdala in post-traumatic stress disorders play a stronger constraining role, such dysfunctions will never become linear efficient causes.[12]

Basically, we can always describe mental disorders in terms of both complementary aspects. Hence, basal vulnerability can equally be viewed as a neurobiological function disorder (e.g., hyperarousal, hypofrontality, and others) and as an experienced sensitivity or incapacity. An impulse control disorder is biologically describable as a disorder of the brain's maturation and serotonin metabolism, but also psychologically as an abnormal relationship structure resulting from continued trauma during childhood. A depression can be seen

---

[12] A linear cause can at best be referred to in the case of lesion-induced failures of function, such as in stroke or dementia. Even these, however, arguably result in multifaceted secondary adjustment and coping processes.

as a serotonergic dysfunction in limbic centers or as a personality-specific reaction to a present experience of loss. The choice of the suitable aspect depends on the relevant question and, not least, on the practical and therapeutic options under consideration (Henningsen & Kirmeyer 2000). However, if we attempt to define the genesis of an illness, an "either–or" of biogenetic or psychogenetic etiology is as inadequate as an additive "multifactorial genesis." In keeping with the principles of neuroplasticity and transformation, both aspects do not remain incommensurable with one another. Instead, we can always enquire into the *circular causality* of biological and psychosocial processes.

The basal vulnerability is already experienced and reacted upon by the individual, namely in the form of implicit compensational or avoidance behavior, as indicated for depression. However, this is also well documented for borderline personality disorder or for basic self-disorders in schizophrenia (Weinberger 1987, Linehan 1993, Herpertz et al. 1997, Sass & Parnas 2003). These modes of behavior lead to the formation of corresponding neuronal schemes. Yet they also influence interaction with reference persons, often in dysfunctional and pathogenic modes. Hence, vulnerability can in no sense be situated at the neurobiological level alone. Circular processes at a vertical and equally horizontal level also describe the triggering and course of manifest disorders as outlined earlier in reference to depression or anxiety disorder. With regard to the interrelations of neurobiological, subjective, and intersubjective components of the process of illness, the brain functions in each case as an organ of transformation.

Asserting this general basic structure does not imply that all mental illnesses need to be considered in the same way. It is by all means necessary to distinguish whether an illness is to be traced back to a comprehensible relation between a person's learning history and experience of the environment (as in the case of anxiety disorders), whether it corresponds to a genetically conditioned neurosystemic dysfunction affecting the constitution of the self (as in the case of schizophrenia), or to a macroscopically identifiable lesion of the brain (as in the case of an apoplectic stroke). Depending on the illness, psychosocial and biological aspects have to be weighted differently.

Intentional and psychosocial explanations remain indispensable for neurotic disorders that are derived from dysfunctional patterns of perception, behavior, and relationships (Henningsen & Kirmayer 2000). Even if dysfunctions (e.g., hyper- or hypoactivation) of neural systems are involved here as well, these are usually epiphenomena, even if they may become independent or chronic secondarily. Neurophysiological approaches are generally more relevant for those disorders that can be seen as defects in ordinary functioning. Yet such psychiatric or neurological defects are always connected to adaptive coping processes that are accessible to intentional modes of understanding and treatment—and this even applies to the formation and psychotherapy of delusions (Solms 2004, Kern et al. 2009).

7.4 Circular processes in therapy

Finally, an ecological conception of mental illness also suggests a pluralistic understanding of treatment. If we consider mental disorders under a person's dual aspect as the living unity and physical organism, then all therapeutic action must be describable *under both aspects*. The dualistic distinction between somatic therapies acting on the brain and psychological therapies having elusive, purely subjective effects is no longer tenable. This means that any therapeutic intervention is of a physiological *as well as* of a psychological nature.

Moreover, as already shown in section 6.4.3, the traditional notion of "psychophysical interactions" is inappropriate and should be abandoned. Granted, we may speak of an "anxiolytic" effect of a tranquilizer. Strictly speaking, however, a drug only modifies basal biochemical conditions in the brain that are *correlated* with the experience of anxiety. Chemical agents have no direct effect on emotional experience, which is the experience of a human being as a whole. Conversely, a calming conversation with an anxious and agitated patient does not mean that our words had a favourable effect on the functioning of his amygdala. We spoke *with the patient*, not with the amygdala. But using appropriate means of brain imaging, we might be able to determine that neuronal resonance patterns in the brain's temporal centers, which corresponded to his intentional understanding of words, have been *transformed* into changed activity patterns within his limbic system—including a decreased activity of the amygdala.

Thus, in the one case, pharmacological effects at the micro-level of brain processes are transformed *bottom-up* into changes of brain activity and organism-environment interaction at higher levels, resulting in altered emotional experience. In the other case, overarching psychosocial interactions based on the intersubjective exchange and understanding of meanings are transformed by implication or *top-down* into altered patterns of neuronal activity at the micro-level. Figure 7.2 represents these interrelations in a schematic diagram, using the examples of psychotherapy and medication.

The dual aspect does not mean a dualism of somatotherapy and psychotherapy. On the contrary, the circular interactions of self, body, brain, and environment may be approached at various levels or turning-points, since any mode of treatment will be transformed by the brain and thus contribute to a holistic effect. Moreover, all treatments are initially *actions, that is, integral and*

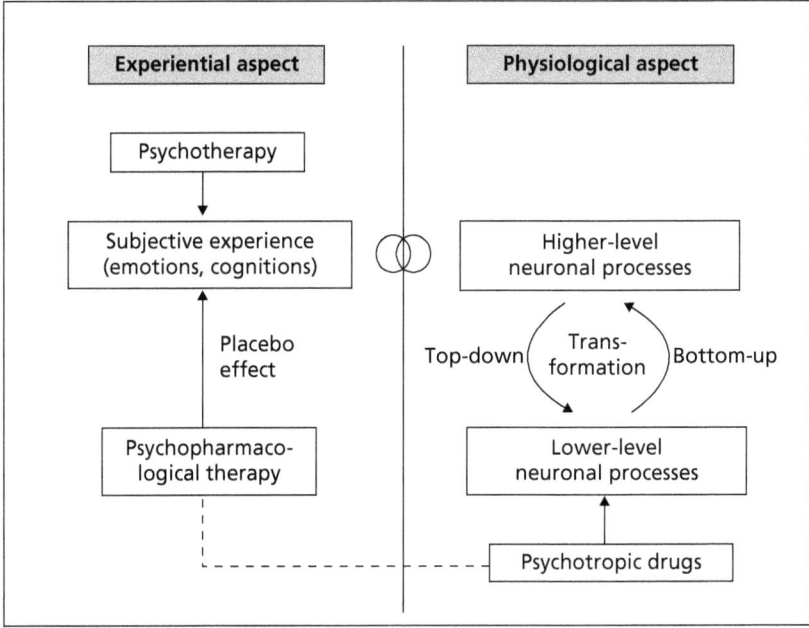

Figure 7.2 Effects of psychotherapy and drug therapy as seen from an experiential aspect (left) and from a physiological aspect (right). The two circles in the middle (∞) signify corresponding or concordant changes within both aspects; there is no "efficient causality" between them.

The effect of psychotherapy on the brain is mediated by participating higher-level neuronal processes being transformed into changes on the micro-level (transmitter metabolism, synaptic weighting, gene expression). This is equivalent to top-down or global-to-local influence.

Conversely, the physiological effect of psychotropic drugs (right) is transformed into higher level changes that correspond to altered subjective experiences (e.g., decreased anxiety). This means a local-to-global influence or bottom-up transformation. However, drugs appear on the left side as well, because they are also efficient within the dimension of subjective experience, expectation, and trust, this being known as placebo effect. Additional explanations are given in the text.

interpersonal forms of communication. Somatotherapy is a form of joint practice embedded in the doctor–patient relationship, even though it is mainly described at an organismic level. Conversely, psychotherapy never occurs on a "purely mental," but always on an embodied level as well. This naturally involves neuronal processes and is therefore proven to be manifested in a modification of neuronal networks (see later). Both forms of therapy are directed at one and the same entity, which may be viewed in terms of two alternative attitudes—the one reflecting a human person who experiences and maintains relationships and the other revealing an organism interacting with its environment. Nevertheless, both approaches suggest markedly different aspects.[13]

7.4.1 Somatic therapy

First, I examine *somatic, in particular psychopharmacological therapy* whose principal effect has already been discussed (6.4.3). Such effects start at a low level, namely, with changes to the transmitter metabolism primarily in subcortical, but also cortical brain regions. These are transformed "bottom-up" into changed higher-level neuronal activity patterns corresponding to an indirect influence on subjective experience (see Figure 7.2).

The goal of the therapy is basically *restitutive*: the aim is to cause the neurodynamic system to "swing back" to its starting position (or at least to a better adjusted state) via pharmacological deflection (or in system-theoretical terms, through "perturbation"). In detail, the mechanisms are admittedly only known at the synaptic level, though usually non-transparent in the wider context of brain processes with their manifold feedback mechanisms. In any case, antidepressants, for example, can be seen as the cause of a reduction of amygdala hyperactivity and a new regulation of the serotonergic and noradrenergic transmitter metabolism. However, the time course of an improvement will be the same for antidepressant therapy as for placebo treatment (Aldenhoff 1997). Hence, we may conclude that the process of recovery is to a large extent the organism's own achievement, namely the restitution of neuroendocrine balance, which is only stimulated by the medication.

> It is even more important to prevent "pharmacological fallacies": just as fever is not caused by a lack of acetylsalicylic acid merely because it drops after taking aspirin, depression cannot be traced back to "deficient serotonin levels" or similar, even if medication raises the concentration of the transmitter in the synaptic cleft, and reveals an antidepressive effect. The existence of an effective medication M to treat a psychological condition P proves *neither* that the cause of P is a deficiency of M, *nor* that the cause was any form of specific brain condition. Psychotropic drugs are only an incentive for the organism's holistic response and for a modification of the organism–environment system through circular causality.

---

[13] I restrict my account to pharmacological and psychotherapeutic approaches in order to analyze them from an embodied and enactive point of view, or under the dual aspect, respectively. Certainly social psychiatric, systemic, and ecological approaches aiming at the social interactions and systems on the macro-level are no less important, but have to be left out of account here for reasons of space. I refer in particular to Willi (1999), Morgan and Bhugra (2010), and Becvar and Becvar (2012).

On the other hand, a patient's and a doctor's subjective attitudes, as well as their relationship and other contextual conditions play a central role in the effectiveness of any treatment. Such factors are generally reflected in negatively loaded terms such as the "placebo effect" or "patient compliance," and thus as a disruptive necessity that should be distinguished from the "actual" effect. However, this attitude highlights a physicalist–dualist view of the organism, which cannot do justice to its integral environmental relations. The "placebo" effect is based on a specific meaningful relation, that is, the subject's interpretation which transforms medication into a symbolic and as such an intrinsically effective object (see Figure 7.2). To this extent, it forms an inherent part of the functional cycles, whereby human persons lend meaning to their world—there is no such thing as a "purely physiological" environmental relationship. Thus, somatic therapy is an interpersonal treatment, not "accidentally" but as a matter of necessity. Accordingly, adherence, placebo effect, and outcome of pharmacotherapy have been shown to be crucially dependent on the therapeutic alliance (Krupnick et al. 1996, Weiss et al. 1997, McCabe & Priebe 2004).[14]

7.4.2 Psychotherapy

Each psychotherapeutic treatment is primarily aimed at the patient as an experiencing, self-conscious, and self-relating subject. However, it is nevertheless an embodied process, which may also be described at a physiological level, and not only reaches the cortical structures, but also deeper layers of the limbic system through manifold transformations. In the same way as for pharmacological treatment, psychotherapy is organismically effective and thus represents an integral treatment. Meanwhile, it is clearly established that psychological interventions have the potential to modify brain function across a range of different psychopathological conditions.

> Firstly, several positron emission tomography (PET) studies could demonstrate that for patients with OCD both cognitive behavioral therapy as well as the antidepressant imipramine lead to a reduction of the typical overactivation in the caudate nucleus after several weeks of treatment (Baxter et al.

---

[14] However, both components can be distinguished in relation to their cerebral mode of processing: Mayberg et al. (2002) used neuroimaging (PET scans) to examine the effect of the antidepressant fluoxetine in comparison to placebo treatment for depressive patients. Overlapping effects were revealed in the brain's metabolism, albeit with more *cortical* emphasis in the case of placebo as contrasted with more *subcortical-limbic* and *brainstem* changes in the case of the antidepressant. This would correspond to a primary cognitive-evaluative mode of processing for the placebo and an additional subcortical—and to that extent subpersonal—effective mode for medication.

1992, Schwartz et al. 1998). For depressive patients, both interpersonal psychotherapy and antidepressant treatment resulted in a similar decrease in prefrontal hyperactivity (Brody et al. 2001, Martin 2001). Since then, similar results referred to the functional and neuroplastic effects of psychotherapy in anxiety disorders, specific phobias, depression, and schizophrenia (for an overview, see Fuchs 2004, Barsaglini et al. 2014).

On the one hand, psychotherapy seems to reverse abnormalities specifically associated with a disorder, or in other words, it is capable of "normalizing" the functions and structures of the brain. This refers to the reduction of metabolism in the orbitofrontal cortex and the caudate nucleus in OCD, to fronto-limbic areas in depression, and to frontotemporal regions in schizophrenia (Barsaglini et al. 2013). On the other hand, symptom remission in panic disorder and post-traumatic stress disorder seems to be associated with compensatory changes in areas not impaired before therapy (Beutel et al. 2010, Mecheli 2010). Generally, it is assumed that the effects are mediated by influencing synaptic plasticity and gene expression in the neurons (Kandel 2001).[15]

Such neurobiologically proven effects of psychotherapy are nonetheless so far only concerned with symptoms. Changes representing the actual goal of psychotherapy affect a person's patterns of perception, behavior, and relationships whose neuronal foundations cannot be mirrored in this way. Such changes are in principle possible through procedural learning on the basis of neuroplasticity. Psychotherapy can provide new experiences of interpersonal relationship that are incorporated as changed neuronal dispositions, and thus alter the

---

[15] Despite similar effectiveness, for psychotherapy and pharmacotherapy the processing routes seem to be different. In a PET study of depressive patients, Goldapple et al. (2004) found differential target areas of successful cognitive behavioral therapy (CBT) vs pharmacotherapy: CBT primarily produced changes in the medial frontal and cingulate cortex, whereas drug treatment changed metabolism in limbic-subcortical regions (brainstem, insula, subgenual cingulate). This corresponds to the clinical observation that psychotherapeutic interventions primarily change dysfunctional patterns of perception and evaluation, leading then to an alleviation of vegetative symptoms and inhibition, while pharmacotherapy rather takes the opposite course.

Accordingly, it has been suggested that psychological treatment may exert its effects through a *top-down* mechanism, targeting mainly dysfunctional thought processes associated with prefrontal activity, whereas pharmacotherapy may produce *bottom-up* changes by disengaging ventral and limbic regions mediating attention to emotional stimuli (Roffman et al. 2005). Transformation thus proceeds in one case from subjective experience to the vegetative level (*top-down*), in the other case from pharmacological effects on the subcortical transmitter metabolism to a change of attitude and cognition (*bottom-up*).

patient's *implicit relational knowledge* (Stern 1998; see 5.1.3). The condition here is not only a cognitive, but also intercorporeal and emotionally experienced therapeutic relationship. Furthermore, unfavourable relationship patterns and unpleasant experiences usually must be activated and experienced in order to be changed and "overwritten" through the therapeutic situation.[16]

In the context of attachment research and theory (see 5.2.1), the psychotherapeutic relationship can also be regarded as a new form of attachment relationship, which helps the patient to better regulate his emotional equilibrium and to restructure his or her implicit relational memory (Amini et al. 1996). In this sense, the core of therapeutic interaction lies in emotional communication, which is also conveyed through bodily resonance, non-verbal signals, and the atmosphere of a meeting as well as symbolic language (Fuchs 2017d). Even if biographical analysis of the person's learning and life history can convey important insights, nonetheless the focus of the therapeutic process is not so much the explicit past, but rather the implicit, presently effective past that shapes a patient's relationships with his social world.

Such implicit behavioral and relationship patterns should not be described with reference to an inner-psychic structure that could then be identified with brain structures. Rather, they are incorporated in the patient's bodily existence, in his attitude, patterns of movement, expression, and behavior, and they are also manifested in the structures of his lived space and relationship sphere. To this extent, body-oriented as well as ecological approaches of psychotherapy are especially suitable for overcoming the still prevailing concept of a psychic or cerebral inner space, and to characterize the patient in terms of his or her concrete and bodily *being-in-the-world* (see Downing 1996, Willi 1999, Fuchs 2006b, 2006c).

7.4.3 Comparison of therapeutic approaches

Let us conclude with a comparative appraisal of somatic therapy and psychotherapy. Both approaches are *interpersonal treatments* which are essentially oriented towards the person as a whole. Furthermore, both treatments can be described in terms of the dual aspect. They are effective not in a linear but a circular fashion, thus establishing "stimuli" for the organism's own dynamic or for the dynamic of the patient's relationships with others. In both processes, the principle of circular causality plays an essential role: in somatic therapy, the direction is predominantly "bottom-up," being primarily effective at the subcortical and synaptic level. For psychotherapy, in contrast, this is predominantly

---

[16] Grawe characterized this as *"processual activation"*: "The therapist has to activate what he wants to eliminate in order to change it" (Grawe 2004, 195–196).

"top-down," beginning with superordinate processes on the interpersonal level, which correspond, in a more focused view, to cortical processes in one individual and finally lead to modifications in brain structure on the micro-level of gene expression and synaptic connections. These changes in turn enable new interactions and experience, thus resulting in a circular relation of process and structure.

As we saw, however, somatic therapy is basically *restitutively or conservatively* predisposed. Through a deflection of the neuronal system (using pharmacological or other means of brain stimulation), it attempts to reinstate a starting condition. This takes effect comparatively quickly, and yet is also essentially reversible: changes to the neurodynamic are usually bound to the lasting presence of the agent (Ulrich 1990). Treatment as such does not essentially anchor a new disposition for experiences and behavior. Nevertheless, the stabilization achieved by the agent can give the patient an opportunity to adopt a certain mode of behavior within the social context that reinforces the effect via feedback.

On the contrary, psychotherapy pursues the goal of *changing* a pre-morbid constitution or structure of the person, which has proved unfavourable and maladaptive, *through learning and maturation processes*. In the course of treatment, new experiences should lead to a less cognitive and more procedural readjustment, which primarily includes new self-evaluations, interaction patterns, and competencies with the superordinate objective of enhancing the person's autonomy in shaping relations to the world. Since the focus is on changing habitual dispositions, this approach takes effect more slowly. On the other hand, it is generally irreversible, even if the accomplished learning processes may prove insufficient under certain circumstances for preventing new crises or relapses.

In summary: our view of the person as a psychophysical unity renders the previous dualist distinction clearly no longer sustainable for somatic therapies only "affecting the brain" and psychological therapies only "exerting subjective effects." This underlines that there is no separation, but rather a circular interaction of psychological and biological processes, and accordingly, no "merely biological" or "merely psychological" treatment. This interaction, however, cannot be expressed in terms such as "the mind acting on the body" or "the brain producing the mind." Instead, the brain acts as a mediator and transformer which may be addressed through input on different hierarchical levels and which converts it in both directions: neurobiochemical changes become mood and other changes on the subjective level, but subjectivity in turn influences the plasticity, structuring, and functioning of the brain. Vertical circular causality allows for both approaches equally.

Thus, both ways of treatment may also interact synergically. On the one hand, beyond a certain point, the neurobiological and endocrine dysfunctions involved in, for example, depression may be too advanced to be accessible to interventions on the psychological level. Pharmacological treatment may then enable the patient to re-engage in his relationships and thus will indirectly further his social well-being. Moreover, medication may not only be used to alleviate symptoms, but also to treat basal dysfunctions such as increased excitability, impulsivity, or emotional instability. On the other hand, psychotherapy can help the individual to change his implicit relational patterns, attitudes, and behavior, and to reframe his beliefs so that they align with the actual nature of events (Glannon 2008). In view of the limited effectiveness of medication, especially in chronic illness, it would be wrong to neglect these "top-down" options of treatment. Nor can we forego (inter-)subjective experience if we want to change the patient's maladaptive cognitive, emotional, and behavioral patterns that have led to his illness.

A biological approach—whether pharmacological or via another form of influencing brain conditions—will only be sufficient in alleviating or suppressing existing symptoms and inhibiting previously anchored neuronal dispositions. It neither produces any new relationship patterns and self-concepts nor can it conjure up any new learning processes. Only conscious, embodied experience is able to correct the corresponding dysfunctional patterns of neural activity. And only repeated interactions with the environment, that is, processes of interpersonal learning can stabilize new attractors of perception and behavior in the brain. Since the neural structures that underlie our personal dispositions are shaped by embodied experience, there will probably never be a way to create new views of the self and the world by brain manipulation directly. Any psychotherapeutic and social approach to psychiatry is thus based on a holistic, ecological view of life.

7.5 Summary: the role of subjectivity

In this chapter, I linked the concept of embodiment and dual aspectivity to the phenomena of mental illness and examined their consequences for psychiatry and psychotherapy. As I have shown, mental disorders crucially affect a person's self-experience and interpersonal relationships. At the same time, however, they are disturbances of embodied existence, including its biological basis. As such, they can only be adequately appreciated under both complementary aspects. A reductionist view of mental disorders as "brain disorders" is therefore ruled out, as are the dualistic separation of "somatic" and "psychic" components and their assumed "interactions."

Since mental disorders affect a person's relation with the self and with others, they are always disorders of the superordinate processes on the macro-level. But persons are also living, embodied beings, and therefore all their psychic processes are biological processes as well—although not limited to the brain. Advances in neurobiology may well have contributed to overcoming dualist ideas about mental disorders. However, such gains would be jeopardized if all forms of disorders were traced back to brain processes in an undifferentiated and linear manner. Seen in isolation, neurophysiologically determinable aberrations usually imply nothing more than a correlative character. They only become etiologically relevant if they are embedded in the overarching circular processes that include the organism–environment system as well as the patient's interpersonal relationships.

A biological psychiatry in the appropriate sense would therefore need an adequate notion of the *biological*, namely as the life process connected with the entire organism and its interactions with the environment. It needs an ecological theory of biology that integrates the social and cultural processes outside the brain, even though these are functionally sedimented in genome and brain structures. Only then is it in the position to adequately grasp the brain as the central mediating organ for the overarching processes. Then the social neurosciences, too, may contribute important components to our understanding of the mechanisms involved (Schilbach et al. 2013, Kotchoubey et al. 2016).

However, to proclaim psychiatry only as clinical neuroscience, and to expect salvation from genomics and proteomics, would be misguided. For the experiences and the relationships of a mentally ill person are the core of his illness, and they may not be identified with neuronal or molecular processes. As I suggested at the beginning of the chapter, we should rather consider psychiatry as a *relational medicine in an encompassing sense*: as a science and practice of biological, psychological, and social relations and their disorders. An ecological concept of the psyche as the superordinate gestalt of the relations between organism and environment, and between person and world, would be a suitable foundation of such a relational medicine. Without doubt, all biological processes involved belong to the domain of psychiatry thus conceived. Its center, however, constitutes the human person in relation to others. For the patient himself ultimately integrates all levels and circular processes, which as psychiatrists we can consider, explore, and within which we can also act and treat.

Integrally viewing mental illnesses as relational dysfunctions is a precondition for treating them adequately. The complexity of the circular processes is not best captured either by an opposition between or a mere summation of various therapeutic approaches. What is called for is rather a polyperspectival approach. Here various, especially somatic, psychotherapeutic and social psychiatric approaches can be combined to influence circular causalities. Psychosocial descriptions and interventions will nonetheless remain indispensable, for a purely neurobiological explanation or treatment of mental illnesses is basically impossible.

This establishes the fundamental importance of subject-oriented psychopathology based on a phenomenological and hermeneutical approach for psychiatry and psychotherapy. For the more objective the account we seek of the patient's experienced condition, by breaking it down into measurable data or physiological symptoms, the more we distance ourselves from the perspective in which the condition is experienced. How does this patient feel his anxiety? What does it feel like to encounter this pain? What is it like to hear voices? Such questions can only be answered from the subjective perspective of the affected individual. Hermeneutical understanding is, however, the method that comes closest to describing this particular perspective. It consists of awakening in oneself a similar mode of experience through attentive perception, empathy, translation, and imagination. True, this is a different form of recognition than conventionally used in objective science—recognition not via analysis, but via involvement in a shared relationship. The patient's subjectivity is primarily accessed through the schooling and differentiation of clinical experience (Fuchs 2010b). As the American psychiatrist Nemiah (1989) once aptly expressed it, as psychiatrists

> we are ourselves the instrument that sounds the depth of the patient's being, reverberates with his emotions, detects his hidden conflicts, and perceives the gestalt of his recurring patterns of behaviour. (p. 465)

No brain scan, no matter how much detail it reveals, could ever be superior to this instrument.

Yet, the matter does not rest here. Also in mental disorders, subjectivity is no mere epiphenomenon of physiological processes, but rather a constantly effective component of the course of the illness. Moreover, unfavourable patterns of perception and reaction, which form the basis of psychiatric disorders, can only be changed by new subjective and intersubjective experiences. The existential dimension of self-recognition, relationship, and meaning, which is crucial for every type of intensive therapy, is beyond the reach of neuroscientific methods. Thus, psychotherapy will never become a branch of applied neurobiology. Its essential grounding sciences remain psychology, hermeneutics, and the social sciences and humanities overall.

Chapter 8

Conclusion

8.1 Brain and person

We have reached the conclusion of our investigation. On the one hand, its result may seem rather sobering. For all its myriad fascinating achievements, the brain is no world creator—no "cosmos in the head." Rather, the brain is primarily an organ of mediation, transformation, and modulation, being embedded in the human organism's relationships with the surrounding world and in interpersonal relationships. The brain appropriates such relationships, acting as their carrier and facilitator, without producing them of its own accord. On the other hand, the brain's high-level plasticity makes it a matrix for human experiences that are sedimented in the neuronal structures of memory as the basis of human capacities. The brain facilitates the appropriation of all the abilities, dispositions, and behavioral modes that comprise a person's essential character traits. Hence, the brain might be called the "organ of potentialities." However, only a living being or human person as a whole can *realize* such potentialities.

The brain does not possess mental states or consciousness as such, for it does not *live*—it merely exists as the *organ* of an animate being or living person. Neither neuronal assemblies nor brains, but only *human persons* can feel, think, perceive, and act. It is erroneous to identify the brain with the human subject and to look inside for what makes up the person. What essentially characterizes a human person is being in relationships. The brain neither produces nor inherently contains such intentional and social relationships with the world. True, a human person's capacities and their realization as conscious acts of life are uniquely linked to brain functions. In this sense, the brain is a primary condition of possibility of personal existence in the world. However, a person is not a localizable part of the body, but is embodied and animate. We do not exist a second time inside ourselves. Human persons have brains, but they are not brains.

All the same, we can only describe this unity as intrinsically mediated, with respect to two complementary aspects. From one aspect, a person appears in his or her integral bodily, emotional, and intellectual acts of life. From the other aspect, a person is a bodily organism—thus, it appears as lived body and physical body, *Leib* and *Körper*. We saw how both aspects can be considered closely aligned with each other and that correlations, isomorphisms, and structural similarities may certainly be identified among them, for they are essentially aspects of one and the same life process. But whenever we want to *explain* their connection, we reach the limits of the aspects. There is no hidden passage leading from one side of the coin to the other.

We may assume we can grasp the effectiveness of intellectual acts in, as it were, palpable neuronal processes. Alternatively, we may suppose we can witness the production of perceptions through brain stimulation; indeed, we may think that we can evidently identify specific physiological processes with conscious experiences. However, on each occasion it is as though we merely accessed the other side through a revolving door, but without still grasping the previous side in turning around. For initially, we considered a human person; now, we are dealing with a complexly organized physical body. We know that both aspects are related to one and the same entity, namely, a living human being, and we understand there must be unity. And yet this insight remains strangely unfulfilled. We cannot fully realize this knowledge *in our thoughts*.[1] The unity of the living person is a *dialectical unity of "monism" and "dualism."* This can neither be resolved in one sense as a clear identity, nor in the other as a clear dualism of both aspects. We have no other choice but to refer to ourselves as animate, embodied beings in two kinds of speech.

All the same, we may also question why this should be so. I sought a response here in terms of Helmuth Plessner's concept of the human person's "eccentric positionality" which inherently contains the dual aspect of subjective and physical body. This concept corresponds to a person's ability to adopt a reflective position in relation to himself as well as his bodily existence, though without being able to shake off his bodily being and place himself in a purely objectivizing standpoint or a *"view from nowhere"* (Nagel 1989). While perceiving and being aware of his perception, a human person can never get *behind* his perceiving body. His self-relationship remains irreducibly ambiguous.

---

[1] Despite his fundamentally different dualist ontology, to this extent Descartes's insight remains valid: "Things that pertain to the union of the soul and body are known only obscurely by the understanding on its own […] *it is in using only life and ordinary conversations and in abstaining from meditating and studying* those things that exercise the imagination that we learn to conceive the union of the soul and the body […] as it does not seem to me that the human mind is capable of conceiving very distinctly, and at the same time, the distinction between the soul and the body and their union, since to do so it is necessary to conceive them as one single thing and at the same time to conceive them as two, which is contradictory." Descartes, letter to Princess Elizabeth, 28 June 1643 (Elisabeth of Bohemia 2007, 69–70; emphasis added).

The same ambiguity recurs in human beings' collective cognitive structures. The human lifeworld and communicative community exist prior to any specialized scientific practice and corresponding extract of reality. The observer perspective presupposes a participant or "we"-perspective and is in no position to overcome this. This explains why the objectivizing view that we may direct on nature and ultimately on ourselves is never able to get sight of its own ground, that is, our lifeworld *as a whole*. Even in shared knowledge, we can never attain a viewpoint *behind* our knowledge. To that extent, even Laplace's nebula or the Big Bang, as Merleau-Ponty argues, is "not behind us, at our remote beginnings, but in front of us in the cultural world" (Merleau-Ponty 1962, 502; see 2.4).

However, the situation is no different for life, consciousness, and the mind itself. These *ground* our perception and recognition, and they are the *media* in which we move. Therefore, they can never completely become *objects* of our knowledge. Contrary to all the relevant book titles, there will never be a complete scientific explanation of human life or consciousness, for all such proposed explanations already presume the explaining subject and thus the *explanandum*. We can only ever state the *conditions* for the life of a living being, or for a hungry cat chasing after a mouse, or a human person feeling and thinking. Moreover, we can *re-enact and comprehend* this to the extent that we ourselves exist as animate beings. Yet just as little as we have explained life by fully deciphering the genome, will we explain the human mind by completely recognizing brain functions—*our understanding would find the mind nowhere*. Life and mind as such are removed from the scientific form of cognition.

If life can only be perceived by the living, and equally human persons can only be perceived by other persons, then a co-involvement is essential for such perceptions: in one case, participation is in a *shared life form*, in the other case, participation is in a *shared mental world*, that is, through the capacity of intentionality whereby human beings are jointly directed towards a third entity. In this way, they also recognize each other as mindful, intentional living beings or as persons. Here, it holds true that one can only know what one maintains a vital relationship with. Accordingly, this gives rise to other, participating modes of knowing. These are cultivated mainly in the hermeneutical humanities, in interpersonally and psychodynamically oriented psychotherapy processes, yet also in Goethean or similar forms of scientific enquiry into the natural world. In this case, the aim is not a kind of knowledge purified from any subjectivity. Rather, the focus is on inner co-involvement or emulating what is perceived via mimesis, empathy, and understanding: "The subject of knowledge is potentially what the object is, and knows in becoming itself what the known is" (Böhme & Böhme 1985, 279; own translation).

Adopting a scientific approach, which systematically suspends such co-involvement, it is undoubtedly possible both for living beings and human persons to investigate certain extracts of their reality and to yield up useful knowledge. For this approach describes the aspect of our existence which actually pertains to us as physical, embodied beings. At least theoretically, it can also give a full account of this aspect, while at the same time remaining alert to the fact that it cannot reconstruct the other aspect, which was previously excluded. As our investigation has repeatedly shown, the aporias of the brain–mind problem are based on short circuits emerging due to the systematic exclusion of the enactment of life, and thus life itself from the study of the human body.

We may prevent such short circuits and erroneous conclusions if we consider conscious enactments of life and physical–organismic processes as complementary aspects of the living human being. In the unity of the person, both aspects are intertwined, thereby being neither identical nor radically different from each other. The human body is alive and thereby *also mindful*; whereas the human mind is alive and thus *also bodily*. As a living entity, the body is also *my lived body*, or the medium of my enactment of life in feeling, thinking, willing, and action. As a living entity, the mind is *embodied*, and all my feeling, thinking, and acting is achieved via physiological and, in particular, neuronal processes. Only to the extent that conscious processes are also of a physical nature are they in a position to determine physical life and have an effect in the world.

Emphasis on the unity of life admittedly does not mean a return to a pre-dualist idyll. It is connected with recognition for the ambivalent, conflict-ridden form of our existence, whereby as bodily organisms we can nevertheless confront our own side of nature and turn our body into an object. Indeed, we frequently experience the body as an opponent and barrier to our enactment of life. In terms of our existence, the dual aspectivity of life invokes the contradictory quality of our self-relationship where spontaneity and reflexivity, freedom and necessity, mind and body, and culture and nature are always also at odds with each other. The lived bodily nature, which we are, and the physical body, which we have, can never be entirely reconciled with each other. The ambiguity of the human person resists all attempts at resolution into a homogeneous unity.

Granted, the urge for clarity is a powerful reality. Ultimately, this urge is also the basis of reductionist efforts at dissolving personhood entirely into objectivized physical reality. The incentive of rendering existence entirely transparent and explicit, and parading life as the ground of one's own being *in front of oneself*, so that it is exposed to the full and unobscured light of the understanding—*clare et distincte*, as Descartes proposed—must surely be one of the prime research motives for the neurosciences. The fascination of images of the brain engaging in activity is rooted in our belief of having shed light on the last *terra incognita* of knowledge, and of capturing the slippery reaches of the mind within the network of neurons, thus suspending the ambiguity of existence once and for all. It is true that, on the basis of dual aspectivity, we possess the possibility of seeing ourselves as complex physical machines. However, there is no option to suspend the dual aspect per se. The images only suggest an ability to grasp what cannot be tangibly grasped. The more precisely we look at the physiological processes, the more inanimate they become. Life is no "clear and distinct perception."

8.2 The scope of neurobiological research

The intrinsic limitations to the form of scientific knowledge have certain consequences for the basic scope of neurobiological research. For the positivist ideal of a unified science based on physics and involving chemistry and biology as well as the social and cultural sciences (Wilson 1998), neurobiology would potentially represent the decisive "*missing link*." This ideal collapses, however, in view of the human person's dual aspect. The brain is only the biological condition for realizing conscious and intentional acts of life, not their cause. A common deception fueled by physicalism is to believe that neurobiological events are the actual reality, while psychological and cultural descriptions of mental processes can only grasp at vacuous concepts rather than concrete facts. But neurobiology is not in a position to infer a single intentional meaning from its main object of study—that is, the brain—which was not previously gleaned from introspection or intersubjective communication and only then attributed to its object. Ontogenetically, too, there is no one-sided, monolinear relationship between "biology" and "culture," but rather a circular connection. Neuronal dispositions are largely traceable to an individual's life experiences, especially if these relate to a person's cultural capacities. As regards their functional description and even more their explanation, recourse is essential to an individual history of *lived meanings*, and thereby also to the integral aspect.

For all the emotional–intellectual capacities, which a child appropriates during the course of socialization, and for all the culturally communicated learning processes, the brain represents the highly plastic matrix. The neuronal structures and couplings thus formed serve as "open loops" for future action, yet they only mirror those connections and relationships which the individual has experienced and incorporated himself. The brain can only receive these semantic, logical, gestalt-like, motivational, and interpersonal interconnections; it cannot produce them. The neuronal processes of a person thinking logically or calculating correctly follow the laws of logic or mathematics, and not the converse. Even if it were possible to give a complete account of the processes and links involved in this case—and the neurosciences are not only far, but endlessly remote from this point—there would be nothing whatsoever to gain. The statement $3 \times 16 = 48$ is not more "correct" because we know how it was expressed in the neuronal circuitry.

The same applies for all structures of meaning and significance imprinted on memory. If I react to an insulting remark in an offended manner, the cortical and limbic structures engaged here are only activated in a specific way because the respective links have formed in previously experienced and comparable situations implying their specific meaning. The intentional vocabulary, which we use to describe the event, is exactly what reflects these meanings—which other description should serve this purpose? However, this means that not a single one of the meaningful connections of mental life becomes more comprehensible by our ability to correlate it with neuronal activity patterns. Neurobiology may certainly supply specific framework conditions (e.g., processing capacity, learning speed, deletion resistance, etc.). However, the *intrinsic* intentional links of experience, emotion, cognition, and behavior can only be appreciated hermeneutically, even if we are becoming more and more precise about recognizing the neurophysiological links formed here. Hence, what we require in psychological, pedagogical, and therapeutic contexts are still excellent psychological, phenomenological, and psychodynamic concepts rather than excellent neurobiological descriptions.

Relationality is a constituent part both of life and mind: the living relationship with the world, emotional ties to others, the intentional relationship to mental contents, and the directedness towards past and future. The same objects, thoughts, and words can mean different things in different contexts and at different times to one and the same individual. The failure of all attempts at naturalizing life, subjectivity, and consciousness is ultimately grounded in this relationality. For this is removed from physicalist description inasmuch as physical objects, processes, and properties always remain external in relation to each other, and, moreover, are strictly localized and compelled to be momentary: brain states as such cannot "anticipate" anything (see 2.3.2). While science otherwise expands its knowledge by advancing from life experience to its physical basis (e.g., heat being traced back to particle movement), in the case of intentional, meaningful, and thus relational phenomena, this reduction does not lead to any advancement of knowledge— but quite the reverse.

The brain is embedded in these relationships, yet it can only mediate them and not produce their meaning. Instead, neurobiological advances precisely open up a new understanding for the brain's dependence on the human lifeworld. The brain can now increasingly be understood as a socially and historically shaped organ, whose functions of transformation and pattern formation enable biographical experiences to be turned into permanent dispositions and capacities. Seen in isolation, the brain is merely a fragment; however, in the context of the organism and its environment it can become a mediator for relational and intentional processes. From this viewpoint, the attempt at a "localization of the mind" through research into brain activities represents no future-oriented research program. An ecological neurobiology is rather obliged to draw on the integrated approaches of dynamic systems theory, psychology, cultural studies, and philosophy. Precisely because the brain is the organ of mediation between different areas of reality, it may not adequately be described by a single paradigm.

8.3 Naturalistic versus personalistic concept of the human being

The dual aspect of the human person corresponds to two basically different attitudes: namely, those we can adopt towards ourselves, and those in relation to others. We described these as personalistic or naturalistic attitudes. Different concepts of the human being derive from these attitudes which are compared and contrasted in this final section of our concluding chapter.

By linking neurobiology and materialist neurophilosophy, we are dealing with an attempt to establish a radical *naturalistic concept of the human person*. Its main implications are explaining mental phenomena as physically describable processes, understanding human cultural development as a continuation of the evolution of nature, and thus comprehending human persons as a particular species of information-processing biological machines. To optimize their chances of survival under the process of natural selection, these machines have developed self-constructs or self-models.[2] The result of this naturalization would be a progressive self-reification of humankind at the same time as the intensification of man's technical abilities for self-modeling through direct manipulation of the person's neuronal substratum.

---

[2] "The phenomenal self can now be regarded as a weapon, developed in a cognitive arms race. Conscious selves are like instruments or abstract organs, invented and constantly optimized by biological systems" (Metzinger 2003, 273). "Besides, it [the conscious self, T. F.] was an important instrument of social cognition. It enables us to take a guess at the aims of our counterpart. For one of its biggest functional advantages is that we can put ourselves in the position of our conspecifics […]—yet only then to deceive them!" (Metzinger 2006, 47; own translation). This corresponds to the basic principle of "Theory of Mind" research, according to which social cognition is essentially a process of "inferring" or "seeing through" others' intentions.

This naturalization attempt is, in itself, nothing new. It is basically a given option with human beings' "eccentric positionality." With advances in sciences since the modern era, this approach has constantly gained ground. Descartes's displacement of subjectivity to an interior space, which he thought could not be touched by physics, was meant to posit the *res cogitans* nonetheless as an independent, sovereign position. Yet, this was only at the expense of imprisoning the subject in the citadel of the brain where physicalism finds it even easier to exploit nowadays.

In contrast, Kant's attempt to preserve subjectivity proved more resistant. Reacting to Hume's already envisaged dissolution of the subject into a "bundle of experiences," Kant countered this with the primacy of the cognitive faculty of the transcendental subject. However, even Kant had to defend himself against the well-meaning efforts of contemporaries to locate the organ of knowledge in the human brain (Hagner 1997, 83–85). However, this bastion of subjectivity was deeply shattered with evolutionary epistemology, which set out to explain the Kantian categories of the understanding as the result of natural selection (Popper 1984, Vollmer 1975, 2005), and with neuroconstructivism, which declares Kant's categories including those of transcendental apperception as neuronal representation and construction achievements.

Even if the naturalization project dates back at least to the early modern era, it has gained an entirely new impetus through research into the human brain. In an unprecedented way, it has moved closer to the correlates and substrata of mental process, thus making these elements appear within our grasp. It changes little that this proximity was long since *too close*, as Lichtenberg already observed (see 2.2.2), and that those phenomena, which form the object of enquiry, are no longer detectable under the intense level of magnification. What matters is that this spyglass at least produces the sufficiently effective *illusion* that the locus of the mind is in the brain. A culture's perception of the human being is not the result of theoretical evaluations, conclusive proofs, or convincing counterarguments, but rather of subtle, and for the most part scarcely reflected changes of perception and attitudes. A "cerebrocentric" concept of the human being need not obtain proof, but merely becomes increasingly plausible. Historically, the current disputes about free will merely represent the continuation of a 2500-year-old debate and provisionally will lead to a stalemate. Yet this should not disguise the fact that it is only the visible expression of an even more fundamental challenge of the traditional concept of the human being whose social and ethical effects cannot yet be forecast.

If the naturalization project is contrasted here with a *personalistic perception of the human being*, this does not mean a rejection or devaluation of the naturalistic attitude as such—the aspect that it highlights for living beings and human persons is undoubtedly something objective about them. Moreover, its eminent practical meaning is undisputed for our civilization and, ultimately, for the practice of medicine. However, I repeatedly emphasized that the naturalistic attitude and research practice are grounded in the lifeworld where we jointly participate in a life form, thus perceiving and acknowledging each other as human beings. This foundation is indispensable not only for practical purposes of life, but also for *epistemology*. All objects of attempts at naturalization—life, consciousness, intentionality, and personhood—presuppose for their identification precisely what should be explained in naturalistic terms, namely conscious life and intersubjectivity. Hence, the personalistic attitude remains the primary one, as it is non-reducible, irrefutable, and can always only ever be temporarily "bracketed" by scientific practice for specifically designated purposes.[3]

However, nor can this foundation be *ethically* refuted. No goal, purpose, or value can be yielded from the naturalization project, since this consists precisely of reducing such anthropomorphic concepts to nature processes. The evolutionary survival benefit of specific modes of behavior does not represent any ethically relevant value, as it is always open to us to reject the result of this blind process of selection (such as a genetically predisposed "altruism") as non-binding for us. No less illusory would be the hope to use the brain as an "orientation organ" that would allow us to deduce conclusions about practical and ethical issues relevant to our life. The brain is the "organ of potentialities," the matrix for our experiences, thoughts, and behavioral patterns. It absorbs and mirrors what we instil in it. But we should not expect the brain to give us *answers* to the question of what we should do, and what the central issues are in life. This would correspond to the first naturalistic fallacy (see 2.4).

A personalistic concept, which comprehends human subjects as a primary psychophysical unity, is, admittedly, nothing new. For example, this viewpoint can be aligned with ideas drafted from Aristotle to Thomas Aquinas and up to twentieth-century philosophical anthropology and phenomenology. However, undoubtedly, the challenge of the naturalism debate rests in finding a new way to explicate and interpret this concept specifically with reference to neurobiological insights. This also appears necessary because the primary experience of the psychophysical unity of the person, as is naturally expressed in all the

---

[3] In this sense, Husserl made it clear that naturalism and personalism are "no two attitudes with equal rights and of the same order, […] but that the naturalistic attitude is in fact subordinated to the personalistic, and that the former only acquires by means of an abstraction, or, rather by means of a kind of self-forgetfulness of the personal Ego, a certain autonomy—whereby it proceeds illegitimately to absolutise its world, i.e. nature" (Husserl 1989, 193).

personal predicates (speaking, laughing, playing, greeting, and feeling pain—see 3.1.3), is nowadays increasingly being undermined.

The virtualization of intercorporeal relations and the ubiquitous presence of information contribute not least to this. These aspects are no longer tied to the bodily present and, hence, to this extent only supply *virtual* information. The world seems literally filled with information: inorganic molecules "recognize" each other in chemical reactions; "messenger molecules" can "read" and pass on genome information; and receptors and nerves supply the brain with information, which it further processes in neuronal networks. At the same time, outside information freely circulates around the world in electronic networks. Where everything appears to be so "filled with mind" and information even informs other information, there appears to be no further need for a mind connected to an animate and lived body that must still *understand* the information. The mind then only becomes an information structure which could basically also be "realized" by other data storage media or "downloaded" onto a new server, irrespective of whether its medium is animate or artificial.

This ubiquity and "idealism" of information is therefore the counterpart to sealing off subjectivity in a mental world of pure cognition. Matter and information relate to each other like "body" and "mind." If this fundamental dichotomy underlying the naturalization project is to be overcome, the human person must be considered as *living* and subjectivity as unconditionally *embodied*. As an animate being, the human person cannot be assembled from body and mind. A new foundation of the concept of life based on the bodily self-experience of the human person is to that extent a key condition for overcoming the naturalistic split of the person into physical and mental properties.

One route to achieve this, as approached in the various steps here, may be to link the perspectives of philosophical biology with phenomenology of the lived body. This is enabled by the fact that the *ecological* structure of the organism–environment relationship and the *phenomenological* structure of bodily "being-towards-the-world" closely correspond to one another. Specifically, the complementary dual aspect of the lived body and physical body can serve as a guideline for the rediscovery of embodied subjectivity. If we detach ourselves from the cerebrocentrism of the neurosciences, the phenomenal bodily space no longer appears as a virtual product of the brain, but rather as coextensive and isomorphic with the organic body. The unity of the bodily subject corresponds to the unity of the living organism.

The central significance of embodiment for the release of the subject from the citadel of the brain was the reason for focusing so closely on the coextension of the space of the lived and physical body in the first chapter. It is true that we could only legitimize this phenomenal space through its congruence with the commonly constituted intersubjective space. It was not incidental that our chosen paradigm for this congruence was the doctor–patient relationship—the patient revealing to the doctor the locus of pain. For what is at stake in this medical context is specifically the dual aspect of the human person as a lived body and as a physical body. Basically, every exchange of objects already presupposes the coincidence of subjective and shared space. To release the subject from the supposed virtuality, it only takes a look at intercorporeality, that is, the concrete bodily encounter of human persons who always acknowledge each other as embodied subjects, and who share their space with one another.

After this release, the lived body can be perceived for what it is: namely, as the medium through which the human person makes his appearance, expresses himself, and inhabits the world. However, this also means that we may not view the physical body as an arbitrary carrier construction for the brain or consciousness, but instead as an organism of precisely this form and function of becoming the lived body of a human person. In terms of its overall biological constitution, the physical body is specifically oriented towards human sociality and mental life: the upright gait, the position and mobility of the human hand, the sensitivity of the skin without hair, intensified sexuality, which is no longer bound to certain seasons, prolonged childhood, the differentiation of facial emotional expressions, the lowered position of the voice box adapted for human speech, and finally even the whites of the eyes, which makes the positions of the pupils and thus the direction of the gaze recognizable for others in most subtle ways—all this provides evidence that only a human body shaped in this form in the course of evolution could become the medium of expression and interaction for the personal subject (see Portman 1969, Maxwell 1984, Aiello & Dean 1990, Boyd & Silk 2003, Hrdy 2011).

The human organism's suitability for embodying the individual person is completed with the brain's faculties. The brain not only manifests a unique plasticity for appropriating culturally communicated patterns of experience, but also resonance systems for perception and communication, which are specifically designed for interaction with peers. Lastly, prefrontally localized functions for *inhibition* are of key importance for the development of the human person. They are at the roots of the particular characteristics of human "eccentric positionality": namely, the capacity to gain distance from bodily and affective motives, for decentring and adopting a superordinate standpoint, and for a personal relationship with the self (see 3.1.2). The phylogenetic and ontogenetic development of the human brain implies ever greater *freedom*, insofar as, on the one hand, the brain acquires increasingly differentiated dispositions or "open loops," thus becoming an "organ of potentialities." On the other hand, the *deployment* of these potentialities is increasingly under the remit of the person via specific functions of inhibition and delay.

By the same token, the embodiment of the person also means that his or her individual development is not only expressed in an "inner" character or personality structure, but equally in his or her bodily nature. Hence, a lived body is always *this* person's unique body. From early childhood onwards, social interactions and experiences are sedimented in the implicit body memory as behavioral repertoires, emotionally interactive schemes, bodily stances, gestures, and practices (Fuchs 2006, 2016b). The lived body is therefore no shell that conceals the person and merely symbolically indicates his or her presence to others. Rather, a person's attitudes, behavioral patterns, and habits are also always stances, motor patterns, and dispositions of the lived body. This extends to the typical and unmistakable style of a person's gait, gestures, facial expressions, articulation, and prosody. Not only "inner," psychic, or mental qualities, but also the individually shaped lived body constitute the human person to the extent that the person's capacities and dispositions are incorporated by bodily practice and can only be realized by the body's mediation.

The human person is formed in and through the lived body, and appears in ever clearer and more individual relief. All of a person's experiences and actions leave behind traces in the organism and thus change their dispositions, skills, and potentialities. A person's being is continually becoming, and this becoming is also increasingly doing. However, human persons shape their development not through direct formation of the self, but in a recurring process of undergoing the function cycles of perception and action, and of interactions with the social environment. They indirectly shape their own becoming through their decisions and actions as well as by the choice of a specific environment, which influences their development. They not only live, but they *lead* their life, and in this way they also form themselves.

The brain is involved in these circular structures as an organ of mediation and relationships and as an organ of the human person. The route to education, promotion, and maturity for the individual is not through direct manipulation of the brain, which can only ever have an inhibiting or modulating, but, non-creative effect. Rather, it is through shaping and organizing the milieu and relationships that influence the brain. The decisive condition for the development of personal freedom is learning from each other, educating and cultivating ourselves in our social interactions. Human persons become at one with themselves not in a mental or neuronal inner world, but in their bodily and inter-bodily "being-in-the-world" and "acting-in-the-world."

To truly become themselves, human persons must become real *for one another*. The unity of the inner and outer, subjectivity and objectivity lies in the intercorporeal interaction of human individuals where their respective own reality and shared reality coincide. *Conversation* is a paradigm for this interaction: the other party becomes real for me through his words and, in turn, I become real for him. This is an intercorporeal and simultaneously intentional relationship, a unity of bodily voice and mental language. The spoken word is equally heard by me and by the other person—it is the "intermediate word" or "*dia-logue.*" Such interaction can only be completed because the object of the exchange, namely, the spoken word, *simultaneously exists for us both*.

Admittedly, the physiological route of receiving the stimulus, of forwarding and processing it, is strictly separate and ongoing in our respective organisms. One might thus suppose that the spoken word merely exists as a construct in each person's brain. However, there can be no dialogue between brains. If the brain indeed produced its own respective world, we could not speak to each other, for instead of listening to animated words, we would merely hear tones with nobody expressing themselves. Rather, we would only be able to interpret these as signs for foreign inner worlds existing beyond our own personal world. However, this is not the case. The brain is only an organ of mediation and resonance, not the creator of our world. In every conversation, the other person's claim is reflected that his words are not heard as mere sounds or external signs, but that *he himself* be understood in his words. This is arguably the most profound reason to regard the conception of the subject as a construction of the brain as nothing else but the human person's depersonalization. For persons are the primordial *phenomenon*: that is, what *shows itself*, and what is present in its very appearing.

I hear the other's thoughts in his words. Grasping his hand, I give *him* my hand. Looking into his eyes, I see *him*. We are not the figments of our brains, but human persons in the flesh.

References

Abbott, A. (2010). The drug deadlock. *Nature* **468**: 158–159.

Abramowitz, J. S., Taylor, S., McKay, D. (2009). Obsessive-compulsive disorder. *The Lancet* **374**: 491–499.

Adolphs, R. (2001). The neurobiology of social cognition. *Current Opinions in Neurobiology* **11**: 231–239.

Adolphs, R. (2009). The social brain: neural basis of social knowledge. *Annual Review of Psychology* **60**: 693–716.

Aiello, L., Dean, C. (1990). *An introduction to human evolutionary anatomy*. London: Elsevier Academic Press.

Akil, H., Brenner, S., Kandel, E., Kendler, K. S., King, M. C., Scolnick, E., Watson, J. D., Zoghbi, H. Y. (2010). The future of psychiatric research: genomes and neural circuits. *Science* **327**: 1580–1581.

Aldenhoff, J. (1997). Überlegungen zur Psychobiologie der Depression. *Der Nervenarzt* **68**: 379–389.

Allison, T., Puce, A., McCarthy, G. (2000). Social perception from visual cues: role of the STS region. *Trends in Cognitive Science* **4**: 267–278.

Amini, F., Lewis, T., Lannon, R., Louie, A., Baumbacher, G., McGuiness, T., Schiff, E. Z. (1996). Affect, attachment, memory: contributions toward psychobiologic integration. *Psychiatry* **59**: 213–239.

Amodio, D. M., Frith, C. D. (2006). Meeting of minds: the medial frontal cortex and social cognition. *Nature Reviews Neuroscience* **7**: 268–277.

Anderson, M. L., Pessoa, L. (2011). Quantifying the diversity of neural activations in individual brain regions. In Carlson, L., Hölscher, C., Shipley, T. (Eds.) *Proceedings of the 33rd Annual Conference of the Cognitive Science Society*, pp. 2421–2426. Austin, TX: Cognitive Science Society.

Anderson, M. L., Richardson, M. J., Chemero, A. (2012). Eroding the boundaries of cognition: implications of embodiment. *Topics in Cognitive Science* **4**: 717–730.

Anderson, S. W., Bechara, A., Damasio, H., Tranel, T., Damasio, A. R. (1999). Impairment of social and moral behavior related to early damage in human prefrontal cortex. *Nature Neuroscience* **2**: 1032–1037.

Andreasen, N. C. (2007). DSM and the death of phenomenology in America: an example of unintended consequences. *Schizophrenia Bulletin* **33**: 108–112.

Aoki, F., Fetz, E. E., Shupe, L., Lettich, E., Ojemann, G. A. (2001). Changes in power and coherence of brain activity in human sensorimotor cortex during performance of visuomotor tasks. *Biosystems* **63**: 89–99.

Appelros, P., Karlsson, G. M., Hennerdal, S. (2007). Anosognosia versus unilateral neglect. Coexistence and their relations to age, stroke severity, lesion site and cognition. *European Journal of Neurology* **14**: 54–59.

Armstrong, D. M. (1968). *A materialist theory of the mind*. London: Routledge.

Assmann, J. (2011). *Cultural memory and early civilization: Writing, remembrance, and political imagination.* Cambridge: Cambridge University Press.

Aurell, C. G. (1989). Man's triune conscious mind. *Perceptual and Motor Skills* **68**: 747–754.

Aziz-Zadeh, L., Wilson, S. M., Rizzolatti, G., Iacoboni, M. (2006). Congruent embodied representations for visually presented actions and linguistic phrases describing actions. *Current Biology* **16**: 1818–1823.

Backenstrass, M., Fiedler, P., Kronmüller, K. T., Reck, C., Hahlweg, K., Mundt, C. (2007). Marital interaction in depression: a comparison of structural analysis of social behavior and the Kategoriensystem für Partnerschaftliche Interaktion. *Psychopathology* **40**: 303–311.

Bailey, A. (2009). Zombies and epiphenomenalism. *Dialogue* **48**: 129–144.

Baldwin, D. A., Baird, J. A. (2001). Discerning intentions in dynamic human action. *Trends in Cognitive Science* **5**: 171–178.

Bangert, M., Altenmüller, E. O. (2003). Mapping perception to action in piano practice: a longitudinal DC-EEG study. *BMC Neuroscience* **4**: 26. http://www.biomedcentral.com/1471-2202/4/26

Bao, S., Chan, V. T., Merzenich, M. M. (2001). Cortical remodeling induced by activity of ventral tegmental dopamine neurons. *Nature* **412**: 79–83.

Baron-Cohen, S., Lombardo, M., Tager-Flusberg, H. (Eds.) (2013). *Understanding other minds: Perspectives from developmental social neuroscience.* Oxford: Oxford University Press.

Barsaglini, A., Sartori, G., Benetti, S., Pettersson-Yeo, W., Mechelli, A. (2014). The effects of psychotherapy on brain function: a systematic and critical review. *Progress in Neurobiology* **114**: 1–14.

Barton, B., Lin, L., Asher, D. E., Brewer, A. A. (2009). Alteration of visuomotor processing following left-right prism adaptation. *Journal of Vision* **9**: 763.

Bateson, G. (1972). *Steps to an ecology of mind: Collected essays in anthropology, psychiatry, evolution, and epistemology.* Chicago, IL: University of Chicago Press.

Baxter, L. R. Jr., Schwartz, J. M., Bergman, K. S., et al. (1992). Caudate glucose metabolic rate changes with both drug and behavior therapy for obsessive-compulsive disorder. *Archives of General Psychiatry* **49**: 681–689.

Becchio, C., Adenzato, M., Bara, B. G. (2006). How the brain understands intention: different neural circuits identify the componential features of motor and prior intentions. *Consciousness and Cognition* **15**: 64–74.

Beck, D. M., Kastner, S. (2009). Top-down and bottom-up mechanisms in biasing competition in the human brain. *Vision Research* **49**: 1154–1165.

Becvar, D. S., Becvar, R. J. (2012). *Family therapy: A systemic integration*, 8th edition. Harlow: Pearson Education Ltd.

Bedi, R. P. (1999). Depression: an inability to adapt to one's perceived life distress? *Journal of Affective Disorders* **54**: 225–234.

Beebe, B., Sterne, D. (1977). Engagement-disengagement and early object experiences. In: N. Freedman, S. Grand (Eds.) *Communicative structures and psychic structures*, pp. 35–55. New York: Plenum Press.

Belin, P., Zatorre, R. J., Ahad, P. (2002). Human temporal-lobe response to vocal sounds. *Cognitive Brain Research* **13**: 17–26.

Bennett, M. R., Hacker, P. M. S. (2003). *Philosophical foundations of neuroscience*. Malden: Blackwell Publishing.
Bergson, H. (1950). *Time and free will: An essay on the immediate data of consciousness*. Trans. F. L. Pogson, 6th edition. London: George Allen & Unwin.
Bertalanffy, L. von (1968). *General system theory*. New York: George Braziller.
Beutel, M. E., Stark, R., Pan, H., Silbersweig, D., Dietrich, S. (2010). Changes of brain activation pre-post short-term psychodynamic inpatient psychotherapy: an fMRI study of panic disorder patients. *Psychiatry Research: Neuroimaging* **184**: 96–104.
Bieri, P. (1981). *Analytische Philosophie des Geistes*. Königstein: Hain.
Björklund, A., Lindvall, O. (2000). Self-repair in the brain. *Nature* **405**: 892–895.
Blakemore, C. (1988). *The mind machine*. London: BBC Publications.
Blakemore, C. (2000). Achievements and challenges of the decade of the brain. *Eurobrain* **2**: 1–4.
Blankenburg, W. (1970). *Der Verlust der natürlichen Selbstverständlichkeit. Ein Beitrag zur Psychopathologie symptomarmer Schizophrenien*. Stuttgart: Enke.
Block, N. J. (1978). Troubles with functionalism. In: W. Savage (Ed.) *Perception and cognition: Minnesota studies in the philosophy of science*, Vol IX, pp. 261–326. Minneapolis, MN: University of Minnesota Press.
Block, N. J. (2005). Review of Alva Noë, Action in perception. *Journal of Philosophy* **102**: 259–272.
Böhme, G. (1992). *Natürlich Natur. Über Natur im Zeitalter ihrer technischen Reproduzierbarkeit*. Frankfurt: Suhrkamp.
Böhme, H., Böhme, G. (1985). *Das Andere der Vernunft. Zur Entwicklung von Rationalitätsstrukturen am Beispiel Kants*. Frankfurt: Suhrkamp.
Botvinik, M, Cohen, J. (1988). Rubber hands feel touch that eyes see. *Nature* **391**: 756.
Bourdieu, P. (1977). *Outline of a theory of practice*. Cambridge: Cambridge University Press.
Bowlby J. (1982). *Attachment and loss. Vol I: Attachment*. New York: Basic Books. [German edition (2006). *Bindung. Vol. 1 Bindung und Verlust*. Munich: Reinhardt.]
Boyd, R., Silk, J. B. (2003). *How humans evolved*, 3rd edition. New York: Norton & Co.
Braun, K., Bogerts, B. (2001). Erfahrungsgesteuerte neuronale Plastizität Bedeutung für Pathogenese und Therapie psychischer Erkrankungen. *Der Nervenarzt* **72**: 3–10.
Broca, P. (1861). Remarks on the seat of the faculty of articulated language, following an observation of aphemia (loss of speech). *Bulletin de la Société Anatomique* **6**: 330–357.
Brodmann, K. (1909). *Vergleichende Lokalisationslehre der Großhirnrinde, in ihren Prinzipien dargestellt auf Grund des Zellenbaues*. Leipzig: Barth.
Brody, A. L., Saxena, S., Stoessel, P., et al. (2001). Regional brain metabolic changes in patients with major depression treated with either paroxetine or interpersonal therapy. *Archives of General Psychiatry* **58**: 631–640.
Brooks, R. A. (1991). Intelligence without representation. *Artificial Intelligence* **47**: 139–159.
Brown, G. W., Harris, T. (1978). *Social origins of depression: A study of psychiatric disorder in women*. New York: Free Press.
Brown, J. (2001). Microgenetic theory: reflections and prospects. *Neuropsychoanalysis* **3**: 61–74.
Bruner, J. (1983). *Child's talk*. New York: Norton & Co.

Bruner, J. S. (1978). Learning how to do things with words. In J. S. Bruner, R. A. Garton (Eds.) *Human growth and development*, pp. 62–84. Oxford: Clarendon Press.

Buccino, G., Binkofski, F., Riggio, L. (2004). The mirror neuron system and action recognition. *Brain and Language* **89**: 370–376.

Buchheim, T. (2006). *Unser Verlangen nach Freiheit. Kein Traum, sondern Drama mit Zukunft.* Hamburg: Meiner.

Bunge, M. (1980). *The mind-body problem: A psychobiological approach.* Oxford: Pergamon Press.

Bunge, M. (2003). *Emergence and convergence: Qualitative novelty and the unity of knowledge.* Toronto: University of Toronto Press.

Burns, D. D., Sayers, S. L., Moras, K. (1994). Intimate relationships and depression: is there a causal connection? *Journal of Consulting and Clinical Psychology* **62**: 1033–1043.

Cacioppo, J. T., Berntson, G. G., Adolphs, R., et al. (2002). *Foundations in Social Neuroscience.* Cambridge, MA: MIT Press.

Cacioppo, J. T., Berntson, G. G., Larsen, J. T., Poehlmann, K. M., Ito, T. A. (2000). The psychophysiology of emotion. In: M. Lewis, J. M. Haviland-Jones (Eds.) *Handbook of emotions*, 2nd edition, pp. 173–191. New York: Guilford Edition.

Caharel, S., Jiang, F., Blanz, V., Rossion, B. (2009). The human brain recognizes individual faces faster from shape than surface reflectance information. *Neuroimage* **47**: 1809–1818.

Calarge, C., Andreasen, N. C., O'Leary, D. S. (2003). Visualizing how one brain understands another: a PET study of theory of mind. *American Journal of Psychiatry* **160**: 1954–1964.

Calder, A. J., Keane, J., Manes, F., Antoun, N., Young, A. W. (2000). Impaired recognition and experience of disgust following brain injury. *Nature Neuroscience* **3**: 1077–1078.

Cameron, O. G. (2001). Interoception: the inside story—a model for psychosomatic processes. *Psychosomatic Medicine* **63**: 697–710.

Campbell, D. (1974). 'Downward causation' in hierarchically organized biological systems. In: F. J. Ayala, T. Dobzhansky (Eds.) *Studies in the philosophy of biology*, pp. 179–186. Berkeley, CA: University of California Press.

Caramazza, A., Anzellotti, S., Strnad, L., Lingnau, A. (2014). Embodied cognition and mirror neurons: a critical assessment. *Annual Review of Neuroscience* **37**: 1–15.

Carlson, S. M., Moses, L. L. (2001). Individual differences in inhibitory control and children's theory of mind. *Child Development* **72**: 1032–1053.

Carpenter, M., Nagell, K., Tomasello, M., Butterworth, G., Moore, C. (1998). Social cognition, joint attention, and communicative competence from 9 to 15 months of age. *Monographs of the Society of Research in Child Development* **63** (4, Serial No. 255).

Carpenter, W. T. (2016). The RDoC controversy: alternate paradigm or dominant paradigm? *American Journal of Psychiatry* **173**: 6.

Carruthers, O., Smith, P.K. (1996). *Theories of theories of mind.* Cambridge: Cambridge University Press.

Carruthers, P. (2005). *Consciousness: Essays from a higher-order perspective.* Oxford: Oxford University Press.

Cartwright, N. (1999). *The dappled world.* Cambridge: Cambridge University Press.

Casey, E. S. (1984). Habitual body and memory in Merleau-Ponty. *Man and World* **17**: 279–297.

Cassidy, J., Shaver, P. R. (Eds.) (2002). *Handbook of attachment: Theory, research, and clinical applications*. New York: Guilford Press.

Catmur, C., Mars, R., Rushworth, M., Heyes, C. (2011). Making mirrors: premotor cortex stimulation enhances mirror and counter-mirror motor facilitation. *Journal of Cognitive Neuroscience* 23: 2352–2362.

Catmur, C., Walsh, V., Heyes, C. (2007). Sensorimotor learning configures the human mirror system. *Current Biology* 17: 1527–1531.

Chalmers, D. J. (1996). *The concious mind: In search of a fundamental theory*. New York: Oxford University Press.

Chalmers, D. J. (2000). What is a neural correlate of consciousness? In: Metzinger, T. (Ed.) *Neural correlates of consciousness: Conceptual and empirical questions*, pp. 17–40. Cambridge, MA: MIT Press.

Chalmers, D. J. (2006). Strong and weak emergence. In: P. Clayton, P. Davies (Eds.) *The re-emergence of emergence: The emergentist hypothesis from science to religion*, pp. 244–256. Oxford: Oxford University Press.

Changizi, M. A., Hsieh, A., Nijhawan, R., Kanai, R., Shimojo, S. (2008). Perceiving the present and a systematization of illusions. *Cognitive Science* 32: 459–503.

Charney, D. S., Barlow, D. H., Botteron, K. et al. (2002). Neuroscience research agenda to guide development of a pathophysiologically based classification system. In: D. J. Kupfer, M. B. First, D. A. Regier (Eds.) *A research agenda for DSM-V*, pp. 31–83. Arlington, VA: American Psychiatric Association.

Chiccetti, D., Carlson, V. (1989). *Child maltreatment: Theory and research on the causes and consequences of child abuse and neglect*. New York: Cambridge University Press.

Chisholm, R. (1976). *Person and object*. La Salle, IL: Open Court.

Chivers, M. L., Seto, M. C., Lalumière, M. L., Laan, E., Grimbos, T. (2010). Agreement of self-reported and genital measures of sexual arousal in men and women: a meta-analysis. *Archives of Sexual Behavior* 39: 5–56.

Churchland, P. M. (1988). *Matter and consciousness*, revised edition. Cambridge, MA: MIT Press.

Churchland, P. S. (2013). *Touching a nerve: The self as brain*. New York: Norton & Co.

Cirulli, F., Berry, A., Alleva, E. (2003). Early disruption of the mother–infant relationship: effects on brain plasticity and implications for psychopathology. *Neuroscience & Biobehavioral Reviews* 27: 73–82.

Clark, A. (2003). *Natural-born cyborgs and the future of human intelligence*. Oxford: Oxford University Press.

Clark, A. (2013). Whatever next? Predictive brains, situated agents, and the future of cognitive science. *Behavioral and Brain Sciences* 36: 181–204.

Clarke, R. (2003). *Libertarian accounts of free will*. Oxford: Oxford University Press.

Colombetti, G. (2010). Enaction, sense-making, and emotion. In: J. Stewart, O. Gapenne, E. Di Paolo (Eds.) *Enaction: Toward a new paradigm for cognitive science*, pp. 145–164. Cambridge, MA: MIT Press.

Colombetti, G. (2013). *The feeling body: Affective science meets the enactive mind*. Cambridge, MA: MIT Press.

Conrad, K. (1947). Über den Begriff der Vorgestalt und seine Bedeutung für die Hirnpathologie. *Nervenarzt* 18: 289–306.

Cosmelli, D. Thompson, E. (2010). Embodiment or envatment? Reflections on the bodily basis of consciousness. In: J. Stewart, O. Gapenne, E. Di Paolo (Eds.) *Enaction: Toward a new paradigm for cognitive science*, pp. 361–386. Cambridge, MA: MIT Press.

Cosmelli, D., Lachaux, J.-P., Thompson, E. (2007). Neurodynamic approaches to consciousness. In: P. D. Zelazo, M. Moscovitch, E. Thompson (Eds.) *The Cambridge handbook of consciousness*, pp. 731–772. Cambridge: Cambridge University Press.

Cox, D. D., Savoy, R. L. (2003). Functional magnetic resonance imaging (fMRI) 'brain reading': detecting and classifying distributed patterns of fMRI activity in human visual cortex. *Neuroimage* 19: 261–270.

Cozolino, L. (2014). *The neuroscience of human relationships: Attachment and the developing social brain*. New York: Norton & Co.

Craig, A. D. (2002). How do you feel? Interoception: the sense of the physiological condition of the body. *Nature Reviews Neuroscience* 3: 655–666.

Craig, A. D. (2003). Interoception: the sense of the physiological condition of the body. *Current Opinion in Neurobiology* 13: 500–505.

Craver, C. F. (2007). *Explaining the brain: Mechanisms and mosaic unity of neuroscience*. Oxford: Clarendon Press.

Craver, C. F., Bechtel, W. (2007). Top-down causation without top-down causes. *Biology & Philosophy* 22: 547–563.

Crick, F. (1994). *The astonishing hypothesis*. New York: Simon & Schuster.

Cross-national Collaborative Group (1992). The changing rate of major depression. *Journal of the American Medical Association* 168: 3098–3105.

Csibra, G., Gergely, G. (2009). Natural pedagogy. *Trends in Cognitive Sciences* 13: 148–153.

Csibra, G., Gergely, G. (2011). Natural pedagogy as evolutionary adaptation. *Philosophical Transactions of the Royal Society B* 366: 1149–1157.

Cuffari, E. C., Di Paolo, E., De Jaegher, H. (2015). From participatory sense-making to language: there and back again. *Phenomenology and the Cognitive Sciences* 14: 1089–1125.

Cumming, M. E. (1995). Security, emotionality, and parental depression: a commentary. *Developmental Psychology* 31: 425–427.

Cuthbert, B. C., Insel, T. R. (2013). Toward the future of psychiatric diagnosis: the seven pillars of RDoC. *BMC Medicine* 11: 126.

Damasio, A. (1995). *Descartes' error: Emotion, reason and the human brain*. London: Picador.

Damasio, A. (1999a). *The feeling of what happens: Body and emotion in the making of consciousness*. New York: Hartcourt Brace & Co.

Damasio, A. (1999b). How the brain creates the mind. *Scientific American* 281: 112–117.

Damasio, A. (2010). *Self comes to mind: Constructing the conscious brain*. New York: Pantheon Books.

Davidson, D. (1980a). *Essays on actions and events*. Oxford: Oxford University Press.

Davidson, D. (1980b). Mental events. In: *Essays on actions and events*, pp. 207–224. Oxford: Oxford University Press. [Republished in: P. Bieri (Ed.) (1993). *Analytische Philosophie des Geistes*, 2nd edition, pp. 73–92. Bodenheim: Athenäum.]

Davidson, D. (2001). The emergence of thought. In: *Subjective, Intersubjective, Objective*, pp. 123–134. Oxford: Oxford University Press.

De Casper, A. J., Fifer, W. P. (1990). Of human bonding. Newborns prefer their mothers' voices. *Science* **208**: 1174–1176.

De Casper, A. J., Spence, M. J. (1986). Prenatal maternal speech influences newborns' perception of speech sounds. *Infant Behaviour and Development* **9**: 133–150.

De Catanzaro, D. A. (1999). *Motivation and emotion*. Upper Saddle River, NJ: Prentice Hall.

De Haan, S. (forthcoming). *Enactive psychiatry*. Cambridge: Cambridge University Press.

De Jaegher, H., Di Paolo, E. (2007). Participatory sense-making: an enactive approach to social cognition. *Phenomenology and the Cognitive Sciences* **6**: 485–507.

De Jaegher, H., Di Paolo, E., Adolphs, R. (2016). What does the interactive brain hypothesis mean for social neuroscience? A dialogue. *Philosophical Transactions of the Royal Society B*, **371**: 20150379.

De Jaegher, H., Di Paolo, E., Gallagher, S. (2010). Can social interaction constitute social cognition? *Trends in Cognitive Sciences* **14**: 441–447.

De Kloet, E. R., Joels, M., Holsboer, F. (2005). Stress and the brain: from adaptation to disease. *Nature Reviews Neurosciences* **6**: 463–475.

De Vignemont, F., Singer, T. (2006). The empathic brain: how, when and why? *Trends in Cognitive Sciences* **10**: 435–441.

de Weerth C., van Hees Y., Buitelaar, J. K. (2003). Prenatal maternal cortisol levels and infant behavior during the first 5 months. *Early Human Development* **74**: 139–151.

Deacon, T. (2006). Emergence: the whole at the weel's hub. In: P. Clayton, P. Davies (Eds.) *The re-emergence of emergence: The emergentist hypothesis from science to religion*, pp. 111–150. Oxford: Oxford University Press.

Decety, J., Ickes, W. (2011). *The social neuroscience of empathy*. Cambridge, MA: MIT Press.

Decety, J., Lamm, C. (2007). The role of the right temporoparietal junction in social interaction: how low-level computational processes contribute to meta-cognition. *The Neuroscientist* **13**: 580–593.

Decety, J., Sommerville, J. A. (2003). Shared representations between self and other: a social cognitive neuroscience view. *Trends in Cognitive Science* **7**: 527–533.

Deecke, L., Scheid, P., Kornhuber, H. H. (1969). Distribution of readiness potential, pre-motion positivity, and motor potential of the human cerebral cortex preceding voluntary finger movements. *Experimental Brain Research* **7**: 158–168.

Dennett, D. (1991). *Consciousness explained*. Boston, MA: Little, Brown.

Depraz, N. (1994). Temporalité et affection dans les manuscrits tardifs sur la temporalité (1929-1935) de Husserl. *Alter* **2**: 63–86.

Deregowsky, J. B. (1973). Illusion and culture. In: R. L. Gregory, E. H. Gombrich (Eds.) *Illusion in nature and art*. London: Duckworth.

Descartes, R. (1993). *Meditations on first philosophy*. Trans. E. S. Haldane, G. R. T. Ross. New York: Routledge.

Descartes, R. (2015). *The passions of the soul, and other late philosophical writings*. Trans. M. Moriarty. Oxford: Oxford University Press.

Dewey, J. (1896). The reflex arc concept in psychology. *Psychological Review* **3**: 357–370.

Di Paolo, D., De Jaegher, H. (2012). The interactive brain hypothesis. *Frontiers in Human Neuroscience* **6**: 163.

Di Paolo, E. (2005). Autopoiesis, adaptivity, teleology, agency. *Phenomenology and the Cognitive Sciences* **4**: 429–452.

Di Paolo, E. (2009). Extended life. *Topoi* **28**: 9–21.
Dierks, T., Linden, D.E., Jandl, M., Formisano, E., Goebel, R., Lanfermann, H., Singer, W. (1999). Activation of Heschl's gyrus during auditory hallucinations. *Neuron* **22**: 615–621.
Dittmann, J. (2002). *Der Spracherwerb des Kindes. Verlauf und Störungen*. München: Beck.
Donald, M. (2001). *A mind so rare: the evolution of human consciousness*. New York: Norton & Co.
Doron, G., Kyrios, M. (2005). Obsessive compulsive disorder: a review of possible specific internal representations within a broader cognitive theory. *Clinical Psychology Review* **25**: 415–432.
Downing, K. L. (2009). Predictive models in the brain. *Connection Science* **21**: 39–74.
Dretske, F. (1995). *Naturalizing the mind*. Cambridge, MA: MIT Press.
Dreyfus, H. L. (2002). Intelligence without representation—Merleau-Ponty's critique of mental representation. *Phenomenology and the Cognitive Sciences* **1**: 367–383.
Dreyfus, H., Taylor, C. (2015). *Retrieving realism*. Cambridge, MA: Harvard University Press.
Dunbar, R. I. (1998). The social brain hypothesis. *Evolutionary Anthropology* **6**: 178–190.
Dunbar, R. I. M. (1993). Coevolution of neocortical size, group size and language in humans. *Behavioral and Brain Sciences* **16**: 681–735.
Dunbar, R. I., Schultz, S. (2007). Understanding primate brain evolution. *Philosophical Transactions of the Royal Society of London, Biological Sciences* **362**: 649–658.
Dunn, B. D., Galton, H. C., Morgan, R., Evans, D., Oliver, C., Meyer, M., Cusack, R., Lawrence, A.D., Dalgleish, T. (2010). Listening to your heart how interoception shapes emotion experience and intuitive decision making. *Psychological Science* **21**: 1835–1844.
Dupré, J. (2001). *Human nature and the limits of science*. Oxford: Oxford University Press.
Durt, C., Fuchs, T., Tewes, C. (Eds.) (2017). *Embodiment, enaction, and culture: Investigating the constitution of the shared world*. Cambridge, MA: MIT Press.
Eagleman, D. (2015). *Incognito: The secret lives of the brain*. New York: Pantheon Books.
Eagleman, D. (2016). *The brain: The story of you*. Edinburgh: Canongate Books.
Edelman, G. M. (1987). *Neural Darwinism: The theory of neuronal group selection*. New York: Basic Books.
Edelman, G. M. (2004). *Wider than the sky: The phenomenal gift of consciousness*. New Haven, CT: Yale University Press.
Edelman, G. M., Tononi, G. (2000). *A universe of consciousness: How matter becomes imagination*. New York: Basic books.
Edelman, S., Grill-Spector, K., Kushnir, T., Malach, R. (1998). Toward direct visualization of the internal shape representation space by fMRI. *Psychobiology* **26**: 309–321.
Editorial (2010). A decade for psychiatric disorders. *Nature* **463**: 9.
Ehrenberg, A. (2010). *The weariness of the self. Diagnosing the history of depression in the contemporary age*. Montreal: McGill-Queen's University Press.
Ehrenfels, C. von (1978). Über 'Gestaltqualitäten' (1890). Reprinted in: F. Weinhandl (Ed.) *Gestalthaftes Sehen*, pp. 11–43. Darmstadt: Wissenschaftliche Buchgesellschaft.
Ehrlich, P. R., Raven, P. H. (1964). Butterflies and plants: a study in co-evolution. *Evolution* **18**: 586–608.

Ehrsson, H. H., Spence, C., Passingham, R. E. (2004). That's my hand! Activity in premotor cortex reflects feeling of ownership of a limb. *Science* 305: 875–877.

Eisenbach, M., Lengeler, J. W., Varon, M., et al. (2004). *Chemotaxis*, Vol. 1. London: Imperial College Press.

Eisenberger, N.I., Lieberman, M.D., Satpute, A.B. (2005). Personality from a controlled processing perspective: an fMRI study of neuroticism, extraversion, and self-consciousness. *Cognitive, Affective, and Behavioral Neuroscience* 5: 169–181.

Elbert, T., Pantev, C., Wienbruch, C., Rockstroh, B. (1995). Increased use of the left hand in string players associated with increased cortical repräsentation of the fingers. *Science* 270: 305–307.

Elbert, T., Rockstroh, B. (2004). Reorganization of human cerebral cortex: the range of changes following use and injury. *Neuroscientist* 10: 129–141.

Elisabeth of Bohemia/ R. Descartes (2007). *The correspondence between Princess Elisabeth of Bohemia and René Descartes*. Trans. L. Shapiro. Chicago, IL: University of Chicago Press.

Ellis, R. D., Newton, N. (2010). *How the mind uses the brain (to move the body and image the universe)*. Chicago, IL: Open Court Publishing.

Engel, G. (1977). The need for a new medical model: a challenge for bio-medicine. *Science* 196: 129–135.

Ennen, E. (2003). Phenomenological coping skills and the striatal memory system. *Phenomenology and the Cognitive Sciences* 4: 299–325.

Etkin, A., Klemenhagen, K.C., Dudman, J.T., Rogan MT, Hen R, Kandel ER, Hirsch J. (2004). Individual differences in trait anxiety predict the response of the basolateral amygdala to unconsciously processed fearful faces. *Neuron* 44: 1043–1055.

Fadiga, L., Fogassi, L., Pavesi, G., Rizzolatti, G. (1995). Motor facilitation during action observation: a magnetic stimulation study. *Journal of Neurophysiology* 73: 2608–2611.

Farroni, T., Csibra, G., Simion, F., Johnson MH. (2002). Eye contact delectation in humans from birth. *Proceedings of the National Academy of Sciences of the United States of America* 99: 9602–9605.

Faw, B. (2003). Pre-frontal executive committee for perception, working memory, attention, long-term memory, motor control, and thinking: a tutorial review. *Consciousness and Cognition* 12: 83–139.

Feigl, H. (1981). Mind-body, not a pseudo-problem [first published in 1960]. In: *Selected Writings*, pp. 342–350. Dordrecht: Reidel.

Feinstein, J. S., Adolphs, R., Damasio, A., Tranel, D. (2011). The human amygdala and the induction and experience of fear. *Current Biology* 21: 34–38.

Feuerbach, L. (1985a). Wider den Dualismus von Leib und Seele, Fleisch und Geist. In: *Anthropologischer Materialismus. Ausgewählte Schriften I* (ed. A. Schmidt), pp. 165–191. Frankfurt: Ullstein.

Feuerbach, L. (1985b). Der Spiritualismus der sogenannten Identitätsphilosophie oder Kritik der Hegelschen Psychologie. In: *Anthropologischer Materialismus. Ausgewählte Schriften I* (ed. A. Schmidt), pp. 192–209. Frankfurt: Ullstein.

Fichte, J. G. (1992). *Foundations of transcendental philosophy (Wissenschaftslehre) nova methodo* [1796–1799]. Trans. D. Breazeale. Ithaca, NY: Cornell University Press.

Field, T. M., Woodson, R., Greenberg, R., Cohen, D. (1982). Discrimination and imitation of facial expressions by neonates. *Science* **218**: 179–181.

Flor, H., Braun, C., Elbert, T., Birbaumer, N. (1997). Extensive reorganisation of primary somatosensory cortx in chronic back pain patients. *Neuroscience Letters* **222**: 5–8.

Fodor, J. A. (1983). *The modularity of mind: An essay on faculty psychology.* Cambridge, MA: MIT press.

Fonagy, P., Target, M. (2003). *Frühe Bindung und psychische Entwicklung. Beiträge aus Psychoanalyse und Bindungsforschung.* Gießen: Psychosozial-Verlag.

Frank, M. (1986). *Die Unhintergehbarkeit von Individualität.* Frankfurt: Suhrkamp.

Frank, M. (1991). *Selbstbewusstsein und Selbsterkenntnis. Essays zur analytischen Philosophie der Subjektivität.* Stuttgart: Reclam.

Freeman, W. (1999). *How brains make up their minds.* London: Weidenfeld & Nicholson.

Freud, S. (1911). Formulierung über die zwei Prinzipien des psychischen Geschehens. In: *Gesammelte Werke,* Vol. 8, pp. 229–238. Frankfurt: Fischer.

Freud, S. (1911/1958). Formulations on the two principles of mental functioning. In: *The Standard Edition of the Complete Psychological Works of Sigmund Freud,* Vol. XII, pp. 213–226. London: Hogarth.

Freud, S. (1917/1963). Introductory lectures on psycho-analysis. Part 3: General theory of the neuroses. In: *The Standard Edition of the Complete Psychological Works of Sigmund Freud,* Vol. 16. Ed. and trans. J. Strachey. London: Hogarth Press.

Friston, K. J. (1994). Functional and effective connectivity in neuroimaging: a synthesis. *Human brain mapping* **2**: 56–78.

Friston, K.J., Harrison, L., Penny, W. (2003). Dynamic causal modelling. *Neuroimage* **19**: 1273–1302.

Fuchs, T. (2000a). *Leib, Raum, Person. Entwurf einer phänomenologischen Anthropologie.* Stuttgart: Klett-Cotta.

Fuchs, T. (2000b). Das Gedächtnis des Leibes. *Phänomenologische Forschungen* **5**: 71–89.

Fuchs, T. (2001). Melancholia as a desynchronization. Towards a psychopathology of interpersonal time. *Psychopathology* **34**: 179–186.

Fuchs, T. (2004). Neurobiology and psychotherapy: an emerging dialogue. *Current Opinions in Psychiatry* **17**: 479–485.

Fuchs, T. (2005). Corporealized and disembodied minds. A phenomenological view of the body in melancholia and schizophrenia. *Philosophy, Psychiatry & Psychology* **12**: 95–107.

Fuchs, T. (2006). Gibt eine leibliche Persönlichkeitsstruktur? Ein phänomenologisch-psychodynamischer Ansatz. *Psychodynamische Psychotherapie* **5**: 109–117.

Fuchs, T. (2007a). Was heißt ‚sich entscheiden'? Zur Phänomenologie der Entscheidung. In: T. Buchheim, T. Pietrek (Eds.) *Freiheit auf Basis von Natur?*, pp. 101–117. Paderborn: Mentis.

Fuchs, T. (2007b). Psychotherapy of the lived space. A phenomenological and ecological concept. *American Journal of Psychotherapy* **61**: 432–439.

Fuchs, T. (2010a). Phenomenology and psychopathology. In: S. Gallagher, D. Schmicking (Eds.) *Handbook of phenomenology and the cognitive sciences,* pp. 547–573. Dordrecht: Springer.

Fuchs, T. (2010b). Subjectivity and intersubjectivity in psychiatric diagnosis. *Psychopathology* **43**: 268–274.

Fuchs, T. (2011). The brain—a mediating organ. *Journal of Consciousness Studies* **18**: 196–221.

Fuchs, T. (2012a). The phenomenology of body memory. In: Koch, S., Fuchs, T., Summa, M., Müller, C. (Eds.) *Body memory, metaphor and movement*, pp. 9–22. Amsterdam: John Benjamins.

Fuchs, T. (2012b). The feeling of being alive. Organic foundations of self-awareness. In: J. Fingerhut, S. Marienberg (Eds.) *Feelings of being alive*, pp. 149–166. Berlin: De Gruyter.

Fuchs, T. (2013a). Depression, intercorporality and interaffectivity. *Journal of Consciousness Studies* **20**: 219–238.

Fuchs, T. (2013b). The phenomenology and development of social perspectives. *Phenomenology and the Cognitive Sciences* **12**: 655–683.

Fuchs, T. (2013c). Brain mythologies. Jaspers' critique of reductionism from a current perspective. In: T. Fuchs, T. Breyer, C. Mundt (Eds.) *Karl Jaspers: Phenomenology and psychopathology*, pp. 75–84. Berlin: Springer.

Fuchs, T. (2015). Schizophrenia, embodiment and intersubjectivity. In: D. S. Stoyanov (Ed.) *Towards a new philosophy of mental health: Perspectives from neuroscience and the humanities*, pp. 269–291. Cambridge: Cambridge Scholars Publishing.

Fuchs, T. (2016a). *Das Gehirn—ein Beziehungsorgan. Eine phänomenologisch-ökologische Konzeption*, 5th edition (1st edition 2008). Stuttgart: Kohlhammer.

Fuchs, T. (2016b). Embodied knowledge—embodied memory. In: S. Rinofner-Kreidl, H. Wiltsche (Eds.) *Analytic and continental philosophy. Methods and perspectives. Proceedings of the 37th International Wittgenstein Symposium*, pp. 215–229. Berlin: De Gruyter.

Fuchs, T. (2016c). The embodied development of language. In: G. Etzelmüller, C. Tewes (Eds.) *Embodiment in evolution and culture*, pp. 107–128. Tübingen: Mohr Siebeck.

Fuchs, T. (2017a). Self across time: the diachronic unity of bodily existence. *Phenomenology and the Cognitive Sciences* **16**: 291–315.

Fuchs, T. (2017b). Delusion, reality and intersubjectivity: a phenomenological and enactive analysis. *Philosophy, Psychiatry, & Psychology* **24** (forthcoming).

Fuchs, T. (2017c). Collective body memories. In: C. Durt, C. Tewes, T. Fuchs (Eds.) *Embodiment, enaction and culture: Investigating the constitution of the shared world*, pp. 333–352. Cambridge, MA: MIT Press.

Fuchs, T. (2017d). Intercorporeality and interaffectivity. In: C. Meyer, J. Streeck, S. Jordan (Eds.) *Intercorporeality: Emerging socialities in interaction*, pp. 3–24. Oxford: Oxford University Press.

Fuchs, T. (2017e). Levels of empathy—primary, extended, and reiterated empathy. In: V. Lux, S. Weigel (Eds.) *Empathy: Epistemic problems and cultural-historical perspectives of a cross-disciplinary concept*, pp. 27–48. Basingstoke: Palgrave Macmillan.

Fuchs, T., De Jaegher, H. (2009). Enactive intersubjectivity: participatory sense-making and mutual incorporation. *Phenomenology and the Cognitive Sciences* **8**: 465–486.

Fuchs, T., Koch, S. (2014). Embodied affectivity: on moving and being moved. *Frontiers in Psychology. Psychology for Clinical Settings* **5**: Art. 508, pp. 1–12.

Fuster, J. M. (2001). The prefrontal cortex—an update: time is of the essence. *Neuron* **30**: 319–333.

Gadamer, H. G. (1996). *The enigma of health: The art of healing in a scientific age*. Stanford, CA: Stanford University Press.

Gallagher, S. (2005). *How the body shapes the mind*. Oxford: Clarendon Press.

Gallagher, S., Zahavi, D. (2008). *The phenomenological mind. An introduction to the philosophy of mind and cognitive science*. London: Routledge.

Gallagher, S., Bower, M. (2017). Making enactivism even more embodied. In: Gallagher, S. (Ed.) *Enactivist interventions: Rethinking the mind*, pp. 150–163. Oxford: Oxford University Press.

Gallese, V. (2001). The'shared manifold' hypothesis. From mirror neurons to empathy. *Journal of Consciousness Studies* 8: 33–50.

Gallese, V. (2002). The roots of empathy: the shared manifold hypothesis and the neural basis of intersubjectivity. *Psychopathology* 36: 171–180.

Gallese, V. (2008). Mirror neurons and the social nature of language: the neural exploitation hypothesis. *Social Neuroscience* 3: 317–333.

Gallese, V., Goldman, A. (1998). Mirror neurons and the simulation theory of mind reading. *Trends in Cognitive Science* 12: 493–501.

Gallese, V., Umiltà, M. A. (2002). From self-modeling to the self model: agency and the representation of the self. *Neuro-Psychoanalysis* 4: 35–40.

Gazzaniga, M. S. (2005). *The ethical brain*. Chicago, IL: University of Chicago Press.

George, L. K., Blazer, D. G., Hughes, D. C., Fowler, N. (1989). Social support and the outcome of major depression. *British Journal of Psychiatry* 154: 478–485.

Gergely, G., Bekkering, H., Kiraly, I. (2002). Rational imitation in preverbal infants. *Nature* 415: 755.

Ghaemi, S. N. (2010). *The rise and fall of the biopsychosocial model: Reconciling art and science in psychiatry*. Baltimore, MD: Johns Hopkins University Press.

Gibbs, R. W., Van Orden, G. C. (2010). Adaptive cognition without massive modularity. *Language and Cognition* 2: 149–176.

Gibson, J. J. (1979). *The ecological approach to visual perception*. Boston, MA: Houghton Mifflin.

Glannon, W. (2002). Depression as a mind-body problem. *Philosophy, Psychiatry & Psychology* 9: 243–254.

Glannon, W. (2008). The blessing and burden of biological psychiatry. *Journal of Ethics and Mental Health* 3: 1–4.

Goldapple, K., Segal, Z., Garson, C., Lau, M., Bieling, P., Kennedy, S., Mayberg, H. (2004). Modulation of cortical-limbic pathways in major depression. Treatment-specific effects of cognitive behavior therapy. *Archives of General Psychiatry* 61: 34–41.

Goldman, A., Gallese, V. (2000). Reply to Schulkin. *Trends in Cognitive Science* 4: 255–256.

Grafton, S. T., Fadiga, L., Arbib, M. A., Rizzolatti, G. (1997). Premotor cortex activation during observation and naming of familiar tools. *Neuroimage* 6: 231–236.

Graham, G. (2013). *The disordered mind: An introduction to philosophy of mind and mental illness*, 2nd edition. New York: Taylor & Francis.

Grawe, K. (2004). *Psychological therapy*. Cambridge, MA: Hogrefe & Huber.

Graybiel, A. M. (1998). The basal ganglia and chunking of action repertoires. *Neurobiology of Learning and Memory* 70: 119–136.

Graybiel, A. M. (2005). The basal ganglia: learning new tricks and loving it. *Current Opinion in Neurobiology* 15: 638–644.

Gregory, R. L. (1966). *Eye and brain: The psychology of seeing*. New York: McGraw Hill.

Grice, P. (1989). *Studies in the Way of Words*. Cambridge, MA: Harvard University Press.

Griesinger, W. (1845). *Die Pathologie und Therapie der psychischen Krankheiten für Ärzte und Studirende*. Vol. 1. Stuttgart: Adolph Krabbe.

Grossmann, K. E., Grossmann, K., Winter, M., Zimmermann, P. (2002). Attachment relationships and appraisal of partnership: from early experience of sensitive support to later relationship representation. In: L. Pulkkinen, A. Caspi (Eds.) *Personality in the life course: paths to successful development*, pp. 73–105. Cambridge: Cambridge University Press.

Gündel, H., O'Connor, M. F., Littrell, L., Fort, C. Lane, R. D. (2003). Functional neuroanatomy of grief: an FMRI study. *American Journal of Psychiatry* **160**: 1946–1953.

Haag, A. (2007). Biomarkers trump behavior in mental illness diagnosis. *Nature medicine* **13**: 3–4.

Habermas, J. (2004). Freiheit und Determinismus. *Deutsche Zeitschrift für Philosophie* **52**: 871–890.

Haggard, P. (2011). Decision time for free will. *Neuron* **69**: 404–406.

Hagner, M. (1997). *Homo cerebralis. Der Wandel vom Seelenorgan zum Gehirn*. Darmstadt: Wissenschaftliche Buchgesellschaft.

Haken, H. (1993). *Advanced synergetics*, 3rd edition. Berlin: Springer.

Haken, H., Stadler, M. (Eds.) (1990). *Synergetics of cognition*. Berlin: Springer.

Halbwachs, M. (1939). *La mémoire collective*. Paris: Presses Universitaires de France.

Hardcastle, V. G., Stewart, C. M. (2009). fMRI: A modern cerebrascope? The case of pain. In J. Bickle (Ed.) *The Oxford handbook of philosophy and neuroscience*, pp. 200–225. Oxford: Oxford University Press.

Hardcastle, V. G., Stewart, C. M. (2002). What do brain data really show? *Philosophy of Science* **69**: S72–S82.

Hari, R., Forss, N., Avikainen, S., Kirveskari, E., Salenius, S., Rizzolatti, G. (1998). Activation of human primary motor cortex during action observation: a neuromagnetic study. *Proceedings of the National Academy of Sciences of the United States of America* **95**: 15061–15065.

Harnad, S. (1995). Why and how we are not zombies? *Journal of Consciousness Studies* **1**: 164–167.

Harris, T. O., Brown, G. W. (1996). Social causes of depression. *Current Opinion in Psychiatry* **9**: 3–10.

Hartmann, D. (1996). *Methodischer Kulturalismus. Zwischen Naturalismus und Postmoderne*. Frankfurt: Suhrkamp.

Hartmann, D. (1998). *Philosophische Grundlagen der Psychologie*. Darmstadt: Wissenschaftliche Buchgesellschaft.

Hatala, A. R. (2012). The status of the 'biopsychosocial' model in health psychology: towards an integrated approach and a critique of cultural conceptions. *Open Journal of Medical Psychology* **1**: 51–62.

Havas, M., Gutowski, K. A., Lucarelli, M. J., Davidson, R. J. Havas, D. A., Glenberg, A. (2010). Cosmetic use of botulinum toxin-A affects processing of emotional language. *Psychological Science* **21**: 95–900.

Heim, C., Binder, E. B. (2012). Current research trends in early life stress and depression: review of human studies on sensitive periods, gene–environment interactions, and epigenetics. *Experimental Neurology* **233**: 102–111.

Held, R., Hein, A. (1963). Movement-produced stimulation in the development of visually guided behavior. *Journal of Comparative Physiology and Psychology* **56**: 872–876.

Henningsen, P., Kirmayer, L.J. (2000). Mind beyond the net: implications of cognitive neuroscience for cultural psychiatry. *Transcultural Psychiatry* **37**: 467–494.

Henrich, D. (1970). Selbstbewusstsein. Kritische Einleitung in eine Theorie. In: R. Bubner (Eds.) *Hermeneutik und Dialektik*, Vol. I, pp. 257–284. Tübingen: Mohr.

Henry, M. (1963). L'Essence de la manifestation. Paris: PUF.

Henry, M. (1975). *Philosophy and phenomenology of the body*. Trans. G. Etzkorn. The Hague: Martinus Nijhoff.

Herbert, B. M., Herbert, C., Pollatos, O. (2011). On the relationship between interoceptive awareness and alexithymia: is interoceptive awareness related to emotional awareness? *Journal of Personality* **79**: 1149–1175.

Herpertz, S.C., Sass, H., Favazza, A. (1997). Impulsivity in self-mutilative behavior: psychometric and biological findings. *Journal of Psychiatric Research* **31**: 451–465.

Herrmann, C. S., Pauen, M., Byoung-Kyong, M., Busch, N. A., Rieger, J. W. (2008). Analysis of a choice-reaction task yields a new interpretation of Libet's experiments. *International Journal of Psychophysiology* **67**: 151–157

Hickok, G. (2014). *The myth of mirror neurons: The real neuroscience of communication and cognition*. New York: Norton & Co.

Highstein, S. M. (1991). The central nervous system efferent control of the organs of balance and equilibrium. *Neuroscience Research* **12**: 13–30.

Hihara, S., Notoya, T., Tanaka, M., Ichinose, S., Ojima, H., Obayashi, S., Fujii, N., Iriki, A. (2006). Extension of corticocortical afferents into the anterior bank of the intraparietal sulcus by tool-use training in adult monkeys. *Neuropsychologia* **44**: 2636–2646.

Hirshberg, L. M., Svejda, M. (1990). When infants look to their parents: I. Infants' social referencing of mothers compared to fathers. *Child Development* **61**: 1175–1186.

Hobson, R. P. (2002). *The cradle of thought*. London: Macmillan

Hodge, C. W., Grant, K. A., Becker, H. C., Besheer J, Crissman AM, Platt DM, Shannon EE, Shelton KL. (2006). Understanding how the brain perceives alcohol: neurobiological basis of ethanol discrimination. *Alcoholism: Clinical and Experimental Research* **30**: 203–213.

Hofer, M. A. (2001). Origins of attachment and regulation of development within early social interactions: from animal to human. In: A. F. Kalverboer, A. Gramsberg (Eds.) *Handbook of brain and behaviour in human development*, pp. 821–840. Dordrecht: Kluwer Academic Publishers.

Hofer, M. A. (1994). Hidden regulators in attachment, separation, and loss. *Monographs of the Society for Research in Child Development* **59**: 192–207.

Hohwy, J. (2013). *The predictive mind*. Oxford: Oxford University Press.

Holst, E. von, Mittelstaedt, H. (1950). Das Reafferenzprinzip. *Naturwissenschaften* **37**: 464–476.

Holtzheimer, P. E., Mayberg, H. S. (2011). Stuck in a rut: rethinking depression and its treatment. *Trends in Neurosciences* **34**: 1–9.

Hornik, R., Risenhoover, N., Gunnar, M. (1987). The effects of maternal positive, neutral, and negative affective communications on infant responses to new toy. *Child Development* **58**: 937–944.

Howe, M. L. (2000). *The fate of early memories*. Washington, DC: American Psychological Association.

Hrdy, S. B. (2011). *Mothers and others: The evolutionary origins of mutual understanding*. Cambridge, MA: Harvard University Press.

Huizink A. C., de Medina P. G., Mulder E. J., Visser G.H., Buitelaar J. K. (2002). Psychological measures of prenatal stress as predictors of infant temperament. *Journal of the American Academy of Child Adolescent Psychiatry* **41**: 1078–1085.

Huizink A. C., Robles de Medina P. G., Mulder E. J., Visser G. H., Buitelaar J. (2003). Stress during pregnancy is associated with developmental outcome in infancy. *Journal of Child Psychological Psychiatry* **44**: 810–818.

Hurley, S. (1998). *Consciousness in action*. Cambridge, MA: Harvard University Press.

Hurley, S., Noë, E. (2003). Neural plasticity and consciousness. *Biology and Philosophy* **18**: 131–168.

Husserl, E. (1950). *Ideen zu einer reinen Phänomenologie und phänomenologischen Psychologie*. Vol. I. Husserliana 3. The Hague: Martinus Nijhoff.

Husserl, E. (1952). *Ideen zu einer reinen Phänomenologie und phänomenologischen Philosophie*. Vol. II. Husserliana 4. The Hague: Nijhoff.

Husserl, E. (1960). *Cartesian meditations: An introduction to phenomenology*. Trans. D. Cairns. The Hague: Martinus Nijhoff.

Husserl, E. (1973). *Ding und Raum*. Husserliana 16. The Hague: Nijhoff.

Husserl, E. (1973a). *Zur Phänomenologie der Intersubjektivität. Erster Teil: 1905–1920*. The Hague: Martinus Nijhoff.

Husserl, E. (1973b). *Zur Phänomenologie der Intersubjektivität. Zweiter Teil: 1921–1928*. The Hague: Martinus Nijhoff.

Husserl, E. (1989). *Ideas pertaining to a pure phenomenology and to phenomenological philosophy II*. Trans. R. Rojcewicz and A. Schuwer. Dordrecht: Kluwer.

Husserl, E. (1991). *On the phenomenology of the consciousness of internal time*. Trans. J. Broug. Dordrecht: Kluwer Academic Publishers.

Husserl, E. (2001). *Analyses concerning passive and active synthesis: Lectures on transcendental logic*. Trans. A. J. Steinbock. Dordrecht: Kluwer Academic Publishers.

Hutchison, W. D., Davis, K. D., Lozano, A. M., Tasker, R. R., Dostrovsky, J. O. (1999). Pain-related neurons in the human cingulate cortex. *Nature Neuroscience* **2**: 403–405.

Hyman, S. E. (2003). Diagnosing disorders. *Scientific American* **289**: 96–103.

Insel, T. R. (2014). Understanding mental disorders as circuit disorders. http://www.brainfacts.org

Insel, T. R., Quirion, R. (2005). Psychiatry as a clinical neuroscience discipline. *Journal of the American Medical Association* **294**: 2221–2224.

Insel, T. R., Young, L. R. (2001). The neurobiology of attachment. *Nature Neuroscience* **2**: 129–136.

Iriki, A. (2006). The neural origins and implications of imitation, mirror neurons and tool use. *Current Opinion in Neurobiology* **16**: 660–667.

Iriki, A., Tanaka, M., Iwamura, Y. (1996). Coding of modified body schema during tool use by macaque postcentral neurones. *Neuroreport* **7**: 2325–2330.

Ito, M. (2008). Control of mental activities by internal models in the cerebellum. *Nature Reviews Neuroscience* **9**: 304–313.

Jackson, J. H. (1958). *Selected writings of John Hughlings Jackson, vol 2: On the evolution and dissolution of the nervous system.* New York: Basic Books.

Jacob, P., Jeannerod, M. (2003). *Ways of seeing. The scope and limits of visual cognition.* Oxford: Oxford University Press.

Jacob, P., Jeannerod, M. (2004). The motor theory of social cognition. A critique. *Trends in Cognitive Science* **9**: 21–25.

James, W. (1884). What is an emotion? *Mind* **9**: 188–205.

James, W. (1890). *The principles of psychology.* New York: Henry Holt and Company.

Janich, P. (1996). *Konstruktivismus und Naturerkenntnis. Auf dem Weg zum Kulturalismus.* Frankfurt: Suhrkamp.

Jasnow, M., Feldstein, S. (1986). Adult-like temporal characteristics of mother-infant vocal interactions. *Child Development* **57**: 754–761.

Jaspers, K. (1997). *General Psychopathology.* Trans. J. Hoenig, M. W. Hamilton. Baltimore, MD: Johns Hopkins University Press.

Jeannerod, M. (1995). Mental imagery in the motor context. *Neuropsychologia* **33**: 1419–1432.

Jeannerod, M. (1997). *The cognitive neuroscience of action.* Oxford: Blackwell.

Jeannerod, M., Decety, J. (1995). Mental motor imagery: a window into the representational stages of action. *Current Opinion in Neurobiology* **5**: 727–732.

Jirak, D., Menz, M. M., Buccino, G., Borghi, A. M., Binkofski, F. (2010). Grasping language —a short story on embodiment. *Consciousness and Cognition* **19**: 711–720.

Johansson, R. S., Westling, G. (1988). Coordinated isometric muscle commands adequately and erroneously programmed for the weight during lifting task with precision grip. *Experimental Brain Research* **71**: 59–71.

Jonas, H. (1968). Biological foundations of individuality. *International Philosophical Quarterly* **8**: 231–251.

Jonas, H. (2001). *The phenomenon of life: Toward a philosophical biology* [1st edition 1966]. Evanston, IL: Northwestern University Press.

Juarrero, A. (1999). *Dynamics in action: Intentional behaviour as a complex system.* Cambridge, MA: MIT Press.

Kaan, E., Swaab, T. Y. (2002). The brain circuitry of syntactic comprehension. *Trends in Cognitive Science* **6**: 350–356.

Kamitani, Y., Tong, F. (2005). Decoding the visual and subjective contents of the human brain. *Nature Neuroscience* **8**: 679–685.

Kandel, E. R. (2001). Psychotherapy and the single synapse. The impact of psychiatric thought on neurobiological research. *Journal of Neuropsychiatry and Clinical Neuroscience* **13**: 290–300.

Kane, R. (Ed.) (2011). *The Oxford handbook of free will*, 2nd edition. Oxford: Oxford University Press.

Kant, I. (1900). *Dreams of a spirit-seer.* [1766] Trans. E. F. Goerwitz. London: Swan Sonnenschein & Co.

Kant, I. (1914). *Critique of judgement.* [1790] Trans. J. H. Bernard. 2nd ed. London: Macmillan.

Kant, I. (1998). *Critique of pure reason* [1781]. *The Cambridge edition of the Works of Immanuel Kant.* Trans. and ed. P. Guyer and A. W. Wood. Cambridge: Cambridge University Press.

Karenberg, A. (2009). Cerebral localization in the eighteenth century—an overview. *Journal of the History of the Neurosciences* **18**: 248–253.
Keil, G. (2007). *Willensfreiheit*. Berlin: De Gruyter.
Kelso, J. A. S. (1995). *Dynamic patterns: The self-organization of brain and behavior*. Cambridge, MA: MIT Press.
Kendal, J. R. (2011). Cultural niche construction and human learning environments: Investigating sociocultural perspectives. *Biological Theory* **6**: 241–250.
Kendler, K. S., Kessler, R. C. (1993). The prediction of major depression in women: toward an integrated etiological model. *American Journal of Psychiatry* **150**: 1139–1148.
Kendler, K. S., MacLean, C., Neale, M., Kessler, R., Heath, A., Eaves, L. (1995). Stressful life events, genetic liability and onset of an episode of major depression in women. *American Journal of Psychiatry* **152**: 833–842.
Kern, R. S., Glynn, S. M., Horan, W. P., Marder, S. R. (2009). Psychosocial treatments to promote functional recovery in schizophrenia. *Schizophrenia Bulletin* **35**: 347–361.
Keysers, C., E. Kohler, M. A. Umiltà, L. Nanetti, L. Fogassi, Gallese, V. (2003). Audiovisual mirror neurons and action recognition. *Experimental Brain Research* **153**: 628–636.
Kilgard, M. P., Merzenich, M. M. (1998). Cortical map reorganization enabled by nucleus basalis activity. *Science* **279**: 1714–1718.
Kim, J. (1993). *Supervenience and mind: Selected philosophical essays*. Cambridge: Cambridge University Press.
Kim, J. (2006). Being realistic about emergence. In: P. Clayton, P. Davies (Eds.) *The re-emergence of emergence: The emergentist hypothesis from science to religion*, pp. 189–202. Oxford: Oxford University Press.
Kircher, T., David, A. S. (2003). Self-consciousness: an integrative approach from philosophy, psychopathology and the neurosciences. In: T. Kircher, A. S. David (Eds.) *The self in neuroscience and psychiatry*, pp. 445–466. Cambridge: Cambridge University Press.
Kirk, R. (2008). The inconceivability of zombies. *Philosophical Studies* **139**: 73–89.
Knoll, J. (2005). *The brain and its self: A neurochemical concept of the innate and acquired drives*. Berlin: Springer.
Koelsch, S. (2005). Ein neurokognitives Modell der Musikperzeption. *Musiktherapeutische Umschau* **26**: 365–381.
Koelsch, S., Fritz, T., Schulze, K., Alsop, D., Schlaug, G. (2005). Adults and children processing music: an fMRI study. *Neuroimage* **25**: 1068–1076.
Kohler, E., Keysers, C., Umiltà, A., Fogassi, L., Gallese, V., Rizzolatti, G. (2002). Hearing sounds, understanding actions: action representation in mirror neurons. *Science* **297**: 846–848.
Kohler, I. (1951). *Über Aufbau und Wandlungen der Wahrnehmungswelt*. Wien: Österreichische Akademie der Wissenschaften.
Kotchoubey, B., Tretter, F., Braun, H. A., et al. (2016). Methodological problems on the way to integrative human neuroscience. *Frontiers in Integrative Neuroscience* **10**: Art. 41, 1–19.
Krach, S., Cohrs, J. C., de Echeverría Loebell, N. C., Kircher, T., Sommer, J., Jansen, A., Paulus, F. M. (2011). Your flaws are my pain: linking empathy to vicarious embarrassment. *PLoS One* **6**: e18675.

Kronmüller, K. T., Backenstrass, M., Kocherscheidt, K., Hunt, A., Fiedler, P., Mundt, C. (2005). Dimensions of the typus melancholicus personality type. *European Archives of Psychiatry and Clinical Neuroscience* **255**: 341–349.

Kronz, F. M., Tiehen, J. T. (2002). Emergence and quantum mechanics. *Philosophy of Science* **69**: 324–347.

Krueger, J. (2013). Ontogenesis of the socially extended mind. *Cognitive Systems Research* **25**: 40–46.

Krupnick, J. L., Sotsky, S. M., Simmens, S., et al. (1996). The role of the therapeutic alliance in psychotherapy and pharmacotherapy outcome: findings in the National Institute of Mental Health Treatment of Depression Collaborative Research Program. *Journal of Consulting and Clinical Psychology* **64**: 532–539.

Krystal, J. H., State, M. W. (2014). Psychiatric disorders: diagnosis to therapy. *Cell* **157**: 201–214.

Kvaale, E. P., Haslam, N., Gottdiener, W. H. (2013). The 'side effects' of medicalization: a meta-analytic review of how biogenetic explanations affect stigma. *Clinical Psychology Review* **33**: 782–794.

Laan, E., Everaerd, W., Velde, J., Geer, J. H. (1995). Determinants of subjective experience of sexual arousal in women: feedback from genital arousal and erotic stimulus content. *Psychophysiology* **32**: 444–451.

Lakoff, G., Johnson, M. (1999). *Philosophy in the flesh: The embodied mind and its challenge to western thought.* New York: Basic Books.

Langleben, D.D., Schroeder, L., Maldjian, J. A., et al. (2002). Brain activity during simulated deception: an event-related functional magnetic resonance study. *Neuroimage* **15**: 727–32.

Lazarus, R. S. (1966). *Psychological stress and the coping process.* New York: McGraw Hill.

Leder, D. (1990). *The absent body.* Chicago, IL: University of Chicago Press.

LeDoux, J. (1998). Fear and the brain: where have we been, and where are we going? *Biological Psychiatry* **44**: 1229–1238.

LeDoux, J. E. (2003). *The synaptic self: How our brains become who we are.* Viking, New York.

Lee, W.-C. A., Huang, H., Feng, G., et al. (2006). Dynamic remodeling of dendritic arbors in GABAergic interneurons of adult visual cortex. *PLoS Biology* **4**: e29.

Legrand, D. (2006). The bodily self: the sensori-motor roots of pre-reflective self-consciousness. *Phenomenology and the Cognitive Sciences* **5**: 89–118.

Lende, D. H., Downey, G. (2012). *The encultured brain: An introduction to neuroanthropology.* Cambridge, MA: MIT press.

Levine, J. (1983). Materialism and qualia: the explanatory gap. *Pacific Philosophical Quarterly* **64**: 354–361.

Levine, S. (2002). Regulation of the hypothalamic-pituitary-adrenal axis in the neonatal rat: the role of maternal behaviour. *Neurotoxicity Research* **4**: 557–564.

Lewis, M. D. (2005). Bridging emotion theory and neurobiology through dynamic systems modelling. *Behavioral and Brain Sciences* **28**: 169–194.

Liberman, M.C., Dodds, L. W., Pierce, S. (1990). Afferent and efferent innervation of the cat cochlea: quantitative analysis with light and electron microscopy. *Journal of Comparative Neurology* **301**: 443–460.

Libet, B. (1985). Unconscious cerebral initiative and the role of conscious will in voluntary action. *Behavioral and Brain Sciences* **8**: 529–566.

Libet, B., Gleason, C. A., Wright, E. W., Pearl, D. K. (1983). Time of conscious intention to act in relation to onset of cerebral activity (readiness-potential). *Brain* **106**: 623–642.

Lichtenberg, G. C. (1973). *Schriften und Briefe*. Ed. W. Promies, Vol. 1, Sudelbücher. 3rd edition. München: Hanser.

Lin, M. (2004). Spinoza's metaphysics of desire. *Archiv für Geschichte der Philosophie* **86**: 21–55.

Linehan, M. M. (1993). *Cognitive-behavioral treatment of borderline personality disorder*. New York: Guilford.

Liszkowski, U., Carpenter, M., Striano, T., Tomasello M. (2006). 12- and 18-month-olds point to provide information for others. *Journal of Cognition and Development* **7**: 173–187.

Locke, J. (1689/1825). *An essay concerning human understanding*. 25th edition. London: Thomas Davison.

Locke, J. (1813). *An essay concerning human understanding*. London: Cummings & Hilliard and J.T. Buckingham.

Lowe, E. J. (2003). Personal agency. In: O'Hear, A. (Ed.) *Minds and persons*, pp. 211–227. Cambridge: Cambridge University Press.

Lycan, W. (1987). *Consciousness*. Cambridge, MA: Bradford Books/MIT Press.

Mack, A., Rock, I. (1998). *Inattentional blindness*. Cambridge, MA: MIT press.

Maguire, E. A., Gadian, D. G., Johnsrude, I. S., Good, C. D., Ashburner, J., Frackowiak, R. S., Frith C. D. (2000). Navigation-related structural change in the hippocampi of taxi drivers. *Proceedings of the National Academy of Sciences of the United States of America* **97**: 4398–4403.

Malloch, S. (1999). Mother and infants and communicative musicality. *Musicæ Scientiæ* (Special issue: Rhythm, musical narrative, and the origins of human commu-nication) **3**: 29–57.

Mampe, B., Friederici, A. D., Christophe, A., Wermke, K. (2009). Newborns' cry melody is shaped by their native language. *Current Biology* **19**: 1994–1997.

Margulis, L., Sagan, D. (1995). *What is life?* New York: Simon & Schuster.

Markowitsch, H., Welzer, H. (2009). *The development of autobiographical memory*. London: Psychology Press.

Martin, S. D., Martin, E., Rai, S. S., et al. (2001). Brain blood flow changes in depressed patients treated with interpersonal psychotherapy or venlafaxine hydrochloride. *Archives of General Psychiatry* **58**: 641–648.

Martinez-Conde, S., Macknik, S. L., Troncoso, X. G., Dyar, T. A. (2006). Microsaccades counteract visual fading during fixation. *Neuron* **49**: 297–305.

Maxwell, M. (1984). *Human evolution: A philosophical anthropology*. New York: Columbia University Press.

Mayberg, H. S., Silva, J. A., Brannan, S. K., et al. (2002). The functional neuroanatomy of the placebo effect. *American Journal of Psychiatry* **159**: 728–737.

Mayr, E. (1979). *Evolution und die Vielfalt des Lebens*. Berlin: Springer.

McCabe, R., Priebe, S. (2004). The therapeutic relationship in the treatment of severe mental illness: a review of methods and findings. *International Journal of Social Psychiatry* **50**: 115–128.

McCulloch, G. (2003). *The life of the mind: An essay on phenomenological externalism.* New York: Routledge.

McEwen, B. (1999). Stress and the brain. In: R. Conlan (Ed.) *States of mind,* pp. 81–101. New York: Wiley.

McLaren, N. (1998). A critical review of the biopsychosocial model. *Australasian Psychiatry* 32: 86–92.

McMullen, E., Saffran, J. R. (2004). Music and language: a developmental comparison. *Music Perception* 21: 289–311.

Mead, G. H. (1934). *Mind, self and society from the standpoint of a social behaviorist.* Chicago, IL: University of Chicago Press.

Meaney, J. M. (2001). Maternal care, gene expression, and the transmission of individual difference in stress reactivity across generations. *Annual Review of Neuroscience* 24: 1161–1192.

Mechelli, A. (2010). Psychoanalysis on the couch: can neuroscience provide the answers? *Medical hypotheses* 75: 594–599.

Mechelli, A., Price, C. J., Friston, K. J., Ishai, A. (2004). Where bottom-up meets top-down: neuronal interactions during perception and imagery. *Cerebral Cortex* 14: 1256–1265.

Meister Eckehart (1958). *Predigten,* ed. J. Quint. Vol. 1. Stuttgart: Kohlhammer.

Melchner, L. von, Pallas, S. L., Sur, M. (2000). Visual behaviour mediated by retinal projections directed to the auditory pathway. *Nature* 404: 871–876.

Melloni, L., Molina, C., Pena, M., Torres, D., Singer, W., Rodriguez, E. (2007). Synchronization of neural activity across cortical areas correlates with conscious perception. *Journal of Neuroscience* 27: 2858–2865.

Meltzoff, A. N., Brooks, R. (2001). 'Like me' as a building block for understanding other minds: Bodily acts, attention, and intention. In: B.F. Malle, L. J. Moses, D. A. Baldwin (Eds.) *Intentions and intentionality: foundations of social cognition,* pp. 171–191. Cambridge, MA: MIT Press.

Meltzoff, A. N., Moore, M. K. (1977). Imitation of facial and manual gestures by human neonates. *Science* 198: 74–78.

Meltzoff, A. N., Moore, M. K. (1989). Imitation in newborn infants: exploring the range of gestures imitated and the underlying mechanisms. *Developmental Psychology* 25: 954–962.

Meltzoff, A. N., Prinz, W. (2002). *The imitative mind: Development, evolution and brain bases.* Cambridge, MA: Cambridge University Press.

Menary, R. (2013). Cognitive integration, enculturated cognition and the socially extended mind. *Cognitive Systems Research* 25: 26–34.

Merleau-Ponty, M. (1960). Le philosophe et son ombre. In: *Signes.* Paris: Gallimard.

Merleau-Ponty, M. (1962). *Phenomenology of perception.* Trans. C. Smith. London: Routledge.

Merleau-Ponty, M. (1968). *The visible and the invisible: Followed by working notes.* Trans. C. Lefort. Evanston, IL: Northwestern University Press.

Merleau-Ponty, M. (2010). *Institution and passivity: Course notes from the Collège de France. (1954–1955).* Evanston, IL: Northwestern University Press.

Mermillaud, M., Vermeulen, N., Droit-Volet, S., Jalenques, I., Durif, F., Niedenthal, P. (2011). Embodying emotional disorders: new hypotheses about possible emotional consequences of motor disorders in Parkinson's disease and Tou-rette's syndrome. *ISRN Neurology*, Article ID 306918, http://dx.doi.org/10.5402/2011/306918

Metzinger, T. (1999). *Subjekt und Selbstmodell. Die Perspektivität phänomenalen Bewußtseins vor dem Hintergrund einer naturalistischen Theorie mentaler Repräsentation*, 2nd edition. Paderborn: Mentis.

Metzinger, T. (2003). *Being no one: The self-model theory of subjectivity*. Cambridge, MA: MIT Press.

Metzinger, T. (2006). Der Preis der Selbsterkenntnis. *Gehirn & Geist* 7–8: 42–49.

Metzinger, T. (2009). *The EGO Tunnel: The science of the mind and the myth of the self*. New York: Basic Books.

Metzinger, T., Gallese, V. (2003). The emergence of a shared action ontology: building blocks for a theory. *Consciousness and Cognition* 12: 549–571.

Meyer-Lindenberg, A., Tost, H. (2012). Neural mechanisms of social risk for psychiatric disorders. *Nature Neuroscience* 15: 663–668.

Meyer-Lindenberg, A., Tost, H. (2012). Neural mechanisms of social risk for psychiatric disorders. *Nature Neuroscience* 15: 663–668.

Meyniel, F., Safra, L., Pessiglione, M. (2014). How the brain decides when to work and when to rest: dissociation of implicit-reactive from explicit-predictive computational processes. *PLoS Computational Biology* 10: e1003584.

Mezzich, J. E. (2007). Psychiatry for the person: articulating medicine's science and humanism. *World Psychiatry* 6: 65–67

Mikkelsen, J. D. (1992). Visualization of efferent retinal projections by immunohistochemical identification of cholera toxin subunit B. *Brain Research Bulletin* 28: 619–629.

Miller, G. (2010). Is pharma running out of brainy ideas? *Science* 329: 502–504.

Millikan, R. (1984). *Language, thought and other biological categories*. Cambridge, MA: MIT Press.

Millikan, R. (1991). Speaking up for Darwin. In: B. Loewer, G. Rey (Eds.) *Meaning in mind: Fodor & his critics*. Oxford: Blackwell.

Milton, J., Solodkin, A., Hlustík, P., Small, S. L. (2007). The mind of expert motor performance is cool and focused. *Neuroimage* 35: 804–813.

Monroe, S. M., Slavich, G. M., Georgiades, K. (2009). The social environment and life stress in depression. In: I. H. Gotlib, C. L. Hammen (Eds.) *Handbook of depression*, 2nd edition, pp. 340–360. New York: Guilford Press.

Moreno, A., Umerez, J. (2000). Downward causation at the core of living organisation. In: P. B. Andersen, C. Emmeche, N. O. Finnemann, P. V. Christiansen (Eds.) *Downward causation: Minds, bodies and matter*, pp. 99–117. Aarhus: Aarhus University Press.

Morgan, C., Bhugra, D. (Eds.) (2010). *Principles of social psychiatry*. Chichester: Wiley-Blackwell.

Morton, P. A. (1997). *A historical introduction to the philosophy of mind: Readings with commentary*. Ontario: Broadview.

Moylan, S., Maes, M., Wray, N. R., Berk, M. (2013). The neuroprogressive nature of major depressive disorder: pathways to disease evolution and resistance, and therapeutic implications. *Molecular Psychiatry* 18: 595–606.

Mukamel, R., Ekstrom, A. D., Kaplan, J., Iacoboni, M., Fried, I. (2010). Single-neuron responses in humans during execution and observation of actions. *Current Biology* 20: 750–756.

Müller, L. E., Schulz, A., Andermann, M., Gäbel, A., Gescher, D. M., Spohn, A., Herpertz, S. C., Bertsch, K. (2015). Cortical representation of afferent bodily signals in borderline personality disorder: neural correlates and relationship to emotional dysregulation. *JAMA Psychiatry* 72: 1077–1086.

Mundt, C., Backenstrass, M., Kronmüller, K. T., Fiedler, P., Kraus, A., Stanghellini, G. (1997). Personality and endogenous/major depression: an empirical approach to typus melancholicus. *Psychopathology* 30: 130–140.

Mundt, C., Kronmüller, K.T., Backenstraß, M., Reck, C., Fiedler, P. (1998). The influence of psychopathology, personality and marital interaction on the short-term course of major depression. *Psychopathology* 31: 29–36.

Münte, T. F., Altenmüller, E., Jäncke, L. (2002). The musician's brain as a model of neuroplasticity. *Nature Reviews Neuroscience* 3: 473–478.

Murphy, N. (2006). Emergence and mental causation. In: P. Clayton, P. Davies (Eds.) *The re-emergence of emergence: The emergentist hypothesis from science to religion*, pp. 228–243. Oxford: Oxford University Press.

Murray, L., Cooper, P. (2003). Intergenerational transmission of affective and cognitive processes associated with depression: infancy and the pre-school years. In: L. Murray, P. Cooper (Eds.) *Unipolar depression: a lifespan perspective*, pp. 17–46. Oxford: Oxford University Press.

Nagel, T. (1974). What is it like to be a bat? *The Philosophical Review* 83: 435–450.

Nagel, T. (1989). *The view from nowhere*. Oxford: Oxford University Press.

Neisser, U. (1988). Five kinds of self-knowledge. *Philosophical Psychology* 1: 35–59.

Nelson, K. (1996). *Language in cognitive development*. Cambridge: Cambridge University Press.

Nemiah, J. C. (1989). The varieties of human experience. *British Journal of Psychiatry* 154: 459–466.

Niedenthal, P. M. (2007). Embodying emotion. *Science* 316: 1002–1005.

Nijhawan, R. (2008). Visual prediction: psychophysics and neurophysiology of compensation for time delays. *Behavioral and Brain Sciences* 31: 179–198.

Nitsan, Z. (2017). *The brain show—behind the scenes: what is going on inside our brain while we are living our life*. CreateSpace Independent Publishing Platform: Kindle Edition.

Noë, A. (2009). *Out of our heads: Why you are not your brain, and other lessons from the biology of consciousness*. New York: Hill & Wang

Noë, A., Thompson, E. (2004). Are there neural correlates of consciousness? *Journal of Consciousness Studies* 11: 3–28.

Northoff, G. (2004a). *Philosophy of the brain: The brain problem*. Amsterdam: John Benjamins.

Northoff, G. (2004b). Why do we need a philosophy of the brain? *Trends in Cognitive Sciences* 8: 484–485.

Northoff, G., Bermpohl, F. (2004). Cortical midline structures and the self. *Trends in Cognitive Science* 8: 102–107.

Northoff, G., Musholt, K. (2006). How can Searle avoid property dualism? Epistemic-ontological inference and autoepistemic limitation. *Philosophical Psychology* 19: 589–605.

Nuechterlein, K. H., Dawson, M. E. (1984). A heuristic vulnerability-stress model of schizophrenic episodes. *Schizophrenia Bulletin* 10: 300–312.

O'Connor, T. G., Rutter, M. (2000). Attachment disorder behaviour following early severe deprivation. Extension and longitudinal follow-up. *Journal of the American Academy of Child and Adolescent Psychiatry* **39**: 703–712.

O'Regan, J. K., Noë, A. (2001). A sensorimotor account of vision and visual consciousness. *Behavioral and Brain Sciences* **24**: 939–1011.

Ochsner, K. N., Beer, J. S., Robertson, E. R., Cooper, J. C., Gabrieli, J. D., Kihsltrom, J. F., D'Esposito, M. (2005). The neural correlates of direct and reflected self-knowledge. *Neuroimage* **28**: 797–814.

Oostenbroek, J., Suddendorf, T., Nielsen, M., Redshaw, J., Kennedy-Costantini, S., Davis, J., Clark, S., Slaughter, V. (2016). Comprehensive longitudinal study challenges the existence of neonatal imitation in humans. *Current Biology* **26**: 1334–1338.

Panksepp, J. (1998a). *Affective neuroscience: The foundations of human and animal emotions*. Oxford: Oxford University Press.

Panksepp, J. (1998b). The periconscious substrates of consciousness: affective states and the evolutionary origins of the self. *Journal of Consciousness Studies* **5**: 566–582.

Panksepp, J. (2003). Damasio's error? *Consciousness and Emotion* **4**: 111–134.

Panksepp, J. (2012). A synopsis of affective neuroscience—naturalizing the mammalian mind. *Journal of Consciousness Studies* **19**: 6–18.

Papousek, H., Papousek, M. (1987). Intuitive parenting: a dialectic counterpart to the infant's integrative competence. In: J. D. Osofsky (Ed.) *Handbook of infant development*, 2nd edition, pp. 699–720. New York: Wiley.

Papousek, H., Papousek, M. (1991). Innate and cultural guidance of infants' integrative competences: China, the United Staates, and Germany. In: M. H. Bornstein (Ed.) *Cultural approaches to parenting*, pp. 23–44. Hillsdale, NJ: Lawrence Erlbaum.

Pascual-Leone, A., Torres, F. (1993). Plasticity of the sensorimotor cortex representation of the reading finger in Braille readers. *Brain* **116**: 39–45.

Patel, A. (2003). Language, music, syntax and the brain. *Nature Neuroscience* **6**: 674–681.

Paulus, M. P., Stein, M. B. (2010). Interoception in anxiety and depression. *Brain Structure and Function* **214**: 451–463.

Peirce, C. S. (1932). *Collected Papers of Charles Sanders Peirce*. Vol. 2, Elements of Logic. Ed. C. Hartshorne, P. Weiss. Cambridge, MA: Belknap.

Penfield, W., Perot, P. (1963). The brain's record of an auditory and visual experience. *Brain* **86**: 595–696.

Pereboom, D. (Ed.) (2009). *Free will*. Indianapolis, IN: Hackett Publishing.

Phelps, E.A., O'Connor, K.J., Cunningham, W.A., Funayama, E. S., Gatenby, J. C., Gore, J. C., Banaji, M. R. (2000). Performance on indirect measures of race evaluation predicts amygdala activity. *Journal of Cognitive Neuroscience* **12**: 1–10.

Pinker, S. (1997). *How the mind works*. New York: Norton.

Plessner, H. (1970). *Philosophische Anthropologie*. Frankfurt: Fischer.

Plessner, H. (1975). *Die Stufen des Organischen und der Mensch* (1st edition 1928). Berlin: De Gruyter.

Ploog, D. (1997). Epilog: Das soziale Gehirn des Menschen. In: H. Meier, D. Ploog (Eds.) *Der Mensch und sein Gehirn*, pp. 235–252. Munich: Piper.

Polanyi, M. (1967). *The tacit dimension*. Garden City, NY: Anchor Books.

Pollatos, O., Kirsch, W., Schandry, R. (2005). On the relationship between interoceptive awareness, emotional experience, and brain processes. *Cognitive Brain Research* **25**: 948–962.

Popper, K. R. (1984). Evolutionary epistemology. In: J. W. Pollard (Ed.) *Evolutionary theory: Paths into the future*. London: John Wiley & Sons Ltd.

Popper, K. R., Eccles, J. (1977). *The self and its brain*. New York: Springer International.

Portmann, A. (1969). *Zoologie und das neue Bild des Menschen*, 3rd edition. Reinbek: Rowohlt.

Posner, M. I., Rothbart, M. K. (1998). Attention, self-regulation and consciousness. *Philosophical Transactions of the Royal Society of London B: Biological Sciences* **353**: 1915–1927.

Prinz, W. (1992). Why don't we perceive our brain states? *European Journal of Cognitive Psychology* **4**: 1–20.

Puccetti, R. (1974). Physicalism and the evolution of consciousness. *Canadian Journal of Philosophy*, **1**: 171–183.

Pulvermüller, F. (2005). Brain mechanisms linking language and action. *Nature Reviews Neuroscience* **6**: 576–582.

Putnam, H. (1967). Psychological predicates. In: W. H. Capitan, D. D. Merrill (Eds.) *Art, Mind, and Religion*, pp. 37–48. Pittsburgh, PA: University of Pittsburgh Press.

Putnam, H. (1979). The meaning of 'meaning'. In: *Mind, language, and reality. Philosophical papers*, Vol. 2, pp. 215–271. Cambridge: Cambridge University Press.

Quine, W. V. O. (1960). *Word and object*. Cambridge, MA: MIT Press.

Raichle, M. E., MacLeod, A. M., Snyder, A. Z., Powers, W. J., Gusnard, D. A., Shulman, G. L. (2001). A default mode of brain function. *Proceedings of the National Academy of Sciences of the United States of America* **98**: 676–682.

Ramachandran, V. S. (1995). Anosognosia in parietal lobe syndrome. *Consciousness and cognition* **4**: 22–51.

Ramachandran, V. S., Blakeslee, S. (1998). *Phantoms in the brain: Probing the mysteries of the human mind*. New York: William Morrow.

Rameson, L. T., Satpute, A. B., Lieberman, M. D. (2010). The neural correlates of implicit and explicit self-relevant processing. *Neuroimage* **50**: 701–708.

Read, J., Bentall, R. P., Fosse, R. (2009). Time to abandon the bio-bio-bio model of psychosis: exploring the epigenetic and psychological mechanisms by which adverse life events lead to psychotic symptoms. *Epidemiologia e psichiatria sociale* **18**: 299–310.

Revonsuo, A. (2003). The contents of phenomenal consciousness. *Psyche* **9**: 2.

Ricks, M.H. (1985). The social transmission of parental behavior: attachment across generations. In: I. Bretherton, E. Waters (Eds.) *Growing points of attachment theory and research*, pp. 211–227. Monographs of the Society for Research in Child Development, Vol. 50 (1–2, Serial No. 209). Chicago, IL: University of Chicago Press.

Ricoeur, P. (1992). *Oneself as another*. Trans. K. Blamey. Chicago, IL: University of Chicago Press.

Rizzolatti, G., Arbib, M. A. (1998). Language within our grasp. *Trends in Neuroscience* **21**: 188–194.

Rizzolatti, G., Fogassi, L., Gallese, V. (2001). Neurophysiological mechanisms underlying the understanding and imitation of action. *Nature Neuroscience* **2**: 661–70.

Rizzolatti, G., Luppino, G., Matelli, M. (1998). The organization of the cortical motor system: new concepts. *Electroencephalography and Clinical Neurophysiology* **106**: 283–296.

Rizzolatti, G., Sinigaglia, C. (2008). *Mirrors in the brain: How our minds share actions and emotions.* New York: Oxford University Press.

Rochat, P. (2009). *Others in mind: Social origins of self-consciousness.* Cambridge: Cambridge University Press.

Rockwell, W. T. (2005). *Neither brain nor ghost. A nondualist alternative to the mind-brain identity theory.* Cambridge, MA: MIT Press.

Rodriguez, E., George, N., Lachaux, J. P., Martinerie, J., Renault, B., Varela, F. J. (1999). Perception's shadow: long-distance synchronization of human brain activity. *Nature* **397**: 430–433.

Roffman, J. L., Marci, C. D., Glick, D. M., Dougherty, D. D., Rauch, S. L. (2005). Neuroimaging and the functional neuroanatomy of psychotherapy. *Psychological Medicine* **35**: 1385–1398.

Rorty, R. (1970). In defense of eliminative materialism. *Review of Metaphysics* **24**: 112–121.

Rosen, R. (1991). *Life itself: A comprehensive inquiry into the nature, origin and fabrication of life.* New York: Columbia University Press.

Rosenthal, D. M. (2005). *Consciousness and mind.* Oxford: Clarendon Press.

Rosner, B. S., Doherty, N. E. (1979). The response of neonates to intra-uterine sounds. *Developmental Medicine & Child Neurology* **21**: 723–729.

Ruby, P., Decety, J. (2001). Effect of subjective perspective taking during simulation of action: a PET investigation of agency. *Nature Neuroscience* **4**: 546–550.

Ruff, C. B., Trinkaus, E., Holliday, T. W. (1997). Body mass and encephalization in Pleistocene Homo. *Nature* **387**: 173–76.

Ryle, G. (1949). *The concept of mind.* London: Hutchison.

Salk, L. (1962). Mothers' heartbeat as an imprinting stimulus. *Transactions of the New York Academy of Sciences* **24**: 753–763.

Sartorius, N., Nielsen, J. A., Strömgren, E. (1989). (Eds.) Change in frequency of mental disorders over time: results of repeated surveys of mental disorders in the general population. *Acta Psychiatrica Scandinavia* **79**(Suppl. 348): 1–189.

Sass, L. A., Parnas, J. (2003). Schizophrenia, consciousness, and the self. *Schizophrenia Bulletin* **29**: 427–444.

Schacter, C. L. (1987). Implicit memory: history and current status. *Journal of Experimental Psychology: Learning, Memory, and Cognition* **13**: 501–518.

Schacter, D. L., Tulving, E. (1994). *Memory systems.* Cambridge, MA: MIT Press.

Schilbach, L., Timmermans, B., Reddy, V., Costall, A., Bente, G., Schlicht, T., Vogeley, K. (2013). Toward a second-person neuroscience. *Behavioral and Brain Sciences* **36**: 393–414.

Schmahmann, J. D., Anderson, C. M., Newton, N., Ellis, R. D. (2001). The function of the cerebellum in cognition, affect and consciousness: empirical support for the embodied mind. *Consciousness & Emotion* **2**: 273–309.

Schmitz, H. (1995). *Der unerschöpfliche Gegenstand. Grundzüge der Philosophie*. 2nd edition. Bonn: Bouvier.

Schomerus, G., Schwahn, C., Holzinger, A., Corrigan, P. W., Grabe, H. J., Carta, M. G., Angermeyer, M. C. (2012). Evolution of public attitudes about mental illness: a systematic review and meta-analysis. *Acta Psychiatr Scandinavia* **125**: 440–452.

Schopenhauer, A. (1966). *The world as will and representation*, Vols. 1–2, E. Payne, Trans. New York: Dover.

Schore, A. N. (1994). *Affect regulation and the origin of the self: The neurobiology of emotional development*. Hillsdale, NJ: Erlbaum.

Schore, A. N. (2001). The effects of a secure attachment relationship on right brain development, affect regulation and infant mental health. *Infant Mental Health Journal* **22**: 1–66.

Schore, A. N. (2003). *Affect dysregulation and disorders of the self*. New York: Norton & Co.

Schultze-Kraft, M., Birman, D., Rusconi, M., Allefeld, C., Görgen, K., Dähne, S., Blankertz, B., Haynes, J. D. (2016). The point of no return in vetoing self-initiated movements. *Proceedings of the National Academy of Sciences of the United States of America* **113**: 1080–1085.

Schulz, A., Köster, S., Beutel, M. E., Schächinger, H., Vögele, C., Rost, S., Rauh, M., Michal, M. (2015). Altered patterns of heartbeat-evoked potentials in depersonalization/derealization disorder: neurophysiological evidence for impaired cortical representation of bodily signals. *Psychosomatic Medicine* **77**: 506–516.

Schwartz, J. M. (1998). Neuroanatomical aspects of cognitive-behavioral therapy response in obsessive-compulsive disorder: an evolving perspective on brain and behavior. *British Journal of Psychiatry Supplement* **35**: 38–44.

Searle, J. (1992). *The Rediscovery of mind*. Cambridge MA: MIT Press.

Searle, J. R. (1980). Minds, brains, and programs. *The Behavioral and Brain Sciences* **3**: 417–457.

Searle, J. R. (1983). *Intentionality: An essay in the philosophy of mind*. Cambridge: Cambridge University Press.

Searle, J. R. (1998). How to study consciousness scientifically. *Philosophical Transactions of the Royal Society of London B: Biological Sciences* **353**: 1935–1942.

Searle, J. R. (2015). *Seeing things as they are: A theory of perception*. Oxford: Oxford University Press.

Segall, M. H., Campbell, D. T., Herskovits, M. J. (1963). Cultural differences in the perception of geometric illusions. *Science* **139**: 769–771.

Seligman, M. E. P. (1974). *Depression and learned helplessness*. Hoboken, NJ: John Wiley & Sons.

Selimbeyoglu, A., Parvizi, J. (2010). Electrical stimulation of the human brain: perceptual and behavioral phenomena reported in the old and new literature. *Frontiers in Human Neuroscience* **4**: 46.

Serres, L. (2001). Morphological changes of the human hippocampal formation from midgestation to early childhood. In: C. A. Nelson, M. Luciana (Eds.) *Handbook of developmental cognitive neuroscience*, pp. 45–58. Cambridge, MA: MIT Press.

Sheets-Johnstone, M. (1999). Emotion and movement. A beginning empirical-phenomenological analysis of their relationship. *Journal of Consciousness Studies* **6**: 259–277.

Sheets-Johnstone, M. (2011). Kinesthetic memory. In: S. Koch, T. Fuchs, M. Summa, C. Müller (Eds.) *Body memory, metaphor and movement*, pp. 43–72. Amsterdam: John Benjamins.

Sherbourne, C. D., Hays, R. D., Wells, K. B. (1995). Personal and psychosocial risk factors for physical and mental health outcomes and course of depression among depressed patients. *Journal of Consulting and Clinical Psychology* 63: 345–355.

Shoemaker, S. S. (1968). Self-reference and self-awareness. *The Journal of Philosophy* 65: 555–567.

Silverstein, M. (2006). In defence of ontological emergence and mental causation. In: P. Clayton, P. Davies (Eds.) *The re-emergence of emergence: The emergentist hypothesis from science to religion*, pp. 203–226. Oxford: Oxford University Press.

Singer, T., Lamm, C. (2009). The social neuroscience of empathy. *Annals of the New York Academy of Sciences* 1156: 81–96.

Singer, T., Seymour, B., O'Doherty, J. P., Kaube, H., Dolan, R. J., Frith, C. D. (2004). Empathy for pain involves the affective but not the sensory components of pain. *Science* 303: 1157–1162.

Singer, W. (1999). Neuronal synchrony: a versatile code for the definition of relations? *Neuron* 24: 49–65.

Singer, W. (2001). Consciousness and the binding problem. *Annals of the New York Academy of Sciences* 929: 123–146.

Singer, W. (2009). Distributed processing and temporal codes in neuronal networks. *Cognitive Neurodynamics* 3: 189–196.

Slaby, J. (2011). Perspektiven einer kritischen Philosophie der Neurowissenschaften. *Deutsche Zeitschrift für Philosophie* 59: 375–390.

Soccio, D. J. (2012). *Archetypes of wisdom: An introduction to philosophy*. London: Wadsworth.

Solms, M. (2004). Is the brain more real than the mind? In: A. Casement (Ed.) *Who owns psychoanalysis?*, pp. 323–342. London: Karnak.

Solms, M. (2013). The conscious Id. *Neuropsychoanalysis: An Interdisciplinary Journal for Psychoanalysis and the Neurosciences* 15: 5–19.

Soon, C. S., Brass, M., Heinze, H. J., Haynes, J. D. (2008). Unconscious determinants of free decisions in the human brain. *Nature Neuroscience* 11: 543–545.

Soon, C. S., He, A. H., Bode, S., Haynes, J. D. (2013). Predicting free choices for abstract intentions. *Proceedings of the National Academy of Sciences of the United States of America* 110: 6217–6222.

Soto-Faraco, S., Spence, C., Kingstone, A. (2004). Cross-modal dynamic capture: congruency effects in the perception of motion across sensory modalities. *Journal of Experimental Psychology: Human Perception and Performance* 30: 330–345.

Sowell, E. R., Thompson, P. M., Holmes, C. J., Jernigan, T. L., Toga, A. W. (1999). In vivo evidence for post-adolescent brain maturation in frontal and striatal regions. *Nature Neuroscience* 2: 859–861.

Spaemann, R. (2006). *Persons: The difference between 'someone' and 'something'*. Trans. O. O'Donovan. Oxford: Oxford University Press.

Sperry, R. (1980). Mind-brain interactionism: mentalism, yes; dualism, no. *Neuroscience* **5**: 195–206.

Spinoza, B. D. (2000). *Ethics*. Trans. G. H. R. Parkinson. Oxford: Oxford University Press.

Spitz, R. A. (1965). *The first year of life: A psychoanalytic study of normal and deviant development of object relations*. Oxford: International Universities Press.

Sporns, O., Tononi, G., Kötter, R. (205). The human connectome: a structural description of the human brain. *PLoS Computational Biology* **1**: e42.

Sprevak, M. (2011). Neural sufficiency, reductionism, and cognitive neuropsychiatry. *Philosophy, Psychiatry, & Psychology* **18**: 339–344.

Stamm, R., Bühler, K. E. (2001). Vulnerabilitätskonzepte bei psychischen Störungen. *Fortschritte der Neurologie· Psychiatrie* **69**: 300–309.

Stephan, A. (1999a). *Emergenz. Von der Unvorhersagbarkeit zur Selbstorganisation*. Dresden: Dresden University Press.

Stephan, A. (1999b). Varieties of emergence. *Evolution and Cognition* **5**: 50–59.

Stephens, G. J., Silbert, L. J., Hasson, U. (2010). Speaker–listener neural coupling underlies successful communication. *Proceedings of the National Academy of Sciences of the United States of America* **107**: 14425–14430.

Sterelny, K. (2010). Minds: extended or scaffolded? *Phenomenology and the Cognitive Sciences* **9**: 465–481.

Stern, D. N. (1985). *The interpersonal world of the infant: A View from Psychoanalysis and Developmental Psychology*. New York: Basic Books.

Stern, D. N. (1998). The process of therapeutic change involving implicit knowledge: Some implications of developmental observations for adult psychotherapy. *Infant Mental Health Journal* **19**: 300–308.

Stewart, J. R., Gapenne, O., Di Paolo, E. A. (2010). *Enaction: Toward a new paradigm for cognitive science*. Cambridge, MA: MIT Press.

Strack, F., Martin, L., Stepper, S. (1988). Inhibiting and facilitating conditions of the human smile: a non-obtrusive test of the facial feedback hypothesis. *Journal of Personality and Social Psychology* **54**: 768–777.

Straus, E. (1956). *Vom Sinn der Sinne*. Springer: Berlin. [Engl. trans. J. Needleman (1963). *The primary world of the senses: A vindication*. Glencoe, IL: Free Press.]

Strawson, G. (1994). *Mental reality*. Cambridge, MA: MIT Press.

Strawson, P. (1959). *Individuals: Introduction to logical theory*. London: Methuen.

Stroud, B. (2000). *The quest for reality: Subjectivism and the metaphysics of colour*. New York: Oxford University Press.

Sur, M., Rubenstein, J. L. (2005). Patterning and plasticity of the cerebral cortex. *Science* **310**: 805–810.

Sutton, J. (2015). Scaffolding memory: themes, taxonomies, puzzles. In: L. Bietti, C. B. Stone (Eds.) *Contextualizing human memory: An interdisciplinary approach to understanding how individuals and groups remember the past*, pp. 187–205. London: Routledge.

Swaab, D. F. (2014). *We are our brains: A neurobiography of the brain, from the womb to Alzheimer's*. London: Penguin.

Taipale, J. (2014). *Phenomenology and embodiment: Husserl and the constitution of subjectivity*. Evanston, IL: Northwestern University Press.

Tangney, J.P., Fischer, K.W. (Hrsg.) (1995). *Self-conscious emotions: The psychology of shame, guilt, embarrassment, and pride*. New York: Guilford.

Taylor, R. (1973). *Action and purpose*. New York: Prentice Hall.

Terhaar, J., Viola, F. C., Bär, K. J., Debener, S. (2012). Heartbeat evoked potentials mirror altered body perception in depressed patients. *Clinical Neurophysiology* **123**: 1950–1957.

Tetens, H. (1994). *Geist, Gehirn, Maschine. Philosophische Versuche über ihren Zusammenhang*. Stuttgart: Reclam.

Thinus-Blanc C., Gaunet, F. (1997). Representation of space in blind persons: vision as a spatial sense? *Psychological Bulletin* **121**: 20–39.

Thomas Aquinas (1953). *Summa Theologiae*. Ottawa: Commissio Piana.

Thompson, E. (1995). *Colour vision: A study in cognitive science and the philosophy of perception*. London: Routledge.

Thompson, E. (2005). Sensorimotor subjectivity and the enactive approach to experience. *Phenomenology and the Cognitive Sciences* **4**: 407–427.

Thompson, E. (2007). *Mind in life: Biology, phenomenology, and the sciences of mind*. Cambridge, MA: Harvard University Press.

Thompson, E., Varela, F. J. (2001). Radical embodiment: neural dynamics and consciousness. *Trends in Cognitive Sciences* **5**: 418–425.

Thompson, E., Zahavi, D. (2007). Philosophical issues: phenomenology. In: P. D. Zelazo, M. Moscovitch, E. Thompson (Eds.) *The Cambridge handbook of consciousness*, pp. 67–87. Cambridge: Cambridge University Press.

Tomasello, M. (1999). *The cultural origins of human cognition*. Cambridge, MA: Harvard University Press.

Tomasello, M. (2000). The social-pragmatic theory of word learning. *Pragmatics* **10**: 401–413.

Tomasello, M. (2008). *The origins of human communication*. Cambridge, MA: MIT Press.

Tomasello, M., Carpenter, M., Call, J., Behne, T., Moll, H. (2005). Understanding and sharing intentions: the origins of cultural cognition. *Behavioral and Brain Sciences* **28**: 675–735.

Tomasello, M., Carpenter, M., Liszkowski, U. (2007). A new look at infant pointing. *Child development* **78**: 705–722.

Trabant, J. (1991). Parlare Cantando: language singing in Vico and Herder. *New Vico Studies* **9**: 1–16.

Trevarthen, C. (1998). Language development: mechanisms in the brain. In: G. Adelman, B. Smith (Eds.) *Encyclopedia of neuroscience*, pp. 1018–1026. Amsterdam: Elsevier.

Trevarthen, C. (2001). The neurobiology of early communication: intersubjective regulations in human brain development. In: A. F. Kalverboer, A. Gramsberg (Eds.) *Handbook of brain and behaviour in human development*, pp. 841–881. Dordrecht: Kluwer Academic Publishers.

Trevarthen, C. (2012). Communicative musicality: the human impulse to create and share music. In: D. Hargreaves, D. Miell, R. MacDonald (Eds.) *Musical imaginations: Multidisciplinary perspectives on creativity, performance, and perception*, pp. 259–284. Oxford: Oxford University Press.

Trevarthen, C., Hubley, P. (1978). Secondary intersubjectivity: confidence, confiding and acts of meaning in the first year. In: A. Lock (Ed.) *Action, gesture and symbol: The emergence of language*, pp. 183–229. London: Academic Press.

Tronick, E., Als, H., Adamson, L., Wise, S., Brazelton, T. B. (1978). The infant's response to entrapment between contradictory messages in face-to-face interaction. *Journal of the American Academy of Child Psychiatry*, 17: 1–13.

Turati, C., Simion, F., Milani, I., Umiltà, C. (2002). Newborns' preference for faces: what is crucial? *Developmental Psychology* 38: 875–881.

Tye, M. (1995). *Ten problems of consciousness: A representational theory of the phenomenal mind.* Cambridge, MA: MIT Press.

Uexküll, J. von (1973). *Theoretische Biologie* (1st edition 1920). Frankfurt: Suhrkamp.

Uexküll, J. von, Kriszat, G. (1956). *Streifzüge durch die Umwelten von Tieren und Menschen. Bedeutungslehre.* Reinbek: Rowohlt.

Uexküll, T. von, Wesiack, W. (1996). Wissenschaftstheorie und Psychosomatische Medizin, ein bio-psycho-soziales Modell. In: T. von Uexküll, R. Adler (Eds.) *Psychosomatische Medizin*, 5th edition, pp. 1–30. Munich: Urban and Schwarzenberg.

Uhlhaas, P., Pipa, G., Lima, B., Melloni, L., Neuenschwander, S., Nikolić, D., Singer, W. (2009). Neural synchrony in cortical networks: history, concept and current status. *Frontiers in Integrative Neuroscience* 3: Art. 17.

Uhlhaas, P. J., Singer, W. (2006). Neural synchrony in brain disorders: relevance for cognitive dysfunctions and pathophysiology. *Neuron* 52: 155–168.

Ulrich, G. (1990). Was unterscheidet und was verbindet Somatotherapie und Psychotherapie? *Fundamenta Psychiatrica* 4: 132–136.

Umiltà, M. A., Kohler, E., Gallese, V. Fogassi, L., Fadiga, L., Keysers, C., Rizzolatti, G. (2001). I know what you are doing: a neurophysiological study. *Neuron* 31: 155–165.

Uttal, W. R. (2001). *The new phrenology: The limits of localizing cognitive processes in the brain.* Cambridge, MA: MIT Press.

Valenza, E., Simion, F., Cassia, V. M., Umiltà, C. (1996). Face preference at birth. *Journal of Experimental Psychology: Human Perception and Performance* 22: 892–903.

Van den Bergh, B. R. H., Marcoen, A. (2004). High antenatal maternal anxiety is related to ADHD symptoms, exernalizing problems, and anxiety in 8- and 9-year-olds. *Child Development* 75: 1085–1097.

Van der Kolk, B. A. (1994). The body keeps the score: memory and the evolving psychobiology of posttraumatic stress. *Harvard Review of Psychiatry* 1: 253–265.

Van Dijk, J., Kerkhofs, R., Van Rooij, I., Haselager, P. (2008). Special section: can there be such a thing as embodied embedded cognitive neuroscience? *Theory & Psychology* 18: 297–316.

Van Orden, G. C., Pennington, B. F., Stone, G. O. (2001). What do double dissociations prove? *Cognitive Science* 25: 111–172.

Varela, F. J. (1979). *Principles of biological autonomy.* New York: Elsevier (North Holland).

Varela, F. J. (1996). Neurophenomenology: a methodological remedy for the hard problem. *Journal of Consciousness Studies* 3: 330–349.

Varela, F. J. (1997). Patterns of life: intertwining identity and cognition. *Brain and Cognition* 34: 72–87.

Varela, F. J. (1999). Present-time consciousness. *Journal of Consciousness Studies* 6: 111–140.

Varela, F., Thompson, E., Rosch, E. (1991). *The embodied mind: Cognitive science and human experience.* Cambridge, MA: MIT Press.

Verwey, G. (1985). *Psychiatry in an anthropological and biomedical context.* Dordrecht: Reidel.

Vogeley, K., Bussfeld, A., Newen, A., et al. (2001). Mind reading: neural mechanisms of theory of mind and self-perspective. *Neuroimage* **14**: 170–181.

Vogeley, K., May, M., Ritzl, A., Falkai, P., Zilles, K., Fink, G. R. (2004). Neural correlates of first-person-perspective as one constituent of human self-consciousness. *Journal of Cognitive Neuroscience* **16**: 817–827.

Vollm, B. A., Tayler, A. N., Richardson, P., Corcoran, R., Stirling, J., McKie, S., Deakin, J. F., Elliott, R. (2006). Neuronal correlates of theory of mind and empathy: a functional magnetic resonance imaging study in a nonverbal task. *Neuroimage* **29**: 90–98.

Vollmer, G. (1975). *Evolutionäre Erkenntnistheorie. Angeborene Erkenntnisstrukturen im Kontext von Biologie, Psychologie, Linguistik, Philosophie und Wissenschaftstheorie.* Stuttgart: Hirzel.

Vollmer, G. (2005). How is it that we can know this world? New arguments in evolutionary epistemology. In: V. Hösle, Illies, C. (Eds.) *Darwinism & Philosophy*, pp. 259–274. Notre Dame, IN: University of Notre Dame Press

von Senden, M. (1960). *Space and sight: The perception of space and shape in the congenitally blind before and after operation.* London: Methuen & Co.

Vrtička, P., Andersson, F., Sander, D., Vuilleumier, P. (2009). Memory for friends or foes: the social context of past encounters with faces modulates their subsequent neural traces in the brain. *Social Neuroscience* **4**: 384–401.

Vrtička, P., Bondolfi, G., Sander, D., Vuilleumier, P. (2012). The neural substrates of social emotion perception and regulation are modulated by adult attachment style. *Social Neuroscience* **7**: 473–493.

Vul, E., Harris, C., Winkielman, P., Pashler, H. (2009). Puzzlingly high correlations in fMRI studies of emotion, personality, and social cognition. *Perspectives on Psychological Science* **4**: 274–290.

Wakefield, J. (2014). Wittgenstein's nightmare: why the RDoC grid needs a conceptual dimension. *World Psychiatry* **13**: 38–40.

Waldenfels, B. (2000). *Das leibliche Selbst. Vorlesungen zur Phänomenologie des Leibes.* Frankfurt: Suhrkamp.

Waldenfels, B. (2002). *Bruchlinien der Efahrung. Phänomenologie, Psychoanalyse, Phänomenotechnik.* Frankfurt: Suhrkamp.

Weber, A., Varela, F. H. (2002). Life after Kant: natural purposes and the autopoietic foundations of biological individuality. *Phenomenology and the Cognitive Sciences* **1**: 97–125.

Weinberger, D. R. (1987). Implications of normal brain development for the pathogenesis of schizophrenia. *Archives of General Psychiatry* **44**: 660–669.

Weiss, M., Gaston, L., Propst, A., Wisebord, S., Zicherman, V. (1997). The role of the alliance in the pharmacologic treatment of depression. *Journal of Clinical Psychiatry* **58**: 196–204.

Weizsäcker, V. von (1986). *Der Gestaltkreis. Theorie der Einheit von Wahrnehmen und Bewegen*, 5th edition. Stuttgart: Thieme.

Wernicke, C. (1874). *Der aphasische Symptomencomplex: eine psychologische Studie auf anatomischer Basis.* Breslau: Cohn.

White, P. D., Rickards, H., Zeman, A. Z. J. (2012). Time to end the distinction between mental and neurological illnesses. *British Medical Journal* 344: e3454.

Wicker, B., Keysers, C., Plailly, J., Royet, J. P., Gallese, V., Rizzolatti, G. (2003). Both of us are disgusted in my insula: the common neural basis of seeing and feeling disgust. *Neuron* 40: 644–655.

Wiens, S. (2005). Interoception in emotional experience. *Current Opinion in Neurology* 18: 442–447.

Willi, J. (1999). *Ecological psychotherapy: Developing by shaping the personal niche*. Seattle, WA: Hogrefe & Huber.

Williams, L. E., Bargh, J. A. (2008). Experiencing physical warmth promotes interpersonal warmth. *Science* 24: 606–607.

Wilson, E. O. (1998). *Consilience: The unity of knowledge*. New York: Random House.

Wimmer, H., Perner, J. (1983). Beliefs about beliefs: representation and constraining function of wrong beliefs in young children's understanding of deception. *Cognition* 13: 103–128.

Windt, J. (2015). *Dreaming: A conceptual framework for philosophy of mind and empirical research*. Cambridge, MA: MIT Press.

Wittgenstein, L. (1958). *The blue and brown books*. Oxford: Blackwell.

Wittgenstein, L. (1961). *Tractatus logico-philosophicus*. Trans. D. F. Pears and B. F. McGuinness. London: Routledge and Kegan Paul.

Wollmer, M. A., de Boer, C., Kalak, N., et al. (2012). Facing depression with botulinum toxin: a randomized controlled trial. *Journal of Psychiatric Research* 46: 574–581.

Wolpert, D. M., Miall, R. C., Kawato, M. (1998). Internal models in the cerebellum. *Trends in Cognitive Sciences* 2: 338–347.

Yim, H., Dennis, S. J., Sloutsky, V. M. (2013). The development of episodic memory items, contexts, and relations. *Psychological Science* 24: 2163–2172.

Zahavi, D. (1996). Husserl's intersubjective transformation of transcendental philosophy. *Journal of the British Society for Phenomenology* 27: 228–245.

Zahavi, D. (1999). *Self-awareness and alterity: A phenomenological investigation*. Evanston, IL: Northwestern University Press.

Zahavi, D. (2003). Intentionality and phenomenality. In: E. Thompson (Ed.) *The problem of consciousness: New essays in the phenomenological philosophy of mind*, pp. 63–92. Calgary: University of Calgary Press.

Zahavi, D. (2005). *Subjectivity and selfhood: Investigating the first-person perspective*. Cambridge, MA: MIT Press.

Zahavi, D. (2017). Thin, thinner, thinnest: defining the minimal self. In: C. Durt, C. Tewes, T. Fuchs (Eds.) *Embodiment, enaction and culture: Investigating the constitution of the shared world*, pp. 193–199. Cambridge, MA: MIT Press.

Zeki, S. (1992). The visual image in mind and brain. *Scientific American* 267: 69–76.

Zhong, C. B., Leonardelli, G. J. (2008). Cold and lonely: does social exclusion feel literally cold? *Psychological Science* 19: 838–842.

Zubin, J., Spring, B. (1977). Vulnerability—a new view of schizophrenia. *Journal of Abnormal Psychology* 86: 103–126

Index

Note: page numbers followed by *n* indicate footnotes.

action prediction 54–5
act of life *see* life acts
adaptivity 86, 224
affect, basic 110, 113–14, 115, 126
affection–resonance–emotion cycle 122–5
affective neuroscience 110–11, 118
affectivity, embodied 120–5
affordance 88, 155
agent causation 241–2
Akil, H. 254
alien limb phenomenon 156
Altenmüller, E.O. 131–2
Alzheimer's disease 246, 253
amygdala 121, 182, 235–6, 246, 261, 268
Anderson, M.L. 50
Andreasen, N.C. 254
angel 230, 230*n*
anosognosia 156
anterior cingulate cortex 149, 184–5, 189
anticipation 55–6, 86, 92, 97*n*, 128, 128*n*, 150
antidepressant 245–6, 270, 271*n*
anxiety/fear 217, 235–6, 235*n*, 244, 246, 258, 266, 268
apes, great 173–4
Aquinas, St Thomas 17*n*
Arbib, M.A. 198
area postrema 112, 112*n*
Aristotle 17*n*, 18*n*, 20*n*, 57*n*, 99–100, 99*n*, 151, 166–7, 175
arousal 124*n*
artificial intelligence machines 214
aspect dualism 80
aspect-duality *see* dual aspectivity
atomism 5
attachment 183–6, 264, 273
attention 98, 117, 148, 149, 152, 153
  direction 203–4
  joint 193–4
audiomotor mirror neurons 198
auto-affection 33, 72, 116*n*
autobiographical memory 101, 180
autobiographical self 114–15
autoepistemic closure 162
autoepistemic limitation 161
automatic and controlled processes 206*n*
autonomous nervous system 122*n*

autonomous regulatory processes 109–10, 116, 120
autonomy 84–5, 90
autopoiesis 84, 85, 89, 247
Aziz-Zadeh, L. 199*n*

baby talk 179
Bangert, M. 131–2
being-towards-the-world 70, 73, 89, 116, 215, 273
Bennett, M.R. 39*n*, 44, 45
Bergson, H. 55
Bieri, P. 209–10, 211, 232
binding problem 20*n*, 47, 52, 165, 225*n*
biological system 41, 82, 84, 183, 244
biopsychosocial model 255
Blakemore, C. 58, 254
Blakeslee, S. 11
blind man's stick 15*n*
bodily self 91, 93
bodily subjectivity 70
body 68, 72*n*
  connection with brain 109–11
  contemporary philosophy 78–9
  dual aspect 74–7
  physical *see* physical body
  subjective (lived) *see* subjective (lived) body
body-as-object 80, 82, 117–18, 211
  see also objective body; physical body
body-as-subject 70–3, 80, 82, 117–18, 211
  see also lived body; subjective body
body image 16
body memory *see* implicit memory
body schema 11, 15, 16, 18*n*, 132, 144–5, 178, 188
BOLD (blood oxygen level-dependent) signal 49
bottom-up (upward) causality 96, 97, 98, 255–6
  effects of somatic/drug therapy 268, 270, 272*n*, 273–4
  psychosomatic interrelations 244
  somatopsychic interrelations 245
botulinum toxin 125
Bowlby, J. 183
Braille 144, 145

brain
    ecological conception xviii, 68, 107–71, 166
    embeddedness 109–26
    functional specialization 46–52
    higher functions 147–8
    lesions/dysfunctions 49, 53, 142, 246–7
    organ of potentialities 234, 249, 279, 289–90
    personalization of 43–5
    size, evolution 173–4
    as subject 29–66, 219
    unity with organism and environment 126–67
brain development 174–5
    early attachment and 184–6
    incorporation of experience 139, 140–2, 142n
    prenatal 176–7
    socially and culturally scaffolded 173–208
brain-in-a-vat 135–6, 135n, 163
brainstem 109, 110, 111–12, 113–14, 118
Broca's area 46, 187n, 197, 198, 199
Bunge, M. 220–1n

Cameron, O.G. 113n, 120
capacity 99–105
Cartesian central module xiv, 155, 218–19
Cartesian dualism 55, 73, 137, 211–12, 230n, 280n
Cartesianism 8, 13–14, 32, 45, 111
Cartesian materialism 26
cataract, early 10
category formation 151, 151n
category mistake 17n, 20, 43–52
Catmur, C. 190n
caudate nucleus 261, 272
causal closure 59, 61, 210, 233n
causality 94–104
    bottom-up (upward) see bottom-up causality
    circular see circular causality
    downward/top-down see top-down causality
    integral 99–101, 231n, 240–3
    linear 127–8
causation, mental 225–6
cavemen 4, 163
central nervous system (CNS) 87, 90–2, 90n
Chalmers, D.J. 37, 220–1
children
    attachment 183–6
    cultural influences 205–8
    language development 194–205
    see also brain development; infants
Chinese Room 38–9
choice reaction task 54
cingulate gyrus/cortex 121, 149, 184–5, 189, 261
circular causality 94–9, 104, 223–4
    horizontal see horizontal circular causality
    integration and formation of capacities 99–104

linear causality versus 127–34
    in mental disorders 255–62
    in pathogenesis 262–8
    in therapy 268–75
    vertical see vertical circular causality
closure 156
co-emergence 225, 248
    dynamic 94, 222
co-evolution 103–4, 223n
coexistence 22, 168
coextension 12–19, 76, 214, 288–9
cognition 91, 108, 281
cognitive behavioral therapy 271–2, 272n
cognitive neuroscience 38, 43, 62, 65–6, 111, 157, 206n
cognitivism 47–8, 108, 182
coherence 149, 156, 167
    intermodal 15, 15n, 144
    optimal 15, 143–4, 145, 155–6, 169
    situational 101
Colombetti, G. 120, 122, 124n
color 23–5, 66, 99, 151, 160
communication 178–80, 194–205, 273
communicative musicality 197–8
compulsions 53
computational theories 110n
computer 40, 53, 55, 110, 135, 139, 157–8
conation 71, 71n, 72, 91, 111, 113, 116, 153
condition humaine, La (Magritte) 6–8
conditioning 146, 224–5, 244
connectionist models 110, 110n
consciousness xiv, 45–6, 281
    basal levels 110, 110n, 111–14, 115, 126
    capacities/dimensions 60, 227–8
    dynamic core hypothesis 47n, 59, 218
    ecological model 127–9, 134–5
    embodied see embodied mind
    emergence 220–5, 248
    extended see extended mind
    hierarchy 116–17
    higher levels 114–20
    idealist conception 5–8
    as integral 119, 134–9, 226–7
    intentionality see intentionality
    localization 46–52, 218–19, 219n
    modularity 46, 47–8
    neural correlates 48–9, 136
    neuroconstructivist thesis 30, 67
    phenomenal 32–6, 57–8
    primary, vital or interoceptive 111–14
    relation to environment 127
    role of 57–61, 227–8
    unity of action 52–7
    see also mind
core self 114–15, 115n, 116, 117
corollary discharge 92
cortex, cerebral 115–16, 141–2, 142n
Cosmelli, D. 48, 135n, 152
Crick, F. 3

culture 143, 175–6, 205–8, 265
cybernetic system 39, 39n, 41
cycle of embodied affectivity 122–5

Dalmatian image 148–9, 150, 151, 152, 155, 160
Damasio, A. 3, 14n, 30, 111–15, 111n, 116, 118–19, 121–2, 121n, 239n
Davidson, D. 159, 193, 221
Deacon, T. 96, 96n, 97n, 222, 224
decision making 53–7, 237–40
default mode 50
delusion 156
Democrit 5
Dennett, D. 29n
depression 246, 254
  circular processes 257, 257n, 258, 260
  pathogenesis 262–3, 264–6, 266–7
  therapy 245–6, 272, 275
Descartes, R. xiv, xv, 5, 14n, 18, 18n, 30, 73, 280n, 282, 286
determinism 53–4, 237n, 242–3
development, brain see brain development
Dewey, J. 129–30
dialogical thought 203
diamond 96n
diencephalon 109–11, 119, 141
disgust 189–90
doctor–patient relationship 12–13, 75, 269, 271, 289
dorsolateral prefrontal cortex 204
double sensations (Doppelempfindungen) 74
downward causality see top-down causality
dreaming 136–7n, 150
Dretske, F. 39n
Dreyfus, H. 134, 166–7
drug therapy 245–6, 268–71, 272n, 273–5
dual aspectivity 73–83, 137–9, 209–49
  downward causality and 225–32
  life 77–83, 282
  psychophysical relations and 232–47
  subjective and physical body 74–7, 279–80
  therapy of mental illness 268–75
dualism, Cartesian 55, 73, 137, 211–12, 230n, 280n
Dunbar, R.I. 174
dynamical systems theory 109, 149, 156, 223–4
dynamic co-emergence 94, 222
dynamic core hypothesis 47n, 59, 218
dýnamis 99–100, 99n, 101

Eagleman, D. 3, 11, 29
eccentric positionality 75–6, 80, 92, 280, 289
Eckhart, Meister 17n
ecological conception
  brain xviii, 68, 107–71, 166
  living organism xix, 83–92, 247
  psychiatry 255–77

ecological model of consciousness 127–9, 134–5
ecological niche 87, 174, 222n, 229n
ecological theory of perception 10n, 165n
Edelman, G.M. 42n, 47n, 58–9, 133n, 140, 150n, 154, 166n, 218, 219n
efference copy principle 92
ego 45, 55
egocentric perspective 204
Ego Tunnel 4, 31
electrical stimulation 50–1, 51n
eliminative materialism xvi, 210n
embeddedness 15, 108, 109–26, 136
embodied affectivity 120–5
embodied freedom 236–43
embodied intersubjectivity 77, 175, 192–4
embodied memory 115
  see also implicit memory
embodied mind 9, 108–9, 282, 289–91
  capacities of action 56–7, 101, 128
  language development 194–205
  mental disorders 255–62
  objections to 119
  phenomenology 69–83
embodied perception 8–19, 26–7
embodied socialization 207
embodied social perception 191
embodied subjectivity 19, 138, 211, 213, 288–9
  emergence 113, 229
  phenomenology 69–83
emergence 219–32, 248
  diachronic 222–3
  morphodynamic 224
  strong 220, 221–5
  teleodynamic 224
  weak 220–1
emotional body memory 122, 239
emotions 91, 115–16, 117, 120–5, 126
  amplifying learning 141
  cognitive or appraisal theories 124n
  embodied 120–5
  mother–child dyad 179
  primary and secondary 121–2
  psychosomatic interrelations 244
  role in perception 149
  social resonance 189–90, 191
empathy 189, 189n, 191
enactivism 83–94, 108–9
  culture 207–8
  language development 200
  mental disorders 255–62
  perception 9, 26, 130, 131
  social cognition 182
environment 68
  circular relations 98–9
  organism's relationships 86–9, 93, 103–4, 127–34
  role in learning 100, 101, 139–45
  unity with brain and organism 126–67

epigenetics 140–1, 174
epileptic aura 50
epiphenomenalism 210n, 221, 229–30
  biological 227, 249
  conception of subjectivity 29, 32, 57, 58n, 59
equipotential theory 46
Erleben xix, 31, 78, 78n, 94
être-au-monde see being-towards-the-world
evolution 58–9, 173–4, 228–9, 229n
evolutionary epistemology 286
executive control 204
experience, incorporation of 139–45, 168–9, 215
explanatory gap 35n
explicit learning 206–7
extended mind 108, 167–8, 207–8
externalism 159n
external world 6–9
eye movement 92n, 130, 130n

face recognition 73, 146
fear *see* anxiety/fear
feeling of being alive 72, 111–14
ferrets, newborn 141–2, 142n
fetus 176–7
Feuerbach, L. 67, 139–40
Fichte, J.G. 6
first-personal experiential givenness 33–4
first-person perspective 33–5, 81–2, 213
flash-lag effect 56n
flight simulator 30
Flourens, P. 46
fMRI *see* functional magnetic resonance imaging
formative causality 95, 141
4e cognition 108n, 109
Freeman, W. 226
free will xiii–xiv, 53–7, 58, 61, 231n, 236–43
Freud, S. 202, 235n
frontal brain injuries 53
frontopolar cortex 204
function, primacy of 221–3, 228, 249
functional cluster 47, 154
functional cycle 88–9, 93, 98–9, 128–9, 131–2, 167–8
functional failure 49, 246–7
functionalism 37, 38–9, 110, 110n
functional magnetic resonance imaging (fMRI) 49–50, 54–5, 199, 236n, 252
functional specialization 46–52, 118

Galilei, G. xv, 5
Gall, F. J. 46
Gallagher, S. 54n, 135n, 154n
Gallese, V. 186, 187n, 190
gaze 92n
Gazzaniga, M.S. 44
genetic factors 264

gestalt 146–7, 169
  formation 148–9, 150–3
  movement 154
  perceptual 21–2, 28, 105, 145
  see also Vorgestalt
ghost in the machine 17, 31
Gibson, J.J. 10n, 165n
global-to-local causality *see* top-down causality
goal-directed movements 188, 198
Goethe, J.W. von 6, 140n
Goldman, A. 190
Graham, G. 257n
Gregory, R.L. 143
Griesinger, W. 251

Hacker, P.M.S. 39n, 44, 45
hemoglobin 84, 96, 96n
Haggard, P. 53
Haller, A. 46
hallucination 261, 261n
Hardcastle, V.G. 236n
Hartmann, D. 62, 63–4
Haynes, J.D. 54–5
Hebb's rule 141, 150
Hegel, G.W.F. 22n, 80, 107
Heidegger, M. 128
Heidelberg School 33n
Hein, A. 9, 10
Held, R. 9, 10
Henry, M. 33, 33n, 72, 116n
hermeneutical approach 235–6, 277, 284
Herrmann, C.S. 54
hippocampus 142, 177, 186
Hobbes, T. 71n
Holtzheimer, P.E. 254
homeostasis 98, 110, 112, 116
homunculus 42, 42n, 160, 161
horizontal circular causality 94, 95, 98–9, 104, 255
  integration and formation of capacities 99–104
  linear relations vs 127–34
  mental disorders 256, 259–60, 261–2
Hume, D. 6, 170
Husserl, E. 12, 20, 21, 55, 74, 75, 76n, 168, 287n
hydranencephaly 113–14
hypothalamus 109, 112

idealism 5–8, 168
identity theory 17n, 80, 216–19, 248
illusion 11, 15n, 21, 143
image schema 115, 128, 149–50
imaging, brain *see* neuroimaging
imitation 177–8, 178n, 188, 189–90, 206
  reverse 195
implicit coupling 146, 147, 150–1, 169
implicit learning 206

implicit (body) memory 115, 144–5, 290
  early infancy 180–1
  emotional 122, 239
  formation of capacities 101–4
  open loop formation 131–2
  relational 181, 273
impulse control 203–4
impulse control disorder 266
infants
  9-month revolution 192–4
  attachment 183–6
  primary intersubjectivity 177–81
  secondary intersubjectivity 192–205
  see also brain development; children
information 156–9, 288
information processing 39, 40, 157–8, 157n, 158n
inhibition 90, 203–5, 238, 289–90
inner milieu 109–11, 112
inner world xiv–xv, 6–8, 9, 11–12, 25
insula 51, 121, 189
insult 235
integral, consciousness as 119, 134–9, 226–7
integral causality 99–101, 231n, 240–3
intentionality 36–43, 132–3, 133n, 217, 281
  determining physiological processes 233–6
  infants' understanding of 193–4
  phenomenal consciousness and 37–8
  representation and 38–43
interaffectivity 177–80, 185, 189
intercorporeality 77, 177–80, 189, 291
intercorporeal memory 180–1
internalization 181, 181n
interoception 112, 113, 113n, 120, 124
interrelations, organ of xviii, 68
intersubjectivity 13, 21, 27
  embodied 77, 175, 192–4
  implicit 170–1
  neurobiology of early 182–92
  open 21, 27
  primary 176–81
  secondary 192–205
intuitive parental competence 179
involuntary action 71
iron 84, 96
isolation 253

Jackson, J. H. 261n
James, W. 55, 72n, 117n, 121, 124
Jaspers, K. 252
Jeannerod, M. 54, 153–4
joint attention 193–4
Jonas, H. 10n, 39n, 59, 89–90, 90n, 157

Kant, I. 6, 17–18, 85, 286
keyboard user see pianist; typist
kinetic melody 154, 155
kittens, newborn 9, 10

Körper see physical body
Kronz, F.M. 222
Krystal, J.H. 254

language 194–205, 215
learned helplessness 262
learning 101–4, 139–45, 188, 206–7, 224–5
Leben xix, 31, 78, 94
Leib see lived body
Leib-Körper problem 83, 211
Lenin, V. 65
Levine, J. 35n
libertarianism 237–43
Libet, B. 53–5, 227n, 237, 242
Lichtenberg, G. C. 52, 286
life xviii, 70–1, 89–90, 281
  dual aspect 77–83, 282
  precariousness 86
life acts 81, 82, 99, 100
  restriction or loss of 246–7
  unity of 210–11, 219
lifeworld xiii–xiv, 25, 62–6, 281
limbic system 101, 110, 116
linear causal model 127–8, 134
lived body (Leib) xix, 70, 72–3, 139, 282, 288–90
  coextensivity 12–19, 76, 288–9
  dual aspect 74–7, 80, 83
  mental illness 257, 257n
  relations with environment 133, 134
  see also body-as-subject; subjective body
lived body–living body problem 83, 211
living being 79–80, 213
  acts of see life acts
  brain as organ of xviii, 68, 107–71
  circular and integral causality 94–105
living body 17, 80, 82–3, 93–4, 100n
living organism
  brain in context of 109–26
  dual character 75, 77–8
  ecological conception xix, 83–92, 247
  environmental relationships 86–9, 93, 103–4, 127–34
  indivisibility 18, 18n
  physical body as 82
  self-organization and autonomy 84–5
  specific causality 94–104
  subjectivity 89–92, 93–4
  unity of 80, 83, 210–12, 218–19
  unity with brain and environment 126–67
localization fallacy 46–52, 246
local-to-global causality see bottom-up causality
Locke, J. 5, 6, 238n
locked-in syndrome 120
long-term potentiation 141

Mach, E. 80
macro-level interactions 255–6

Magritte, R. 6–8, 9
materialism xv, xvi, 8, 29
  Cartesian 26
  eliminative xvi, 210*n*
Mayberg, H.S. 254
McCulloch, G. 159
meaning 36*n*, 37–8, 158, 159
  attributed to environment 88, 90, 99, 100–1, 158
  coupling of 146, 244
  gestalt 145, 146, 149
medial prefrontal cortex 182
mediated immediacy 22, 144, 146, 164, 167–71
mediation, organ of xvii–xviii, 68, 131
memory
  autobiographical 101, 180
  implicit (body) *see* implicit memory
  intercorporeal 180–1
  of past states 224
  procedural 101, 145
  representational view 42, 42*n*, 102
  sequence 132
mental causation 225–6
mental disorders xiii, 124, 251–77
  circular causality in pathogenesis 262–8
  as circular processes 255–62
  circular processes in therapy 268–75
  neuroreductionist views 251–5
  role of subjectivity 275–7
mentalizing 179–80
mental representation 38, 40–1
mental state 36, 45, 217
mereological fallacy 43–6, 55, 64
mereology 44*n*
Merleau-Ponty, M. 16, 63*n*, 64*n*, 70, 72, 72*n*, 73*n*, 74–5*n*, 77*n*, 116*n*, 132, 178, 188, 191, 200, 281
meso-level interactions 255, 256
metabolism 86, 98
metaphysical realism 62–3
Metzinger, T. xiv, 4, 23, 30–2, 40–1, 41*n*, 162–4, 164*n*, 285*n*
Meynert, T. 252
micro-level brain processes 243, 255–6
Millikan, R. 41*n*
mind
  brain as organ of 207–8
  contemporary philosophy of 78–9
  embodied *see* embodied mind
  extended 108, 167–8, 207–8
  indivisibility 18, 18*n*
  see also consciousness
mind–body problem 78, 82–3, 209–11
mind-reading 179–80
mirror neurons 186–92, 198–200
modularity 46, 47–8
molecular processes 96, 243, 252

Mooney figures 152*n*
mother–infant dyad 176–9, 183–6, 193
  see also infants
motivation 110, 113, 116, 149
motor action 153–5
motor habitual learning 101, 144–5
motor imagery 54, 153–4
movement 53–7, 94–5, 98–9, 154–5
  mirror neurons and 186–92, 198
  perception and 8–12, 90
Müller–Lyer illusion 143
music 131–2, 144–5, 197–8, 197*n*

Nagel, T. 33, 280
naive realism 4–5, 25, 162–3
National Institute of Mental Health 254
naturalistic (objectifying) attitude 75, 138, 213–14, 247, 285–91
  see also third-person perspective
naturalistic fallacy, second 63–5, 63*n*
natural pedagogy 193
negation 203
neglect phenomena 156
Nemiah, J.C. 277
neocortex 110, 141–2, 173
networks, neural 44, 51–2, 51*n*, 141, 175
  artificial computer-based 110*n*, 150*n*
  emergent properties 221*n*, 224–5
  functional clusters 47*n*, 154
neural correlates of consciousness 48–9, 136
neural Darwinism 140
neural resonance system 187–92, 198–200
neurobiology xiii–xiv, xv, xvii
  development of early intersubjectivity 182–92
  functional specialization 46–52
  idealistic legacy 5–8
  language development 196–201
  mental disorders 251–5
  problems posed by 62–6
  scope of research 283–5
  see also reductionism
neurocognitivism 38*n*, 43, 61, 160, 165
neuroconstructivism 3–4, 26, 30, 67, 137, 286
  critique 9–25, 26–8, 170
  idealistic theory 5–8
neuroimaging 3, 46, 47, 49–50, 252
neuronal representation 8, 37, 38, 40, 42, 127
neuroplasticity 49, 131–2, 139–45, 168–9, 196–7
neurosolipsism 13, 163
newborn babies 176–9
Nietzsche, F. 6
Noë, A. 15*n*, 48, 130–1, 159
Northoff, G. 44, 161
nucleus accumbens 101
nucleus parabrachialis 112, 116
nucleus tractus solitarii 112, 116

object-directed movement 188
objectifying attitude *see* naturalistic attitude
objective body 12, 75*n*, 82
  see also body-as-object; physical body
observer perspective *see* third-person perspective
obsessive-compulsive disorder (OCD) 261, 266, 271–2
open loops 102, 147, 149, 167–9
  capacities as 101, 105
  forming functional cycles 129, 131–2
orbitofrontal cortex 182, 184, 185, 204
O'Regan, J.K. 130–1
organic body *see* physical body
organism *see* living organism
orphans 185–6

pain 76, 98, 142*n*
  embodied perception 11, 12–13, 14–17, 14*n*
  fMRI studies 236*n*
  functional definition 37
  identity theory 17*n*
  localization 51–2
  social resonance 189
  subjectivity 34–5
Panksepp, J. 111, 113, 114, 118
paraplegia 120, 122*n*
parietal lobe 11, 121, 132, 156, 189
Parkinson's disease 124–5
participant perspective *see* second-person perspective
Parvizi, J. 50, 51, 51*n*
pattern 43, 148–9
  formation 147, 150–1
  resonance 43, 148, 149, 165, 169
Pavlov, I. 146, 224, 244
Peirce, C.S. 40*n*
Penfield, W. 50–1
perception
  circular causality 98–9
  color 23–5
  conscious 148–53, 154–5
  embodied 8–19, 26–7
  functional cycle 128–9
  idealistic theory 5–8
  implicit knowledge and 145
  intermodal 20, 20*n*
  localization 46, 48–9
  motion and 8–12, 90
  objectivity 20–2, 170–1
  reality 3–5
  reformulation of Aristotelian theory 167
  sensorimotor theory 130–1
  space of 19–20
periaqueductal gray 112, 113, 116
peripheral nervous system 14, 19, 100, 120, 136
periphery/peripheral body 14, 14*n*, 19, 51, 87, 93, 98–100, 109, 113, 118, 126, 134–6, 152, 154, 164

person 215–16, 279–83
  organ of 68, 173–208
personalistic attitude 75, 138, 213–14, 247, 285–91
  see also first-person perspective; second-person perspective
personal predicates 81
perspective-taking 174, 202–5
Pessoa, L. 50
phantom limb 11, 16
phantom pain 12, 16, 142*n*
pharmacological therapy *see* drug therapy
phenomenology xviii, 18, 118, 215, 288–9
  consciousness 32–6, 58–9
  decision-making 237–40
  embodied subjectivity 69–83, 118
  intentionality 37–8
  perception 19–22
  primary intersubjectivity 176–81, 191–2
  psychopathology 277
  reality 3–5
  secondary intersubjectivity 192–6
  transparency 146, 148, 161–2, 164, 191
phenospace 12–13
phrenology 46, 51*n*
phylogenetic history 229
physical, defined 77*n*
physical body *(Körper)* xix, 11–19, 81–2, 139, 289
  coextensivity 12–19, 76, 288–9
  dual aspect 74–7, 82–3
  see also body-as-object; objective body
physicalism 17, 24–5, 28, 58, 62–3, 65, 283
phýsis 82, 83, 83*n*
pianist 131–2, 144–5
placebo effect 271, 271*n*
plants 86, 87
plasticity *see* neuroplasticity
Plato 4, 163, 203
Plessner, H. 22*n*, 75–6, 170, 280
pointing 193, 195, 196, 198–9, 200
point of conversion (body) 74, 75
Polanyi, M. 124, 146
positron emission tomography (PET) 271–2, 271*n*, 272*n*
potentialities, organ of 234, 249, 279, 289–90
powerless subject 52–61
preconception *see* Vorgestalt
predictive brain 61, 97*n*
predictive coding 152*n*
prefrontal cortex 121–2, 175, 177, 204–5
pre-gestalt *see* Vorgestalt
premotor cortex 131, 154, 187, 198, 201*n*
prenatal development 176–7
pre-reflective self-awareness 33, 33*n*, 34*n*, 70*n*, 115
presentation 55–6, 133
primary intersubjectivity 176–81

Prinz, W. 19–20
prism goggles 143–4, 144*n*
process and structure, reciprocity of 139–40, 215, 225
programme
  genetic 64, 100
  neural 58, 100, 134, 154
protention 55–6, 60, 97, 97*n*
protoself 111–12, 114, 115*n*, 116, 117, 118–19
psychiatry 251–77
psychoanalysis 181
psychological medicine 251–77
psychosomatic interrelations 243–7
psychotherapy 269, 271–5
psychotropic drug therapy *see* drug therapy

qualia 35, 35*n*, 37

Ramachandran, V.S. 11, 156
readiness potential 53–4, 55, 57, 154
reading 145
reality 3–5, 9, 26–7, 52–3
reductionism xiv–xvii, 5, 29–32, 282–3
  cognition 108
  color perception 24–5
  consciousness 52–3, 57, 64–5, 73
  mental disorders 251–5, 265
  subjectivity problem 32–43
  see also neurobiology
reflection 170, 203, 238, 241
reflex arc 129–30
reification 253
relationality 284
representandum 37
representation 108, 111*n*, 118–19, 156, 160–4, 165–6
  functional-biological perspective 133–4, 137
  idealistic theory 5–8
  intentionality and 38–43
  memory 42, 42*n*, 102
  mirror neuron system 190–1
  see also mental representation; neuronal representation
representatum 37
Research Domain Criteria 254
resonance 118–19, 126, 134–5, 147
  emotions and 122, 123–5, 124*n*, 244
  external 152, 165
  intercorporeal 178, 189
  internal 152, 165
  language acquisition 198–200, 215
  loss of 124–5, 260
  organ of 145–56, 166
  pattern 43, 148, 149, 165, 169
  patterns and 164–7
  social (neural) 186–92, 198–200
  visual perception 149, 151–2, 167–8

retention 55–6, 60
Ricoeur, P. 35*n*
Rizzolatti, G. 186, 187*n*, 198
Rockwell, W.T. 26, 51–2
Rosen, R. 84
rubber hand illusion 11, 15, 15*n*
Rubenstein, J.L. 141–2, 142*n*
Ryle, G. 17, 31, 212*n*

sandwich model 127–8
schizophrenia 53, 156, 254, 257, 257*n*
Schmitz, H. 34*n*
Schopenhauer, A. 6, 161*n*
Schore, A.N. 184–5
Searle, J. xvi, 16, 36*n*, 38–9, 45, 135*n*, 168, 220, 223
secondary intersubjectivity 192–205
second-person (participant) perspective 63, 65–6, 81, 213–14, 281
  see also personalistic attitude
seeking system 113, 149
self 30–2, 85, 89–90, 219
  autobiographical 114–15
  bodily 91, 93
  core 114–15, 115*n*, 116, 117
  embodied 32*n*
  phenomenal 285*n*
self-affection 33–4, 60, 112, 116, 120
self-as-object 117*n*
self-as-subject 117*n*
self-awareness 31–2, 34, 60, 218–19
  see also pre-reflective self-awareness
self-experience 257–8
self-model theory 30, 162, 163, 164*n*, 285, 285*n*
self-movement 10*n*, 71
self-organization 84–5, 222–3, 247
self-relationship 257–8, 259
self-withdrawal 70
Seligman, M.E.P. 262
Selimbeyoglu, A. 50, 51, 51*n*
semantics 36*n*, 38–9
sense-making 26, 86, 91, 131
sensorimotor cycle 90–1, 128, 142*n*
sensorimotor interaction 9, 27, 115, 130–4
sensorimotor interface 87–8
sensorimotor resonance 188, 198
sensory cortex *see* somatosensory cortex
sensus communis 20, 20*n*, 60, 151, 167
separation behavior 183, 185
sexual arousal 120
shame 123, 244
Shannon–Weaver model 157*n*
sign 40, 40*n*
simulation theory 190
Singer, W. 97*n*, 150, 151–2, 225*n*
smiling, inhibited 125
Smith, A. 197*n*
social brain hypothesis 174

social desynchronization 263
social interaction xviii, 26–7, 173–208
  mental disorders 260, 264–5
  neurobiology 182–92
  primary intersubjectivity 176–81
  secondary intersubjectivity 192–205
socialization, embodied 207
social (cognitive) neuroscience 65–6, 175, 182–92
social referencing 193–4
social resonance system 186–92, 198–200
Soemmerring, S.T. von 52
Solms, M. 111, 114, 116
somatic markers 122, 239n
somatic therapy 269, 270–1, 273–5
somatopsychic interrelations 243–7
somatosensory cortex 16, 51, 121–2, 132, 144, 189
somatotopy 201n
Spaemann, R. 213, 214, 215, 216n
spatial perception 9–11
speech 96–7, 146–7, 291
  acquisition 196, 197–201
  dual aspectivity 212, 213
  performative 35–6
Sperry, R. 225–6
Spinoza, B. 71, 71n, 80, 216n
spiral-shaped relationship 103, 139
Spitz, R. 185
State, M.W. 254
Stern, D.N. 179, 180–1, 206
Stewart, C.M. 236n
still face experiment 180, 185
stimulus–cognition–reaction 127–8
stimulus–reaction 129–30
Strack, F. 125
Straus, E. 45
Strawson, G. 37, 81
stream of consciousness 48, 50
stress 177, 262–3, 264
subjective body 11, 71–3, 81–2, 83
  dual aspect 74–7, 80
  extension 15–16, 18, 18n
  see also body-as-subject; lived body
subjective fact 34–5, 34n
subjectivity xiv–xvi, 29–66, 112, 286
  category mistakes 43–52
  embodied see embodied subjectivity
  irreducibility 32–43
  living organism 89–92, 93–4
  mental disorders 275–7
  powerlessness and 52–61
  reductionist viewpoint 29–32
  role of consciousness 57–61
suicide 259, 259n
superior temporal sulcus 182, 189
supervenience 159, 221
supplementary motor cortex 43, 154, 187n

Sur, M. 141–2, 142n
Swaab, D.F. 53, 221n
synaptic connection 140–1, 174–5
synchronization, neural 151–2, 152n, 154
syntopy 12, 13, 16

tacit knowledge 146
Taipale, J. 74n
taxi driver 142
Taylor, C. 166–7
teleodynamic emergence 224
teleofunctionalism 41n
teleology 43
temporal integration 55–6, 60, 217
temporal lobe 51, 182, 187n, 189, 261
tennis player 133–4
thalamus 113, 115, 142
Theory of Mind 77, 179–80, 190, 203, 285n
therapy 268–75
third-person (observer) perspective 29, 43, 64–5, 281
  dual aspect of life 81, 82
  phenomenal consciousness and 33, 34–5
  see also naturalistic attitude
Thompson, E. 48, 94, 135n, 162, 222, 223
thought 202–5, 211, 212, 234
Tiehen, J.T. 222
Tomasello, M. 192, 193, 195, 205
Tononi, G. 42n, 47n, 150n, 154, 166n, 219n
tool 15, 16, 132
top-down (downward) causality 95, 96, 98
  conscious attention 152
  dual aspectivity and 225–32
  effects of psychotherapy 255–6, 268, 272n, 274
  emergence of consciousness 223–5
  neuroplasticity 141
  psychosomatic interrelations 244
torpedo 39
touch 11–12, 15, 124, 144
Tourette syndrome 53
transformation 145–56, 169
  brain as organ of 68, 98, 99
transparency 145–56, 161–4, 170, 191
trauma, early life 186
tree rings 39–40, 39n, 42
Trevarthen, C. 174, 177, 179, 192
Tye, M. 39n
typist 102–3, 144–5, 144n

Uexküll, J. von 26, 87–9, 191
Umwelt 26, 88, 89, 91, 92
upward causality see bottom-up causality

Varela, F. 84–5, 152, 223
vertical circular causality 94–8, 104, 122, 244, 255
  integration and formation of capacities 99–104
  mental disorders 256–9, 260–1

Vico, G. 197*n*
violinist 142
vision 9–11, 23–5, 48–9, 99
visual cortex 48, 141–2, 142*n*, 143–4, 151–2
vocal gestures 196, 198–9, 200
voluntary action 53–7
Vorgestalt (preconception) 129, 131, 133–4, 148, 150
vulnerability 263–6, 267

walking stick 15, 15*n*
we-perspective *see* second-person perspective
Wernicke, C. 46, 252
Wernicke's area 144, 197, 199
what-is-it-likeness 33*n*, 59
Wittgenstein, L. 34*n*, 202, 212*n*
writing 100–1, 128, 138, 154, 233–4

Zahavi, D. 33–4, 33*n*, 34*n*, 135*n*
zombie 230, 230*n*